CLASSIFICATION AND HUMAN EVOLUTION

CLASSIFICATION
AND
HUMAN
EVOLUTION

EDITED BY SHERWOOD L. WASHBURN

AldineTransaction
A Division of Transaction Publishers
New Brunswick (U.S.A.) and London (U.K.)

Second paperback printing 2009
Copyright © 1963 by Wenner-Gren Foundation for Anthropological Research, Inc.

This book is printed on acid-free paper that meets the American National Standard for Permanence of Paper for Printed Library Materials.

Library of Congress Catalog Number: 2006051143
ISBN: 978-0-202-30935-4

Printed in the United States of America

Library of Congress Cataloging-in-Publication Data

Classification and human evolution / Sherwood L. Washburn, editor.
    p. cm.
Originally published: Chicago : Aldine Pub. Co., 1963. In series: Viking fund publications in anthropology ; no. 37.
Includes bibliographical references and index.
ISBN 978-0-202-30935-4 (alk. paper)
  1. Human evolution. 2. Primates. 3. Human beings—Origin. I. Washburn, S. L. (Sherwood Larned), 1911-2000.

GN281.C54   2007
 599.93'8—dc22   2006051143

PREFACE

TWO YEARS AGO at a meeting of the American Institute for Human Paleontology there was a discussion of the classification of fossil men. Several of those present thought that there had been progress since the days when nearly every find was given a new generic name. It was also suggested that there were not as many kinds of australopithecoids as the abundance of names suggested. Dr. Fejos encouraged several of those present to continue the discussions and a meeting was held at the Wenner-Gren Foundation and a symposium at the American Anthropological Association meetings in Philadelphia in 1961. As a result of these preliminary meetings, an international symposium was planned for the summer of 1962. The eighteen participants met at Burg Wartenstein, the European conference center of the Wenner-Gren Foundation. The results of the conference are presented in this volume. No record of the discussions has been included, but the participants revised their papers in the light of the conference.

All of the participants urged me to thank Dr. Fejos and the staff of the Wenner-Gren Foundation for their helpful hospitality. We also wish to thank the National Science Foundation for paying the travel expenses of the American participants. I wish to thank all the participants for the time and effort devoted to this project and, particularly, Dr. G. G. Simpson for the addition of the notes on the conference and his latest ideas on the classification of apes and man.

S. L. Washburn
Berkeley, California

CONTENTS

CONTENTS

CLASSIFICATION AND HUMAN EVOLUTION

THE MEANING OF TAXONOMIC STATEMENTS

GEORGE GAYLORD SIMPSON

INTRODUCTION

EVERYONE WHO DEALS with evolution has occasion to use and to understand statements in the special language of taxonomy and classification. Communication is impeded by the facts that not all who use that language speak it fluently and that those fluent in it do not all speak the same dialect. In our conference on classification in relationship to human evolution we were talking this language much of the time. The main function of this contribution was to discuss the grammar and semantics of a reasonably standard dialect of the language. Centering the discussion on hominoid classification brings up and may clarify certain crucial points. This chapter is not, however, concerned with expressing opinions about human classification and evolution, but with discussing how such opinions are or should be expressed. I have recently covered theoretical aspects of animal taxonomy in some detail (Simpson, 1961), and mere repetition of parts of that book is here avoided.

CLASSIFICATION, TERMINOLOGY, AND NOMENCLATURE

Taxonomic language involves not only a very large number of different designative words (names, terms) but also several different *kinds* of designations. The things or concepts designated by these words, technically their referents,[1] are also of different kinds, and the meanings or semantic implications are likewise diverse. It is therefore essential that they be clearly distinguished. One way to do this is to consider the main operations involved in classification and the points or levels where special designations are required, as shown schematically in Figure 1.

The process starts with observation of the specimens in hand, the objective materials. The specimens studied and believed to be related in some biologically relevant way are a *sample*. If they are believed to represent a definite taxon (as determined at another level of inference), they constitute a *hypodigm*. Unequivocal designations of the specimens must refer to them as concrete, discrete objects; they are not designated by any name of the population or taxon to which they are supposed to belong. The ideal designation, practically universal in zoology

---

1. A psycholinguistic term also useful in zoological taxonomy. See Brown (1958).

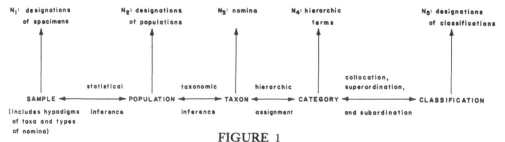

FIGURE 1

Schema of processes (arrows), name sets (N), and referents (capitals) in taxonomy. Vertical arrows all represent the process of designation or symbolization. The processes represented by horizontal arrows proceed logically from left to right, but in practice no one operation can be carried out without reference to the others. These arrows are therefore drawn pointing both ways.

but unfortunately not in anthropology, is by a collection or repository symbol and a catalogue number uniquely associated with each specimen. This is one kind of designation, one set of names (symbols of some sort, not necessarily or usually in words), and may be called the N_1 naming set.[2]

Observations and specimens, no matter how numerous, have no scientific significance purely per se. They acquire significance only when they are considered as representative of a larger group, or population, of possible observations or of individuals united by some common principle or relationship. The population may be abstract, for instance as symbolized in the equation for gravitation, applicable to a potentially infinite number of events but derived from a finite series of experimental observations. In zoological taxonomy the population is finite and concrete: a set of organisms existing (now or formerly) in nature. The existence and characteristics of that population are inferred from the sample drawn (we hope at random) from the population. The methods of inference are statistical by definition, which does not mean that any particular procedure of mathematical statistics is necessarily used although, of course, that is often appropriate and useful. A population is obviously not the same as the specimens actually studied, a sample drawn from the population.

At the next step in the process, all populations belong to taxa and all taxa are composed of populations. However, the two are not necessarily coextensive. It is often necessary to recognize and designate a local population that is a part of a taxon but does not in itself comprise a whole taxon. For some populations a different set of names or symbols, N_2, may therefore be required. Populations are in fact sometimes given distinct designations in zoological systematics, commonly by specification of their geographic location, but there is no established and uniform system. It may be sufficient to designate a population either as that

2. Recognizing different sets of taxonomic designations and distinguishing them in this way is due to Gregg (1954), although I do not follow him in detail.

from which a given sample was drawn (hence by extension of an N_1 designation) or as identical with that of a given taxon (hence by an N_3 designation).

A taxon is a group of real organisms recognized as a formal unit at any level of a hierarchic classification (Simpson, 1961, which see also for definitions and more extended discussions of hypodigms, categories, and hierarchies). A taxon is therefore a population, although the over-all population of one taxon may include many distinct populations of lesser scope. A taxon is created by inference that a population (itself statistically inferred from a sample which now becomes a hypodigm) meets a definition adopted for units in an author's classification. The set of designations for taxa, N_3 names, are those of formal, technical zoological nomenclature, e.g. *Homo*, the name of a taxon in primate classification. The word "name" is used in many different ways, both in the vernacular and in technical discussion, and this has engendered confusion. I propose that technical Neo-Linnaean names in the N_3 set be called *nomina* (singular, *nomen*). Vernacular names ("lion," "monkey," "Neanderthal man") are in the N_3 set if they designate taxa, but they are not nomina.

Each taxon is assigned to (considered as a member of) a category, which has a defined rank in hierarchic classification. A category is a set, the members of which are all the taxa placed at a given level in such a classification. Categories are distinct from taxa, do not have populations as members, and are not represented by samples. They have their own set of names, N_4, which are the relatively few terms applied to levels of the Neo-Linnaean system: basically phylum, class, order, family, genus, and species, with various combinations in super-, sub-, and infra-, and occasionally such additional terms as cohort or tribe.

Finally the various taxa of assigned categorical rank are collocated, superordinated, and subordinated among themselves and so form a hierarchic classification. This is done in terms of the N_3 (nomina) and N_4 (hierarchic terms) names. The added implications are conveyed less by nomenclatural than by topological means, primarily by arrangement and not consistently by verbal or related symbolization. Designations of classifications, N_5, are normally bibliographic references to their authors and places of publication.

What, now, are the meanings or implications of the various sets of designations? N_1 designations refer to particular objects. They imply only that a given specimen exists. They assure that when the same designation is used, the same object is meant. N_2 and N_3 designations both refer to groups that are considered to be populations related in some way. An author using such designations must make clear, explicitly or implicitly, the kind of relationship he has in mind. In modern zoology unless some other usage is definitely stated, it is generally understood that the relationship is genetic, that is, that it reflects evolutionary relationships. Concepts of what constitutes evolutionary relationships, how they are to be determined, and how reflected in classification become difficult and complex, but that is a different point.

Besides the implication that a population, usually genetic in relationship, is designated, nomina, N_3 names, further imply that the unit designated is given a definite rank in classification, that it is associated with an N_4 term. Under the International Code the forms of some nomina reflect the categorical rank of the corresponding taxa. For example, nomina ending in -idae (e.g. Hominidae) name families, and italicized, capitalized single words (e.g. *Homo*) name genera.

Most nomina, however, lack implications as to superordination, and *none have any implications beyond those mentioned*. For instance, nomina have no implications as to relationships among taxa at the same categorical level (e.g. *Homo* and *Tarsius*) or among taxa at any levels with etymologically distinct names (e.g. *Gorilla* and Pongidae). Further implications, which may be numerous and intricate as will be illustrated later, are inherent in the arrangement of nomina in a classification and not in the nomina themselves.

Discussion at the conference repeatedly illustrated the need for employing and distinguishing the different naming sets. The ambiguity and clumsiness of usual references to particular specimens and populations were especially evident. For example no clear and simple way was found for designating the various specimens from Olduvai Bed I that are believed not to belong to the taxon called *Zinjanthropus boisei* by Leakey. Presumably they will eventually be placed in taxa with distinct nomina (in the N_3 set), but that will not solve the problem of referring to the specimens themselves or to the populations inferred from them without ambiguity and without prejudice as to their taxonomic interpretation. As another example, no one maintains that *Telanthropus* is a valid taxon at the generic level, but no way has been found to refer to the specimens in question except as *Telanthropus*, an N_3 designation that necessarily implies a taxonomic conclusion agreed to be incorrect.

It must be emphasized that one of the greatest linguistic needs in this field is for clear, uniform, and distinct sets of N_1 and N_2 designations, applied to specimens and to local populations as distinct from taxa. Just what form such designations should take is a matter for proposal and agreement among those directly concerned with the specimens and their interpretation. It suffices here to stress that they *must not* have the form of Neo-Linnaean nomina. (The catalogue now being compiled by Oakley and Campbell may opportunely provide designations for specimens of fossil Hominidae.)

THE CHAOS OF ANTHROPOLOGICAL NOMENCLATURE

Men and all recent and fossil organisms pertinent to their affinities are animals, and the appropriate language for discussing their classification and relationships is that of animal taxonomy. When anthropologists have special purposes for which zoological taxonomic language is not appropriate, they should devise a separate language that does not duplicate any of the functions of this one and that does not permit confusion with its forms. There is, I believe, no reason for use of an additional language when what is being discussed is in fact the taxonomy of organisms. This language has been developed over a period of hundreds of years by cumulative experience and thought and has been thoroughly tested in non-anthropological use. It is admittedly imperfect, but for its purpose it is the best instrument available. Its imperfections call rather for improvement than for replacement. The most important needed improvement, with particular reference to anthropology, is that all those who use it should speak it well and in accordance with the best established current usages.

It is notorious that hominid nomenclature, particularly, has become chaotic. It is ironical that some of those who have most complained of the chaos have been leading contributors to it. A recent proposal that an international commission be formed to deal with the chaos refuses to recognize the appropriate code and the appropriate commission already set up. The author then proceeds to compound the confusion that he condemns.

Insofar as the chaos is merely formal or grammatical, it could be cleared up by knowledge of and adherence to the International Code of Zoological Nomenclature (Stoll et al., 1961), supplemented, if necessary, by whatever action might be proposed to and endorsed by the International Commission for Zoological Nomenclature. Much of the complexity and lack of agreement in nomenclature in this field does not, however, stem from ignorance or flouting of formal procedures but from differences of opinion that cannot be settled by rule or fiat. For example, when Leakey inferred from an Olduvai specimen (which he made a hypodigm) the existence of a taxon that he called *Zinjanthropus boisei* he was using correct taxonomic grammar to express the opinion that the taxon was distinct at both specific and generic categorical levels from any previously named. In equally grammatical expression of other opinions many other nomina, such as *Paranthropus boisei*, *Australopithecus robustus boisei*, or *Homo africanus boisei*, might have been proposed and might now be used. Or the specimen might have been and might now be referred to (or added to the hypodigm of) some previously named taxon such as *Paranthropus crassidens*. Any of those alternatives accord equally with the Code and would have equal status before the Commission. Decision among them is a zoological, not a nomenclatural or linguistic question, and it will be made by an eventual consensus of zoologists qualified in this special field.

Insofar as the chaos is due to faulty linguistics rather than to zoological disagreements, it stems either from ignorance or from refusal to follow rules and usages.[3] This must be almost the only field of science in which those who do not know and follow the established norms have so frequently had the temerity and opportunity to publish research that is, in this respect, incompetent.

An overt reason sometimes given for refusal to follow known nomenclatural norms is that some nomen is, in the opinion of a particular author, inappropriate. For example, some choose to rename *Australopithecus* as *Australanthropus*, thus adding another objective synonym to the chaos, on the grounds that the Greek

---

3. Mayr, Linsley, and Usinger (1953) provide an excellent introduction to the rules and basic taxonomic usages. The promulgation of a later code (Stoll *et al.*, 1961) must, however, now be taken in account.

anthropos more nearly expresses their opinion as to the affinities of the genus than does *pithekos*. The argument is completely irrelevant. *Australopithecus* does not mean "southern ape." Its meaning (defined by its referent) is simply the taxon to which the nomen was first attached and to which it was the first nomen attached. *Palaeolumbricus* or *Jitu* would have served just as well. The generic nomen does not, in itself, express any opinion as to the affinities of the taxon, and if nomina were changed in accord with every shade of opinion on affinities the chaos would be even worse than it is.[4]

Another reason for the chaos is the previously mentioned failure to develop and use consistently different designations for specimens, populations, and taxa, that is, distinct N_1, N_2, and N_3 name sets. A truly eminent anthropologist insisted on using the (N_3) nomen *Sinanthropus pekinensis* for specimens and a population although he concluded that this nomen does not designate a *taxon* specifically distinct from *Pithecanthropus erectus* or indeed from *Homo sapiens*. The example is far from unique.

Probably no one has ever admitted this, but it seems almost obvious that nomina (N_3) have sometimes been given to single specimens just to emphasize the importance of a discovery that could and should have been designated merely by a catalogue number (N_1). Of course no two specimens are alike, and it is always possible to fulfill the formal requirement that ostensible definition of a taxon must accompany proposal of a nomen. However, and again I would say obviously, the "definition" has often been only a description of an individual "type" with no regard for or even apparent consciousness of the fact that taxa are *populations*. This is not just a matter of exaggerating the taxonomic difference between specimens. It is a much more fundamental misunderstanding of what taxonomy is all about, of what nomina actually name. It is a relapse into pre-evolutionary typology, from which (I must confess) even the nonanthropological zoologists have not yet entirely freed themselves. Nomina have types, but not in the old typological sense. The types are not the referents of the N_3 nomina but are among the referents of N_1 designations. The referents of nomina are taxa—certain kinds of populations.

It is of course also true that the significance of differences between any two specimens has almost invariably come to be enormously exaggerated by one authority or another in this field. Here the fault is not so much lack of taxonomic grammar as lack of taxonomic common sense or experience. Many fossil hominids have been described and named by workers with no other experience in taxonomy. They have inevitably lacked the sense of balance and the interpretive skill of zoologists who have worked extensively on larger groups of animals. It must, however, be sadly noted that even broadly equipped zoologists often seem

4. It is true that when the system was being developed, from 200 to 250 years ago, the then relatively few nomina were usually intended to be etymologically descriptive. The experience of two centuries has, however, conclusively demonstrated that as a general principle this is absolutely unworkable. Except for the occasional mnemonic value, it is unfortunate that nomina do often have ostensible etymological meanings in addition to their real, taxonomic meanings.

to lose their judgment if they work on hominids. Here factors of prestige, of personal involvement, of emotional investment rarely fail to affect the fully human scientist, although they hardly trouble the workers on, say, angleworms or dung beetles.

It is not really my intention to read an admonitory sermon to the anthropologists. You are all well aware of these shortcomings—in the work of others. I must pass on to matters more positive in value.

SPECIES AND GENERA

The undue proliferation of specific and generic nomina is in part a semantic problem. The proposal of such nomina is rarely accompanied by an appropriate definition of the categories (as distinct from the taxa) involved, but ascribing specific or generic status to slightly variant specimens can be rationalized only on a typological basis. Whether consciously or not, taxa are evidently being defined as morphological types and statistical-taxonomic inferences from hypodigm to population to taxon (see Fig. 1) are being omitted. But in modern biology taxa are populations and the following two nonconflicting definitions of the species are widely accepted:

Species are groups of actually or potentially interbreeding populations, which are reproductively isolated from other such groups.

An evolutionary species is a lineage (an ancestral-descendant sequence of populations) evolving separately from others and with its own unitary evolutionary role and tendencies. (Quoted from Simpson, 1961, where sources are cited and the definitions are further discussed.)

The naming of a species either should imply that the taxon is believed to correspond with one or both of those definitions or should be accompanied by the author's own equally clear alternative definition.

Evidence that the definition is met is largely morphological in most cases, especially for fossils. The most widely available and acceptable evidence is demonstration of a sufficient level of statistical confidence that a discontinuity exists *not* between specimens in hand but *between the populations inferred from those specimens.* The import of such evidence and the semantic implication of the word "species" are that populations placed in separate species are either

(1) in separate lineages (contemporaneous or not) between which significant interbreeding does not occur, or

(2) at successive stages in one lineage but with intervening evolutionary change of such magnitude that populations differ about as much as do contemporaneous species.

In dealing with the incomplete fossil record the information at hand commonly cannot establish the original presence or absence of a discontinuity. Allowance must be made for probabilities that further discovery will confirm or confute the existence of an ostensible discontinuity. Those probabilities depend on various circumstances. If populations are approximately contemporaneous, only moderately distinctive, and separated by a large geographic area from which no comparable specimens are known, there is considerable possibility that discovery of intervening populations would eliminate discontinuity. That is, for example, the situation regarding the original hypodigms of *Pithecanthropus erectus* and *Atlanthropus mauritanicus*. In my opinion the possibility that the Trinil population and the Ternifine population belong to the same species is such that different specific (a fortiori, generic) nomina are not justified at present.

If, on the other hand, populations being compared are of markedly different ages, decision to give them different specific nomina should depend on judgment whether such nomina would be justified if it turned out that they belong in successive segments of the same lineage. That would apply, for example, to the Mauer population as compared with the late Pleistocene European neanderthaloid population, and I should think would justify different specific nomina in this example.[5] Still a third situation arises when samples indicate populations that were approximately contemporaneous and living in the same region (synchronous and sympatric) as may be true, at least in part, for the Kromdrai, Swartkrans, Makapan, and Sterkfontein populations. In such cases allowance hardly has to be made for possible discoveries of populations living at other times and in different places. The degree of statistical confidence generated by the samples actually in hand may be taken as definitive of the probability of an original discontinuity, for instance between *Australopithecus africanus* and *A. robustus*.

The category genus is necessarily more arbitrary and less precise in definition than the species. A genus is a group of species believed to be more closely related among themselves than to any species placed in other genera. Pertinent morphological evidence is provided when a species differs less from another in the same genus than from any in another genus. When in fact only one species of a genus is known, that criterion is not available, and judgment may be based on differences comparable to those between accepted genera in the same general zoological group. There is no absolute criterion for the degree of difference to be called generic, and it is particularly here that experience and common sense are required.

It must be kept in mind that a genus is a *different* category from a species and that it is in principle a *group* of species. Much of the chaos in anthropological nomenclature has arisen from giving a different generic nomen to every supposed species, even some clearly not meriting specific rank. In effect no semantic distinction has been made between genus and species, and indeed the number of proposed generic nomina for hominids is much greater than the number of validly definable species. Monotypic genera are justified when, and only when, a single, isolated known species is so distinctive that the probability is that it belongs to a generic group of otherwise unknown ancestral, collateral, or de-

5. I am not suggesting what those nomina should be. Among many other possibilities they might be *Homo heidelbergensis* and *Homo neanderthalensis*, or *Homo erectus* and *Homo sapiens*.

scendent species. No one can reasonably doubt that this is true, for example, of *Oreopithecus bambolii* and that in this case the (at present) monotypic genus is justified. It is, however, hard to see how the application of more than one generic name to the various presently known australopithecine populations can possibly be justified, whatever the specific status of those populations may be.

PHYLOGENY AND RESEMBLANCE

As most biologists understand modern taxonomic language, its implications are primarily evolutionary, but there is some persisting confusion even among professional taxonomists. It is not possible for classification directly to *express*, in all detail, opinions either as to phylogenetic relationships or as to degrees of resemblance. As a rule with important exceptions, degrees of resemblance tend to be correlated with degrees of evolutionary affinity. Resemblance provides important, but *not the only*, evidence of affinity. Classification can be made consistent with, even though not directly or fully expressive of, evolutionary affinity, and its language then has appropriate and understandable genetic implications. Classification cannot, at least in some cases, be made fully consistent with resemblance, and any implications as to resemblance are secondary and not necessarily reliable. These relationships can be explored by consideration of several hypothetical models or examples, set up so as to be simplified parallels of real problems in the use of taxonomic langauge to discuss human origins and relationships.

Classification and taxonomic discussion of related but distinct contemporaneous groups, such as the living apes and living men, involves a pattern of evolutionary divergence. That will first be discussed by means of a model. Discovery of related fossils almost always complicates the picture by revealing other groups divergent from both of those primarily concerned. It may, however, also reveal forms that are ancestral or that are close enough to the ancestry to strengthen inferences about the common ancestor and the course of evolution in the diverging lineages. In general the characters of two contemporaneous groups as compared with their common ancestry will tend to fall into the following classes, exemplified by characters of recent Pongidae and Hominidae:

A. Ancestral characters retained in both descendent groups. E.g. absence of external tail, pentadactylism, dental formula.
B. Ancestral characters retained in the first descendent group but divergently evolved in the second. E.g. quadrupedalism, grasping pes.
C. Ancestral characters retained in the second but divergent in the first group. E.g. undifferentiated lower premolars.
D. Characters divergently specialized in both. E.g. brachiation versus bipedalism.
E. Characters progressive but parallel in both. E.g. increase in average body size.
F. Convergent characters. I know of none between pongids and hominids, a fact which (if it is a fact) greatly simplifies judgment as to their affinites.

Different numbers of characters will fall into different categories. For instance in pongid-hominid comparison there are certainly many more A characters than any others and more B than C characters. (The given example of a C character is dubious.) Many characters do not simply and absolutely fall into one category or other. Retention of ancestral characters is usually relative and not absolute; some changes generally occur and "retained" usually means only "less changed." In constructing the simplest possible model on this basis, further simplifying postulates are that characters evolve at constant rates and that characters in the same group (e.g. D or E) evolve at the same rates. Those postulates are certainly never true in real phylogenies, and more realistic but also much more complicated models can be constructed by taking varying rates of evolution into account. The simplest possible limiting case, although unrealistic in detail, nevertheless more clearly illustrates valid and pertinent matters of principle. Such a model, analogous to pongid-hominid divergence, is illustrated in Fig. 2. Numbers preceding the category designations symbolize relative numbers of characters in the corresponding categories. Exponents symbolize progressive change: a-b-c, or in a different

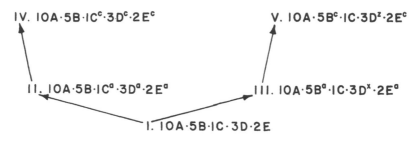

IV. $10A \cdot 5B \cdot 1C^c \cdot 3D^c \cdot 2E^c$ V. $10A \cdot 5B^c \cdot 1C \cdot 3D^z \cdot 2E^c$

II. $10A \cdot 5B \cdot 1C^a \cdot 3D^a \cdot 2E^a$ III. $10A \cdot 5B^a \cdot 1C \cdot 3D^x \cdot 2E^a$

I. $10A \cdot 5B \cdot 1C \cdot 3D \cdot 2E$

FIGURE 2
A model of simple evolutionary divergence.
Symbols are explained in the text.

direction x-y-z. It is assumed that in this example there are no F (convergent) characters. Roman numerals represent taxa: IV and V the two contemporaneous groups being compared, and I their common ancestry, ancestral to IV through II and to V through III.

From such data a comparison matrix can be formed. More sophisticated ways of doing this are exemplified in Campbell's contribution to this symposium, but for present purposes a simpler and sufficient method is to tabulate step differences between taxa. Change from C to C^a, for instance, is one step and from C^a to C^c is two more. These are multiplied by the number of characters in the category, 1 for C characters. The matrix for the model in Fig. 2 is given in Table 1. In this form of comparison, the smaller the number the greater the similarity. In this model I and II are most and IV and V least alike.

Let us suppose now that classification were to be based *entirely* on degrees of resemblance, as has been proposed by some taxonomists, and that classificatory language was therefore understood to be directly and solely expressive of re-

TABLE 1
COMPARISON MATRIX FOR DATA OF THE MODEL IN FIG. 2

	I	II	III	IV	V
I	0	6	9	18	29
II	6	0	12	12	32
III	9	12	0	24	19
IV	18	12	24	0	36
V	29	32	19	36	0

semblance. In building up higher taxa one would of course start by uniting I and II. If I and II are species, they would be placed in one genus; if genera, in one family. The maximum difference within the higher taxon would be 6. If no greater difference were allowed, all other lower taxa, III, IV, and V, would have to be placed in separate, monotypic higher taxa, an arrangement with nearly minimal significance, indicating no more than the close resemblance of I and II. If a difference of 12 were allowed in the higher taxon, II would be united with I and II, but IV should now also go with II, from which its difference is also 12. However, a taxon including IV and I would have to allow a difference of 18 and one including IV and III a difference of 24. But now V must also be added, for its difference from III is only 19. Thus *all* the lower taxa must go in a single taxon of next higher rank, an arrangement that indicates nothing of resemblances or relationships among any of those taxa. Insertion, or in actual examples discovery, of additional taxa, say between II and IV, would only compound the difficulties and lead still more inevitably to equally unsatisfactory alternatives.

I believe that the conclusion from the model is quite general for analogous real cases. In such situations the use of classificatory language as direct expression of degrees of resemblance commonly tends to produce one of two extremely inexpressive results: (1) one higher taxon includes the two most similar lower taxa and all other higher taxa are monotypic; or (2) one higher taxon includes all the lower taxa, no matter how numerous.[6]

Now let us agree that classificatory language is to have primarily evolutionary significance. For the moment degrees of resemblance need not be considered at all. It is clear from consideration of characters in categories B, C, and D that II can be ancestral to IV but not to III or V, and that III can be ancestral to V but not to II or IV. In actual instances the conclusions are neither so simple nor so obvious, but probabilities are readily established by the same categories of evidence. On this basis, I and IV can be placed in one higher taxon and III and V in another of the same categorical rank. That arrangement expresses the

6. Even extremists who would classify by resemblance *only*, usually admit that the biological significance of such classification may be confused by differential rates of evolution and by convergence. Note that in our simplified model both of those admitted sources of confusion have been eliminated by postulation, and that biologically significant classification from the numerical data alone still is impossible.

opinion, postulated as true in the model, that II and IV are phylogenetically related to each other and that III and V are also related in more or less the same way and degree. The arrangement is also consistent with but does not express the opinion, also postulated as true, that II is ancestral to IV and III to V.

In completion of this arrangement there are two alternatives as regards I. It could be placed in a third higher taxon ancestral to both of the two already formed, or it could be placed in the same higher taxon as II and IV, because it is phylogenetically closer to II than to III. Degree of resemblance here enters in as evidence for the latter inference.

Those are not the only classifications that would be consistent with the postulated evolutionary history. It would also be consistent to put I, II, and III in one higher taxon and IV and V in two others, or I, II, III, and IV in one and V in another. The implications on affinity would be somewhat different in each case but not conflicting: all are consistent with the postulates of the model. Choice would depend in part on what implications one wanted especially to bring out, since not all can be expressed in one classification. It would also depend on other considerations such as not changing previous classifications unnecessarily and conveying as much significant information as possible. (The last alternative mentioned above is the least informative.)

The model also illustrates the tendency, which is open to exception, for degree of resemblance to correlate with nearness of common ancestry. II and III are nearer their common ancestry than are IV and V, and they resemble each other more closely. The same is true of III and IV or of II and V as against IV and V. Such relationships are not directly implicit in the classification, but they are important in arriving at the judgments of affinity that are implicit in it.

Another important point illustrated in the model is that II and III resemble each other much more closely than III resembles its descendent V. It is realistic to expect an early—say Miocene—ancestor of *Homo* to be more like an ancestral ape than like modern man. It is unrealistic to expect the Miocene ancestors of either (or both) groups *necessarily* to have any of the specialized features that are diagnostic between *recent* members of the two families.

TAXONOMIC LANGUAGE: HOMINIDAE AS EXAMPLE

The Hominidae may be taken as an example of different principles (e.g. typological versus evolutionary) of classification and of the classificatory implications of different interpretations of data. For purposes of exemplification the data are postulated to be as in Figure 3A. Postulated ranges of variation of known specimens are indicated by the stippled areas, and in order to simplify the subsequent diagrams parts of those ranges are labeled A-F. X is a postulated individual specimen to be classified; it does not represent a specimen actually known. This arrangement is not presented as a realistic summary of what is, in fact, known. It is greatly simplified in several respects. Some known fossils do not fit clearly

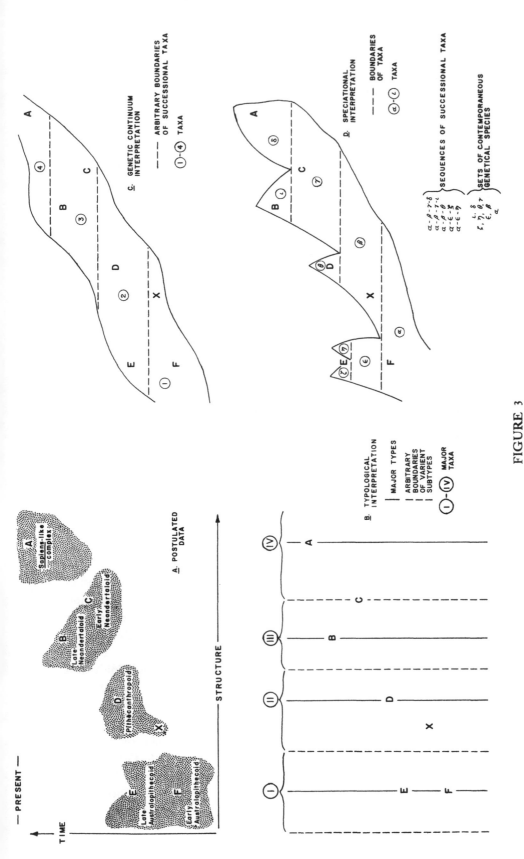

FIGURE 3

Postulated data (simplified and generalized) and three possible taxonomic interpretations of known hominide. Further explanation in text.

into the stippled areas, and some parts of those areas are not clearly represented by known fossils. Structure does not, in fact, follow a linear, one-dimensional scale and could be realistically indicated only by a (quite impractical) *n*-dimensional scale. Nevertheless this is the general *kind* of pattern, however grossly oversimplified, that the data do present.

Typological interpretation, Figure 3B, takes into account morphology only. It ignores temporal sequence and makes no phylogenetic interpretations. It abstracts an arbitrary number of fixed, distinctive types in the morphogenetic field and exercises subjective judgment as to whether a given, concrete specimen belongs to one type or another. Types may be hierarchically divided into subtypes, but variation is then ignored in the sub-types, and genetical or evolutionary considerations do not enter in at any categorical level. As previously mentioned, this basis for classification has been largely abandoned in modern taxonomy. Nevertheless hominid classification started out on this basis, and even some of the most recent work in that field is at least covertly typological. The classificatory and nomenclatural expression of the typological arrangement in the diagram could take several forms, depending on the categorical level assigned to differences between types, for example:

I	*Homo africanus*	*Australopithecus africanus*
II	*Homo erectus*	*Pithecanthropus erectus*
III	*Homo neanderthalensis*	*Homo neanderthalensis*
IV	*Homo sapiens*	*Homo sapiens*

Figures 3C and D represent two possible kinds of phylogenetic interpretations of the same postulated data. Here temporal and genetic relationships are taken into account, and classification is based in principle on inferences as to evolutionary affinity. The differences between Figures 3C and D do not involve any difference in taxonomic principle but only in opinion as to probable evolutionary relationships. Both kinds of interpretation (with many differences in detail as to the particular placing of actual specimens) are currently supported by different students. Choice between them will depend on the accumulation of more data, and the ultimate arrangement will probably not be entirely of either kind in a clear-cut way but will involve elements of both. It will certainly be more complex than either of my diagrams.

Figure 3C diagrammatically represents the interpretation that hominids have been represented by only one interbreeding population since the early Pleistocene, at least. On this interpretation there is only one lineage or evolutionary species and only one genetical species at any one time. In that case, the species would have been highly variable, and even more so during much of past time than *Homo sapiens* is at present. At some time around the middle Pleistocene it might have varied all the way from what in purely morphological (or typological) terms could be called marginal australopithecoid through pithecanthropoid to marginal neanderthaloid. Such variation would be improbable within a single deme or local population. It would be less improbable among geographically separate (allopatric) populations or sub-species. Such geographic semi-isolates would of course be variable in themselves, but some might, for instance, vary about a more australopithecoid modal morphology and others about a more neanderthaloid mode. Discovery that fossil hominids fall into such modally distinct, synchronous but allopatric groups would favor this interpretation. Whether current data do or do not tend to follow such a pattern I leave to the specialists in such matters.

The over-all ranges and modes of morphology change greatly from earlier to later parts of the phylogeny as postulated in Figure 3C, as they also do from early Pleistocene to now in the data actually known. It is useful, if not absolutely necessary, to take this into account in classification. The only possible way to do this (adhering to evolutionary taxonomic principles and accepting the interpretation as one genetic continuum) is to divide the lineage arbitrarily into successional taxa, as also exemplified in the diagram. The placing of the arbitrary boundaries and the ranks given the taxa will depend on judgment as to categorization of morphological differences and also, in practice, on where incomplete knowledge happens to make a morphological gap coincide more or less with a time line, as occurs between C and D in my postulated data. Again several different classifications would be consistent with the given phylogenetic interpretation, among them these two:

1. *Homo africanus*	*Australopithecus africanus*
2. *Homo erectus*	*Pithecanthropus erectus*
3. *Homo neanderthalensis*	*Homo neanderthalensis*
4. *Homo sapiens*	*Homo sapiens*

The same nomina are used here as in the typological interpretation, but the diagrams show that their meanings are different in the two. That is further shown by the fact that specimen X here falls into *africanus* but in Figure 3B into *erectus*.

Figure 3D represents an interpretation with speciation occurring within the Pleistocene hominid group so that there is not a single lineage but successive branching giving rise to two or more distinct, contemporaneous species, of which only one of the two last to arise has survived. The sets of contemporaneous species are separated by natural gaps (noninterbreeding) and are not arbitrary. Successive species in a lineage like α-β-γ-δ are arbitrary as regards that lineage, alone, but in the whole pattern their boundaries are also fixed by the (hypothetically) nonarbitrary points of splitting of the lineage into two species. The probability of this kind of pattern would be supported by discovery of contemporaneous (synchronous) populations with overlapping geographic distribution (sympatric) that did not intergrade and hence were probably not interbreeding significantly. (The existence of two or more distinct species does not, however, depend on their being sympatric.) Again I leave to the appropriate specialists whether data actually in hand do support such an interpretation.

One of several possible nomenclatures consistent with this pattern would be:

> ζ *Australopithecus robustus*
> η *Australopithecus africanus*
> θ *Pithecanthropus erectus*
> ι *Homo neanderthalensis*
> δ *Homo sapiens*

It is not clear what actual specimens might fall into the hypothetical species α, β, γ, and ε, and I therefore suggest no nomenclature for them. It is clear, in any event, that some, at least, of the same nomina as used under the interpretations in Figure 3B and C are also applicable in D but again have different significance and contents. Specimen X, for example, is now neither in *africanus* nor in *erectus* but in unnamed hypothetical species β.

If identical nomina in Figures 3C and D referred to the same populations, there would be no ambiguity. Unfortunately, however, this is not likely to be the case. Population C in Figure 3C would probably be placed in the same taxon and referred to by the same specific nomen as population B. In Figure 3D population C would probably be placed in a different taxon and given a different nomen from population B. The ambiguity resides not in the taxonomic system but in the imperfection of our data and lack of agreement in their zoological interpretation. When such ambiguity persists, clarity demands that an author specify the populations included in his taxa, for example by adequate designation of their hypodigms. In the present example a possible clarifying device (if it accorded with an accepted zoological interpretation) would be to place populations B and C in separate subspecies. The placing of those subspecies in species would then clearly show different placing of the corresponding populations by different students.

TAXONOMIC LANGUAGE: *OREOPITHECUS* AS EXAMPLE

The currently debated classification of *Oreopithecus* may be taken as another example of the use of classificatory language and its implications for phylogeny. Simply for purposes of the example, the following postulates are accepted:

1. Pongidae and Hominidae are distinct families of common ancestry and are united in the superfamily Hominoidea. This is now the usual conclusion and classification.
2. *Oreopithecus* had a common ancestry with both Pongidae and Hominidae at a time when the hominoid ancestry was distinct from that of any other recognized superfamily (e.g. Cercopithecoidea). This is not established, but some, probably most, recent students consider it probable.
3. *Oreopithecus bambolii* is at least generically distinct from any other known species. This is universally accepted.

Table 2 gives four phylogenetic opinions (not the only ones possible) consistent with these postulates and gives for each two classifications consistent with that opinion and also consistent with further opinion as to the lesser or greater difference between *Oreopithecus* and forms considered allied to it according to the respective phylogenetic opinions.

TABLE 2

Some Possible Opinions as to the Phylogeny and Distinctiveness of Oreopithecus, and Some Classifications Consistent with Those Opinions

Opinions as to phylogeny of Oreopithecus	Opinions as to distinctiveness of Oreopithecus	
	a. Lesser	b. Greater
I. In or near hominine ancestry.	A. Hominoidea Hominidae Homininae *Oreopithecus*	B. Hominoidea Hominidae Oreopithecinae *Oreopithecus*
II. Divergent from early hominids after separation from pongids.	B. Hominoidea Hominidae Oreopithecinae *Oreopithecus*	C. Hominoidea Oreopithecidae *Oreopithecus*
III. Divergent from early pongids after separation from hominids.	D. Hominoidea Pongidae Oreopithecinae *Oreopithecus*	C. Hominoidea Oreopithecidae *Oreopithecus*
IV. Divergent from common stem of pongids and hominids.	C. Hominoidea Oreopithecidae *Oreopithecus*	E. Oreopithecoidea Oreopithecidae *Oreopithecus*

Classifications B and C appear more than once in the table. Each is consistent with more than one opinion as to phylogeny. That is also true of all the other classifications, which are consistent with different phylogenetic opinions not distinguished in the table. For instance, A is consistent both with the view that *Oreopithecus* is directly ancestral to *Homo* and that it is a little-differentiated side branch from the direct ancestry. C appears three times in the table, but even that is not exhaustive: C is also consistent with several other possible opinions as to phylogeny. This forcefully illustrates the principle that classification is not intended to be an adequate expression of phylogeny but only to be consistent with conclusions as to evolutionary affinities.

Nevertheless each classification does have quite definite implications as to phylogeny; it is consistent with some and not with other opinions. Classification B, for instance, is definitely inconsistent with phylogenetic opinion III. In this sense a classification does express opinions on phylogeny in a broad or somewhat loose way. B does not express opinion as to whether *Oreopithecus* is a remote ancestor of *Homo*, an ancient hominid sidebranch, or a rapidly evolved hominine sidebranch, but it does express the opinion that *Oreopithecus* had a common ancestry with the hominids after hominids and pongids were distinct. C, which is the arrangement I personally prefer on present evidence, expresses the opinion that my postulate 2, above, is probable, and further that *Oreopithecus* is markedly differentiated from both pongids and hominids, but that the degree of distinctiveness is not so great as to warrant categorization as a superfamily. The basis for this preference is outlined in a following comment written after the conference, and most of the evidence is summarized in Straus's paper in this volume.

CONCLUSION

Classification is not an exact science and is not likely soon to become one. In order for its language to become completely unambiguous and uniform it would be necessary to have adequate samples of all pertinent populations, to reach universal agreement as to their affinities, and similarly to agree on just how to translate those affinities into formal classification. We do not have adequate samples and in the present field probably never will have. Even if the objective data were complete—indeed *especially* if they were complete—classification would have to have many arbitrary elements. Complete agreement as to what is to be said is thus chimerical and indeed may not even be desirable. We can, however, agree as to how to say what we mean. Taxonomic language as an instrument of communication has the failings but also the strengths of any other living language. Linguistic usages determine the clarity, not the content, of expression. The important thing is to use a language grammatically, to be quite sure of the implications of its words, and to make sure that it conveys our intended meanings unambiguously to others using the language.

PART TWO: ADDED NOTES

ASPECTS OF DEFINITION AND DIAGNOSIS

Attention has been given to the definition of man and his distinction, at various taxonomic levels, from his nonhuman or less fully human relatives. It may clarify matters to point out that the problem of definition is a complex with several different aspects. All are related and all intergrade, but ambiguity is introduced if no distinction is made among these aspects. The most essential distinctions may be illustrated by comparison of *Homo sapiens* and *Pan troglodytes*, the phylogenetic relationships of which are represented in Figure 4 as a simple splitting of one ancestral lineage into two and the subsequent divergent evolution of the latter into the present terminal species named. That there were also other splittings and other lineages is a complication that can be ignored for purposes of the example.

The terminal species are completely discrete and, as living populations, completely accessible biological and taxonomic units. The diagnosis of living man with respect to the chimpanzee, symbolized by A in the diagram, is simple and unambiguous, and it may be made in almost any terms we like, for example by the presence of different proteins in the blood, by the nonoverlapping frequency distributions of premolar structure or brain size, by the differences in locomotion, or by the absence in one and presence in the other of true language. Diagnosis of *Homo sapiens* with respect to his own ancestry, represented by B in the diagram, will theoretically involve some but not all of the same criteria as the A-diagnosis. It will involve such diagnostic characters as evolved within the human ancestry

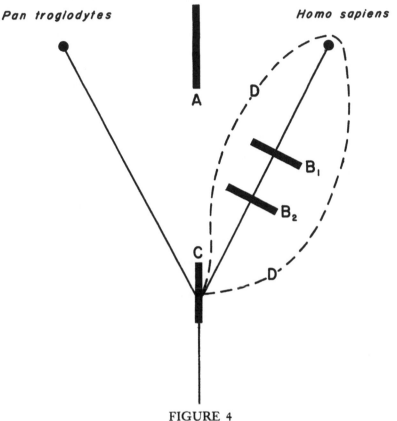

Pan troglodytes   *Homo sapiens*

FIGURE 4

Diagram of different aspects of diagnosis between chimpanzee and man, and between living and fossil man. Further explanation in text.

since the separation from the chimpanzee lineage, but will exclude characters of the common ancestry preserved in *Homo* and not in *Pan*. An operational B-diagnosis must also exclude characters not determinable in the available fossils, because it is the fossils, only, to which the diagnosis must be applied in practice. The origin of such A-diagnosis characters as language, self-awareness, human social structure and the like is of greatest evolutionary interest, but simply is not pertinent to the practical classification of actual fossil specimens.

In principle a B-diagnosis involves drawing a more or less arbitrary line across what was in nature a continuum. In practice these lines are commonly placed where there are gaps in knowledge. There are at present no known specimens unequivocally recognizable as intermediate between *Homo sapiens* and *H. erectus* or between *Homo* and *Australopithecus*, but it is reasonably certain that such creatures existed and probable that specimens of them will eventually be found. In the meantime our ignorance of them permits unambiguous diagnoses between those specific and generic taxa. Campbell (this book) has pointed out that successive taxa may be naturally separated at peaks of evolutionary rates, and it is also likely that such peaks may correspond with gaps in present knowledge. At present, however, and for purposes of practical classification we cannot definitely place such peaks or even be quite sure that they did exist, and in any event a *precise* diagnostic boundary would still be an arbitrary line in a continuum.

It is clear that different characters and character-complexes (e.g. the locomotory, masticatory, and central nervous system complexes) have evolved at different rates and accelerated at different times. The possible B-diagnoses are therefore multiple and will involve different characters or complexes at different times, as symbolized by the separation of B_1 and B_2 in the diagram. As total change is gradual and cumulative, it will also be ranked at different categorical levels at different points. Such ranking is not completely arbitrary because it can be more or less coordinated with current categorical ranking of taxa separated by A-diagnosis. That agreement on such ranking can be reached is evident by unanimity in the contributions to this book that *Homo sapiens* and *H. erectus* differ only specifically but that *Australopithecus africanus* is generically distinct from both. Some disagreement exists at higher levels, but only by one hierarchic substep, e.g. as between superfamily and family or family and subfamily.

Another aspect of definition is symbolized by C in the diagram, representing the difference between the ancestors of *Pan* and *Homo* when they first separated. At the very time of the separation, this distinction was (with high probability if not complete certainty) at the specific level only. As Dobzhansky pointed out, it is further likely that the initially differentiating characters of *both* groups occurred together as simple variants in a single species ancestral to the two. The actual characters involved are likely to be (in simplest, incipient form) some but certainly not all of those eventually involved in the A-diagnosis. I can see no reliable way of judging a priori just which of those characters do in fact stem from the C-differentiation, and indeed it is possible in principle that the C-diagnosis will be by characters not present in either *Pan* or *Homo*. The C-diagnosis does not depend on the characters entirely in themselves or on the magnitude of the difference at the time of separation, but on our knowledge (or opinion) as to the taxonomic significance of the *subsequent* divergence of the lineages involved. The categorical level is assigned *ex post facto*: in the Miocene our ancestor was probably one of several species of a genus of Pongidae but now that same species, if definitely identified, might properly be classified in a quite distinct genus of Hominidae.

Finally, it is desirable that the whole evolutionary unit enclosed in the broken line marked D in the diagram should also be recognized and named as a taxon. Its definition would ideally involve all the differentiating aspects of A, B, and C.

The Dentition of Oreopithecus

The dentition of *Oreopithecus*, which is completely known from a large number of specimens and which I have been able to study in detail through the courtesy of Dr. J. Hürzeler, is extremely distinctive. It differs more from either pongid or hominid dentitions than living pongids and hominids, together, differ among themselves. The proportions of canines and premolars are indeed hominid-like, but the morphology of those teeth, which I take to be far more significant than a simple matter of proportions, is decidedly nonhominid, as is also that of all the other teeth. I completely agree with Dr. Hürzeler's conclusion, strongly reinforced by the study presented in this book by Dr. William L. Straus, Jr., that *Oreopithecus* is not a cercopithecoid and that it is morphologically a hominoid. I cannot, however, support reference to the Hominidae. This negative conclusion is based not only on the great differences of the teeth of *Oreopithecus* from those of any sure hominid (the genera *Homo* and *Australopithecus*, both *sensu lato*, or known Homininae and Australopithecinae) but also on the following considerations:

1. There are known forms (notably *Ramapithecus* and *Kenyapithecus*) approximately contemporaneous with *Oreopithecus* that have much more hominid-like dentitions.

2. The peculiarities of the *Oreopithecus* cheek teeth (e.g., specialized protoconule region and course of the crista obliqua on the upper molars; presence of a mesoconid and of a peculiar paraconid or, more likely, pseudoparaconid on the lower molars) are all present in incipient or fully developed form in *Apidium*, an early Oligocene genus. (Dr. E. L. Simons, and earlier Dr. W. K. Gregory have pointed out this resemblance, which is also mentioned in Straus's paper in this book. Dr. Simons has several very important new specimens of *Apidium* which have not yet been described in print but which I have seen through his courtesy.)

3. No Anthropoidea have yet been definitely identified before the early Oligocene. Comparison of all known Anthropoidea and of conceivably ancestral earlier prosiminians strongly suggests that the anthropoid ancestry in general and the hominoid ancestry in particular lacked peculiarities of dentition already present in *Apidium* and emphasized in *Oreopithecus*.

These considerations seem to me to indicate that the lineage leading to *Oreopithecus* was already distinct near the very base of hominoid differentiation and that its dental differences from both pongids and hominids are not on the whole primitive but are at least in large part divergent with respect to the ancestry of those groups. If that is correct, the placing of *Oreopithecus* (and *Apidium*) in a family Oreopithecidae of the superfamily Hominoidea, for which a preference is expressed above, becomes almost obligatory. It is further highly improbable that the important skeletal resemblances of *Oreopithecus* to the Hominidae, as listed by Straus, are primitive for the Hominoidea. They would therefore appear to be the result of parallelism or convergence to a degree that is noteworthy but not inherently improbable. The total result is a peculiar adaptive type quite unlike that of either the Pongidae or the Hominidae.

This example also illustrates two points of general taxonomic principle. First, the varying significance of different characters and character complexes for classification requires that they be weighted in the light of the whole biological and evolutionary picture as far as it is known. Second, there are often, as in this case, considerations highly pertinent to biological classification and yet difficult or impossible to reduce to simple numerical form and to include in a computer program for obtaining a (likewise highly pertinent) coefficient of similarity or distance function.

Affinities and Classification of the Hominoidea

Within the last few years data for classification of the hominoids have been greatly enriched in breadth and depth: discovery of new pertinent fossils; continued anatomical investigation; studies of serology, hemoglobins, and chromosomes; more detailed behavioral observations of nonhuman hominoids, particularly under natural conditions. Since almost every conceivable view (along with some rather inconceivable ones) has been upheld at one time or another, this new information is useful not so much in giving us a new pattern of affinities as in enabling us to choose more surely among the many already proposed and to gain more confidence in various points of detail. Without reviewing evidence or arguments, in this note I shall briefly state how the probabilities look to me now. I shall also briefly sketch some of the main, different family-group (superfamily, family, subfamily) classifications that would be consistent with those probabilities and indicate my own preference among them.

Gibbons.—On fossil evidence, the gibbon ancestry was probably distinct from that of other apes when hominoids first appear in the fossil record, early Oligocene, and was certainly so in the Miocene. Recent karyological and serological evidence, presented in this book by Klinger and by Goodman, respectively, also indicates strong divergence of living gibbons from all other living hominoids. No evidence suggests special affinities witih orangs, on one hand, or with the chimpanzee–gorilla group on the other. Special affinity with *Homo* is out of the question. It is probable that these three groups did not diverge among themselves until after the gibbon ancestry had already split off. On the other hand, the gibbons have not diverged radically from other apes either morphologically or adaptively. What is distinctive in their facies is largely due to their having remained smaller than other hominoids and to their specialized locomotion, which in turn seems to require the first peculiarity. Miocene fossils (demonstrated to members of the 1962 conference by Professor Zapfe in Vienna) suggest that the locomotory specialization evolved comparatively late. Although gibbons, strictly speaking, and siamangs are usually placed in different genera, they are manifestly very closely related and I now prefer to place them all in *Hylobates*.

Orangs.—Morphologically and to some extent also adaptively *Pongo* is not markedly unlike the living chimpanzees and, to less extent, gorillas. This has long, although not quite unanimously, been considered evidence of rather close relationship. Schultz, whose knowledge of orangs is unexcelled, continued to uphold that view in the conference. On the other hand, karyological (Klinger) and serological (Goodman) evidence seems to separate *Pongo* from *Pan* (and *Gorilla*) as sharply as *Hylobates*. Fossil orangs have not been identified before the Pleistocene, but there is no evident reason why the ancestry of *Pongo* may not be found near that of *Pan* in the dryopithecine complex. On balance, it still seems probable that *Pongo* is especially related to the African apes, but that the split was far enough back to permit considerable, more or less clandestine molecular and chromosomal divergence. Morphological divergence has been less, probably because of retention of somewhat similar adaptation.

Gigantopithecus.—When known only from isolated molars, this Chinese Pleistocene genus was claimed to be a hominid. Later finds of lower jaws and dentitions, not yet adequately described as far as I know, seem clearly to exclude it from the Hominidae. It seems to be a terminal specialization not very close to any living form. During the 1962 conference Leakey suggested special affinity with *Pongo*, and that is a possibility. On present very inadequate evidence I would, however, prefer to place it only as Pongidae *incertae sedis*, and I omit it from further consideration.

African apes.—A consensus has always considered gorillas and chimpanzees as especially and rather closely related, and all the recent evidence, including that of serology and karyology, confirms that view. They are of course sharply distinct species, at a point of divergence where experienced taxonomists may well waver between giving only specific or also generic weight to that divergence. Merely listing characters that demonstrate the self-evident fact of their distinctness does not necessarily suffice to maintain the time-honored generic separation, and at present I prefer to consider both chimpanzees and gorillas as species of *Pan*. Whether *P. paniscus* is a valid third species, closer to *P. troglodytes* than to *P. gorilla*, is still moot. Placing all the African apes in *Pan* permits classification to express the clear fact that they are much more closely related to each other than to any species of other genera, and henceforth I shall use the nomen *Pan* in this sense.

It has long been the virtually universal opinion that *Pan* is anatomically and adaptively rather close first to *Pongo* and then to *Hylobates*. Recent studies, while also confirming that these are quite distinct groups well separated at a generic level, at least, agree with the old conclusion that the three genera belong together in a natural taxon at some higher level. Nevertheless, as noted above, newer sub-

anatomical evidence suggests that separation of the ancestors of the three genera within that higher taxon is ancient. The situation is complicated only by comparisons with the Hominidae, summarized below.

No explicit and particular connection of *Pan* with a Tertiary ancestry has yet been found or, at least, clearly recognized. It is, however, probable that in a general sense the ancestry occurred somewhere in known or unknown members of what is here called the dryopithecine complex.

The dryopithecine complex.—The Miocene and Pliocene of Africa, Europe, and Asia have produced many specimens clearly apelike and distinct from contemporaneous closer relatives of *Hylobates* (notably *Pliopithecus* and the closely allied, perhaps not generically distinct, *Limnopithecus*). They are otherwise highly diverse and clearly represent a greater number of lineages than the four or perhaps five recent species that might possibly have arisen from this complex (*Pongo pygmaens, Pan troglodytes*, perhaps *Pan paniscus, Pan gorilla*, and *Homo sapiens*). Many or most of the dryopithecine-complex lineages have therefore become extinct, and it is the opposite of surprising to find that some of them (e.g. *Proconsul*) have combinations of characters not found in taxa as diagnosed primarily on the basis of living species.

With the sole exception of *Proconsul*, the members of this complex are known only from very incomplete remains, largely single teeth or unassociated upper and lower jaws with partial dentitions. Their classification and nomenclature are unsatisfactory and almost chaotic within the group. This situation could surely be improved by a revision even of the already known fragments, and I consider such revision plus a really systematic search for better specimens the greatest desideratum of primate paleontology at present. Some of these forms, such as *Dryopithecus* itself, may be rather near the ancestry of the living great apes, *Pan* and perhaps also *Pongo*. Others, such as *Ramapithecus*, and possibly *Kenyapithecus*, may belong near the ancestry of *Homo*. Still others, as already mentioned, are doubtless more or less terminal in lineages not close to any living forms. If or when more probable affinities with later groups are established, it should be possible to place some of the dryopithecine-complex species and genera in taxa, e.g. in subfamilies, currently based primarily on living species. It is, however, my opinion that the present unsatisfactory stage of study and incompleteness of sampling do not establish such connections at a sufficient level of probability.

Oreopithecus.—Elsewhere in this chapter I have sufficiently expressed the opinion that *Oreopithecus* probably represents a lineage separate from near the base of hominoid differentiation, with limited parallelism with the Hominidae, but culminating in an extinct terminal form adaptively very unlike either the great apes or any hominids.

Australopithecus.—Despite earlier polemics, it is now perfectly clear that among other sufficiently known genera *Australopithecus* (*sensu lato*, including *Paranthropus* and *Zinjanthropus*) is most closely allied to *Homo*. Late *Australopithecus*, at least, is almost certainly contemporaneous with early *Homo* (including *Pithecan-*

thropus, etc.) and hence not ancestral to it. Present evidence does not exclude, and may be taken to favor, the possibility that early *Australopithecus*, or an unknown genus close to it, was such a direct ancestor. Although *Australopithecus* greatly strengthens the opinion that *Homo* had an apelike ancestor in common with the living great apes, it does not at present seem to me to give additional clear evidence as to precisely which apes, among living forms or the dryopithecine complex, are most nearly related to *Homo*.

Homo.—Since the 19th century it has been the usual, although by no means the universal, opinion that among living mammals *Homo* is most closely allied to *Pan*. That conclusion, based originally on classical anatomical grounds, is strongly supported by all the new evidence, anatomical, karyological, biochemical, and behavioral, presented at recent conferences. It now seems to me so probable that other alternatives need no longer be seriously considered. It should, however, be strongly emphasized that *Homo* represents an anatomical and adaptive complex very radically different from that of any other known animal and (with the partial exception of *Homo*'s close ally *Australopithecus*) differing far more from other living or adequately known fossil hominoids than they differ among themselves. Seemingly contradictory evidence (e.g. that of the haemoglobins as reported by Zuckerkandl in this book) indicates merely that in *certain* characters *Homo* and its allies retain ancestral resemblances and that *these* are not the characters involved in their otherwise radical divergence—a common and indeed universal phenomenon of evolution.

Affinities, adaptive radiation, and phylogeny.—Figure 5 shows, by combination of a dendrogram and an adaptive grid, my present views as to the affinities and the adaptive or structural-functional relationships of the living hominoids. Interpretation of probable closeness of genetic connection is indicated by depth of branching, although it is to be emphasized that such a diagram is not a phylogenetic tree and has no time dimension. Adaptive or ecological (and corresponding structural-functional-behavioral) resemblances and differences are approximated by horizontal distances between the terminal points.

Figure 6 shows in schematic form the combined phylogenetic inferences reviewed above for various groups of hominoids.

Classification.—Evolutionary classification takes into account: degrees of homologous resemblance in *all* available respects; the most probable phylogenetic inferences from all data (including the foregoing resemblances plus evolutionary analysis and weighting of the various characteristics); and also the practical needs of discussion and communication.

It now seems perfectly clear and is all but universally recognized that the animals here called hominoids (in anticipation of a conclusion) form a natural evolutionary unit that should be recognized and named as a taxon. When the whole order Primates is taken into account, the categorical level of this taxon should clearly be no higher than infraorder and no lower than family. A case could be made out for either extreme of those rankings, but in my opinion the

ADAPTIVE AND STRUCTURAL-FUNCTIONAL ZONES

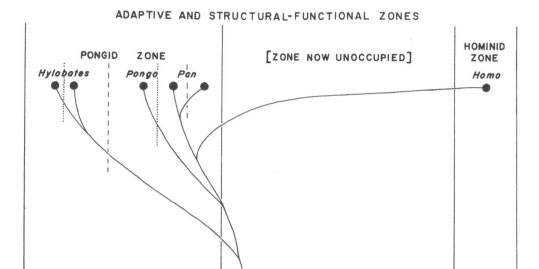

FIGURE 5

Dendrogram of probable affinities of recent hominoids in relationship to their radiation into adaptive-structural-functional zones. The two major adaptive zones are bordered by solid lines. Pongid radiation into sub- and sub-sub-zones is schematically suggested by broken and dotted lines. A dendrogram of this sort has no time dimension and does not indicate lineages, but it is probable that divergences of lines showing affinities are topologically similar to the phylogenetic lineage pattern.

intermediate ranking, that of superfamily, is best in balance and convenience. It also accords with the recent consensus, and thus with the principle that communication is best served if nomenclature is not changed unnecessarily. The current and nomenclaturally correct nomen for this superfamily is Hominoidea.

At the next lower level, I have already expressed the opinion that the apparent lineage *Apidium-Oreopithecus* is at present best ranked as a family, because of its ancient separation plus its marked divergence from any other group now usually given family rank. In view especially of Straus's analysis (in this volume), the only reasonable alternative is to rank this lineage as a subfamily of Hominidae on balance of anatomical resemblance alone. That is not wholly excluded by present evidence, but its phylogenetic implications seem to me extremely improbable. (Unless *Pongo*, *Pan*, and all the dryopithecine complex were also placed in the Hominidae it would be definitely inconsistent with the phylogeny of Figure 6.)

Because of the clear and ancient separation of *Hylobates* and its fossil allies from other hominoids, those forms are now frequently, probably usually, given family status. On the other hand, *Hylobates* almost certainly had a common hominoid ancestry with *Pongo* and *Pan*, and its evolutionary divergence from those genera and their fossil allies is decidedly less than that of either *Homo* or *Oreopithecus*. That would justify placing the *Hylobates* group as a subfamily of a family also containing *Pongo, Pan*, and some, at least, of the dryopithecine complex. Both arrangements are consistent with reasonable interpretations of the available data, and choice becomes a matter of personal judgment and convenience. I continue to prefer the second alternative, partly as a matter of linguistic convenience. One frequently wants to distinguish humans and apes (plus or including gibbons) and this is most conveniently done at the family level. The secondary distinction between gibbons and (other) apes is convenient at the subfamily level.

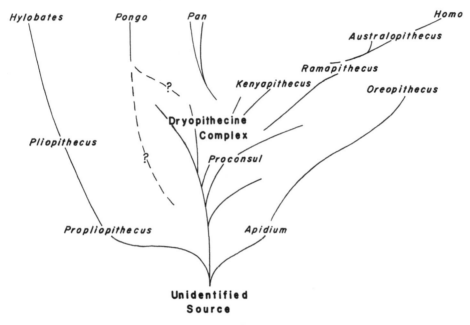

FIGURE 6

Tentative and schematic phylogeny of the Hominoidea. Most of the individual fossil genera are omitted, and lineages as drawn are meant to be impressionistic and diagrammatic (especially in and around the drypopithecine complex) rather than representing all or particular generic or specific lines.

If the gibbons are given family rank, an analogous argument can be made for also giving *Pongo* separate family rank, since it cannot be *demonstrated* to have split off more recently from the *Pan* ancestry and since it now proves to be serologically about equally distinct. This is nevertheless still largely an argument from ignorance, and the most extensive positive evidence we have, that of anatomy, still suggests closer affinities between *Pongo* and *Pan*. Certainly if *Hylobates* and *Pan* are in one family, *Pongo* belongs in the same family and the alternatives involve subfamilies. Of the five possible subfamily arrangements of the three genera (and their fossil allies), only placing *Hylobates* in one subfamily and *Pongo* and *Pan* in another or placing each in a separate subfamily seem worthy of consideration on present knowledge. Both can be defended, but I continue tentatively to favor the former, because I believe that *Pongo* and *Pan* probably are more closely related than *Hylobates* and *Pongo* and because I think monotypic subfamilies should be as few as possible.

For the dryopithecine complex, there are three possibilities: (1) all could be put in the same subfamily as *Pongo* and *Pan;* (2) they could all be placed in a subfamily or family of their own; (3) those clearly related to *Pongo, Pan,* or *Homo* could be put in family-group taxa with those genera and the others in one or more separate subfamilies or families. The first could be justified on grounds of general resemblance and of probability that *part* of this complex is near the ancestry of *Pan,* perhaps also of *Pongo.* The second arrangement makes a horizontal grouping that is phyletically complex and to some extent artificial, but is justifiable *faute de mieux* in our present lack of almost any good knowledge of detailed relationships in this group, and pending withdrawal of particular genera if their affinities with other established taxa are later demonstrated. The third arrangement is definitely preferable and should ultimately be adopted, but as indicated above present knowledge seems inadequate to follow it with sufficient probability. I now waver between (1) and (2), but hesitantly continue to follow (2) simply because there seem to be insufficient grounds for changing it until this complex is better understood.

It is now virtually established that the affinities of *Homo* with *Pan* are closer than with *Pongo* or *Hylobates.* That suggests the possible desirability of placing *Pan* in the Hominidae and *Pongo* and *Hylobates* in one or two other families, an arrangement supported by Goodman and by Klinger at the conference. In fact, as noted above, this view as to affinities is an old one and has been held for two or three generations by students who nevertheless all excluded *Pan* from the Hominidae. The new data increase confidence in the conclusion as to affinities but do not, in my opinion, either require or justify the proposed change in classification. The question involves the whole complex of taxonomic principles and cannot be argued in detail here. The following, in briefest form, are among the principal reasons for continuing to exclude *Pan* from the Hominidae:

    1. *Pan* is the terminus of a conservative lineage, retaining in a general way an anatomical and adaptive facies common to all recent hominoids *except Homo* (and probably to all adequately known fossil ones except *Australopithecus* and the very different *Oreopithecus*). *Homo* is both anatomically and adaptively the most radically distinctive of all hominoids, divergent to a degree considered familial by all primatologists.

    2. *Pan* is obviously not ancestral to *Homo.* The common ancestor was almost certainly more *Pan*-like than *Homo*-like, which suggests not that *Pan* should go in the Hominidae but that the common ancestor should be in a separate family with *Pan* (or in still another ancestral family).

    3. When a younger family arises from an older, the situation is frequently or

usually similar to that of the *Pan* and *Homo* lineages: one of several lineages of the older family splits into two or more, *one* of which diverges (and/or diversifies) until its descendents warrant family status. If the lineages that did *not* diverge are also placed in the later family on the basis of more recent common ancestry, carrying this process on down will eventually require inclusion of all descendents of earlier splittings also in the latest family—eventually the whole animal kingdom would be in the Hominidae on this principle. An arbitrary division must be made in practical classification, and the obvious place to make it is where the lineage *later* reaching family status split off—in this instance where the ancestors of *Homo* split from those of *Pan* and other pongids. Classification cannot be based on recency of common ancestry *alone*.

4. Both arrangements are equally consistent with our present understanding of hominid phylogeny, but the proposed new arrangement is less consistent with other evolutionary considerations, notably that of adaptive divergence. Therefore the change is neither required nor warranted. That radical change of nomenclatural usage would also create great confusion in discussion of hominoid relationships.

Subfamily separation of *Australopithecus* and *Homo* became usual at the time when placing of *Australopithecus* in the Pongidae (as here used) or Hominidae (also of present usage) was disputed and the australopithecines were claimed to include several genera. Now that there is essential agreement that *Australopithecus* belongs in the Hominidae, I see no sufficient reason for having two subfamilies, especially as each has only one known genus as I and, I believe, most others now define the genera. "Australopithecine" and "hominine" may still be used as strictly vernacular terms for structural levels, although there is little need for such terms as long as we know only one genus at each level.

Finally, Leakey suggested at the conference a possible alternative classification that requires comment. (I do not know whether he himself considers the alternative preferable and proposes to use it.) He pointed out that the Hominoidea of my classification could be reduced to family rank and that my families and subfamilies, with some reassignment of genera, could be considered subfamilies (plus another subfamily for *Proconsul*, which I give only generic rank). The supporting phylogeny looked rather different from mine, but was topologically almost identical. Aside from minor points, either classification was consistent with either phylogeny. Without going into detail, I here only say that at the very best I do not think his classification an improvement justifying so many departures from current usage. The reasons are mostly implicit or explicit in the preceding comments.

The following is the outline classification that I now favor, with all accepted recent genera included but a number of fossil genera omitted:

Superfamily Hominoidea
    Family Pongidae
        Subfamily Hylobatinae
           Pliopithecus
           Hylobates

Subfamily Dryopithecinae
   Dryopithecus and other genera of the dryopithecine complex, *tentatively* including *Proconsul*, *Ramapithecus*, and *Kenyapithecus*.
Subfamily Ponginae
   Pongo
   Pan
Family Hominidae
   Australopithecus
   Homo
Family Oreopithecidae
   Apidium
   Oreopithecus

(Placing Oreopithecidae at the end and next to Hominidae has no special significance; as Darwin noted over a century ago, you cannot put organisms in one linear evolutionary sequence or show their true affinities on a sheet of paper. Putting *Australopithecus* in the Hominidae and adding the Oreopithecidae are the only essential changes from my 1945 classification, which I believe *in general* to be consistent with the mass of more recent information. Unfortunately that is less true of some of the nonhominoid primates.)

BIBLIOGRAPHY

BROWN, R.
   1958. *Words and things*. Glencoe, Ill.: Free Press.
CLARK, W. E. LE G.
   1959. *The antecedents of man*. Chicago: Quadrangle Books.
COLD SPRING HARBOR SYMPOSIA ON QUANTITATIVE ZOOLOGY
   1950. *Origin and evolution of man*. C. S. H. Symposia Vol. XV. [Especially Mayr, pp. 109–118.]
GENOVÉS, T. S.
   1960. "Primate taxonomy and *Oreopithecus*." *Science*, 133:760–761. [A recent example of misunderstanding of taxonomic language, corrected by Straus (1960).]
GREGG, J. R.
   1954. *The language of taxonomy*. New York: Columbia Univ. Press.
HEBERER, G.
   1961. "Abstammung des Menschen." In, Bertalanffy and Gessner, *Handbuch der Biologie*. [Incomplete; classification, disagreeing with my concepts of taxonomic language, especially in Lieferung 117/118, pp. 287 and 307.]

MAYR, E., ED.

1957. *The species problem*. Amer. Assoc. Adv. Sci., Pub. No. 50.

MAYR, E., E. G. LINSLEY, and R. L. USINGER

1953. *Methods and principles of systematic zoology*. New York: McGraw-Hill.

SIMONS, E. L.

1959. An anthropoid frontal bone from the Fayum Oligocene of Egypt: the oldest skull fragment of a higher primate. Amer. Mus. Novitates, No. 1976.

SIMPSON, G. G.

1945. *The principles of classification and a classification of mammals*. Bull. Amer. Mus. Nat. Hist., 85:i–xvi, 1–350. [Primate classification now requiring much updating but still illustrative of the method.]

1961. *Principles of animal taxonomy*. New York: Columbia Univ. Press.

1962. *Primate taxonomy and recent studies of nonhuman primates*. New York Acad. Sci., Conference on Relatives of Man. [Contains material pertinent tõ the present topic and not repeated here.]

STOLL, N. R., *et al.*

1961. *International code of zoological nomenclature*. London: Internat. Trust for Zool. Nomencl.

STRAUS, W. L., JR.

1960. "Primate taxonomy and *Oreopithecus*." *Science*, 133:760–761. [Correction of misuse of taxonomic language by Genovés (1960).]

EAST AFRICAN FOSSIL HOMINOIDEA AND

THE CLASSIFICATION WITHIN

THIS SUPER-FAMILY[1]

L. S. B. LEAKEY

INTRODUCTION

EAST AFRICA has contributed very greatly to our knowledge of fossil members of the Hominoidea. These come from deposits ranging from the Lower Miocene through the Lower Pliocene and the Villafranchian to the Middle Pleistocene, as well as from the Upper Pleistocene. It seems to me that if these finds are to be fitted, adequately, into the picture of the evolution of the Hominoidea, a certain amount of revision of some of our earlier ideas becomes an urgent necessity.

Before discussing the problem, let me summarize the available East African material which must be taken into consideration.

Lower Miocene (From sites in Kenya and Uganda)	
Proconsul africanus	Hopwood
Proconsul nyanzae	Le Gros Clark and Leakey
Proconsul major	–do–
Limnopithecus legetet	Hopwood
Limnopithecus macinnesi	Le Gros Clark and Leakey
A small pongid which has not yet been described	
A very large pongid [2]	
"Sivapithecus" africanus	Le Gros Clark and Leakey
Lower Pliocene or Upper Miocene (From a site in Kenya)	
Kenyapithecus wickeri	Leakey
Proconsul sp.	

1. The title chosen for this symposium is, in itself, an example of the muddle that scientists have got into in relation to primate taxonomy. I feel bound to ask just what the word "human" means in this context? The adjective "human" seems to me to be even less well defined than the noun "man." Is "human" used here to signify the Hominoidea or the Hominidae, or the Hominini, or none of these terms?

2. A preliminary account of this has been given by Allcock and Bishop who consider it to be *Proconsul major*. I have seen the material and in my opinion it is not *Proconsul* at all. It is pongid, possibly ancestral to Gorilla.

Lower Pleistocene   (From sites in Kenya and Tanganyika)
  The Kanam mandible which
    Leakey named *Homo kanamensis*
  Zinjanthropus boisei   Leakey
  The "pre-*Zinjanthropus*" hominid
    child and adult from Bed I, Olduvai
Early Middle Pleistocene   (From sites in Tanganyika)
  (a) The milk teeth from BK II, Olduvai,
         found with a Chellean stage
         of culture
  (b) The skull found with Chellean
         stage 3 at LLK II, Olduvai
End Middle Pleistocene   (From sites in Kenya)
  The Kanjera skulls
Upper Pleistocene   (From sites in Kenya and Tanganyika)
  The Eyasi skull   found by Dr. Kohl-Larsen
  Various *Homo sapiens* skulls,
    such as those from Gamble's
    Cave II, Naivasha, etc.

THE CLASSIFICATION OF THE HOMINOIDEA

Until recent times most of those who have been concerned with the study of fossil members of the Hominoidea have been content to try to fit all new discoveries into the very simple classification which allowed for a superfamily called Hominoidea, in turn divided into two, and only two, families—the Pongidae or "apes," and the Hominidae or "men."

Various workers have, from time to time, made use of sub-family divisions within these two families, placing, for example, the gibbons and their fossil relations in a sub-family called Hylobatinae. Or, again, the South African hominid fossils from the limestone caves of the Transvaal have been placed in a sub-family Australopithecinae, to show that while they were admitted to be members of the Hominidae they were regarded as fundamentally distinct. Even *"Telanthropus"* has been classified within the Australopithecinae, although I feel sure, after studying the original material of this creature, that it differs greatly from all members of this sub-family and is a more evolved member of the Hominidae, probably *Homo.*

Almost without exception all fossil members of the Hominoidea found in deposits which are geologically earlier than the Lower Pleistocene have been classified as Pongidae, although such a classification has clearly worried some authors. Lewis did indeed suggest (tentatively) that perhaps *Ramapithecus* was really a primitive member of the Hominidae and recently Simons has supported this claim, while Huerzler and Trevor actually put *Oreopithecus* into the Hominidae in spite of its having many characters which pointed strongly away from such a conclusion. They did this, of course, because they felt that it was clearly even less acceptable to place *Oreopithecus* in the family Pongidae, and perhaps they lacked the courage to set up a full scale new family to receive this aberrant member of the Hominoidea.

The difficulty concerning the correct placing of *Oreopithecus* within the accepted classification pattern of the Hominoidea has forced several authors to consider the setting up of a full family to accommodate *Oreopithecus*. I fully agree that this is essential and proposed such a procedure formally two years ago, but I doubt if such action is enough by itself to sort out our classification problems.

Two years ago I made the proposal that the time had come to recognize not less than five full families within the super-family Hominoidea, and while I am prepared to defend and argue the case for such action as a possible solution to our difficulties, I will suggest an alternative and perhaps more rational way of dealing with the problem later in this paper.

Let us first of all consider what are the basic reasons for setting up a taxonomic classification at all and clarify what it is we are aiming to achieve by making the attempt.

REASONS FOR CLASSIFICATION

The major object of our symposium is to consider "classification" in relation to "human evolution," and we are all agreed that man is simply a mammal belonging to the super-family Hominoidea and that he must have reached his present "human" status through the normal processes of mammalian evolution. We must fully expect, therefore, that the general pattern of evolution within the Hominoidea will have followed the same general lines which we know were characteristic of other mammalian super-families.

In all the well-known mammalian groups the evolutionary picture presented by the fossil evidence is a complex one. We can be certain, therefore, that the more fossil members of the Hominoidea we discover, from year to year, the more complex will our picture of "human" evolution become.

The whole object of attempting systematic taxonomy is, after all, to apply a purely arbitrary set of labels which will enable us to convey our ideas as to the relationship of one group to another, or of one fossil species to another, to our colleagues in other centres of learning, who will often not have the opportunity of examining our original material.

Since the names which we apply, *at any and every level in the taxonomic sequence* are inevitably arbitrary and artificial, it does not, I believe, matter what we decide to do, provided only that the majority of those who are concerned in the classification, at any given time, are agreed as to how they will use the classification system that is set up and provided they are clear as to what they mean by the different names that are applied. While complete unanimity is too much to hope for, we can, at least, aim at a majority agreement.

During the past thirty years there have been so many important discoveries of fossil primates belonging to the super-family Hominoidea that the situation is now very confused and there are scarcely any two workers in this field who employ even the simplest terms with identical meaning and value.

Clearly, then, the time has come to try to achieve a greater measure of uniformity and agreement; and clearly this will only be achieved if those who are concerned with the taxonomy of this group are willing to practice a measure of give and take, rather than force their points of view as the only valid one.[3]

None of us likes giving up our own special ideas in favor of those of one of our colleagues with whom we disagree but in this matter of *terminology* it should not be as difficult as in other fields since, as we have stated already, all names and terms are, after all, only "arbitrary labels" which are made for the convenience of ourselves and of our colleagues.

Provided that the majority can agree on the definition of the labels which we use at any given taxonomic level, the actual nature of these labels is relatively immaterial.

I fully realize that some of my personal views may not be acceptable to some of my colleagues any more than some of theirs are acceptable to me, but in order to achieve agreement on the meaning of our terms, I shall be prepared to give way to the majority view on this question of definitions and I hope others will approach this problem in the same spirit.

What I would like to see is agreement, at this symposium, on matters of *general principle* in connection with the classification of the Hominoidea. I would further like to see a small *ad hoc* committee set up and requested to go into matters of detail with a view to reporting their recommendations on these points at some later stage.[4]

Any agreement achieved now will, of course, only be made in the light of the present state of our knowledge and, therefore, as our knowledge grows it may need further revision.

If the *ad hoc* committee, which I envisage, achieves a good measure of agreement on detail, then its recommendations could be placed before all workers in the field of classification of the Hominoidea through the medium of a journal like "Current Anthropology," or at some international meeting, and there might then be a reasonable hope of achieving at least a majority agreement.

The overriding consideration in any new classification that we propose as a result of our symposium, must be that it is realistic in relation to the present state of our knowledge of fossil and living Hominoidea, and not based upon outmoded concepts.

Since, moreover, there are far more extinct genera and species of the Hominoidea

3. As a result of the give-and-take process, the. members of the conference did reach agreement on some important matters, after much discussion.

4. This proposal was not acceptable to my colleagues.

than there are living ones, we must clearly try to set up a classification which takes this fact into account, even though the extinct representatives are less well-known.

One of the biggest difficulties in taxonomy for those of us who have only fossil material to work upon, is the problem of assessing the value of characters which can be seen in the teeth, in the skulls and in the post cranial skeletal material, as guides to the taxonomic status of our specimens in terms of species, genus, tribe, sub-family and family.

So far as I am concerned, the criteria which I try to use in respect of fossil classifications are based upon the following principles; if the mammalian zoologists have reached general agreement in respect of two living, but allied, groups that the overall differences justify generic distinction, then in dealing with fossils that appear to fall within the same family, sub-family or tribe, I will classify as belonging to distinct genera only those fossils which seem to me to show a comparable degree of difference *in those parts which are preserved in the fossils*, as are to be seen in the corresponding parts in the living genera.

Let me take an example from the Bovidae. Within the tribe Alcelaphini, (or the hartebeests) zoologists recognize the genera *Damaliscus*, *Alcelaphus* and *Beatragus*. If I find a fossil skull which is clearly alcelaphine in its overall picture, but which differs from the skulls of each and all of the recognized species within each of these three genera as much as the species of one genus differ from the species of another, then I feel justified in setting up a new genus.[5]

If, on the other hand, the differences in my fossils are not as great as this, then I must decide to which of the known living species, of one or other of the known living genera, the new specimen stands closest. After that I shall still have to decide as to whether the differences are at the specific, or only the sub-specific level.

In all of this I must be guided by what is known of the overall differences in the bones and teeth and skulls of the various living species within any living genus, even though the main deciding characters, when a particular genus or species of living mammal was first set up, were probably linked with skin, habitat and the characters of the soft parts of the body, criteria which are unfortunately denied to the paleontologist.

If we carry this idea into the field of the Hominoidea, it is generally agreed (except by a very few) that the gorilla and the chimpanzee, while both clearly members of the family Pongidae, are sufficiently different, in the total picture which they present, to justify generic separation.[6]

This total picture has been built up, not only from teeth and skulls and the

---

5. As a result of the conference we decided to reduce the generic status of certain hominoids to subgeneric level. I have therefore prepared an additional note which now follows this paper, since I feel it is important.

6. I am aware that Dr. Mayr, one of my colleagues at the symposium, believes that gorilla and chimpanzee are of the same genus, and I am fully prepared to defend the opposite view.

skeletal structure, but also the anatomy of the soft parts, the psychological differences, the behavior pattern differences and other characters. No single character alone suffices, but the total picture seems adequate for generic distinction to those of us who know these animals in their wild state.

In dealing with fossil Hominoidea we cannot, unfortunately, ever obtain such a total picture and our only line of approach is to say to ourselves, "do the observed differences in the teeth, bones and skulls of this or that group of fossils differ as much from those of some other group, as do the bones, teeth and skulls of, shall we say, gorilla and chimpanzee?" If the answer is in the affirmative, we are justified in assuming that the total differences, if we knew them, would be as great and we may, therefore, tentatively award different generic rank to our fossils.

The greatest difficulty, of course, lies in the fact that we often only have incomplete parts of one or two individual fossil specimens to study, at the time when we have to make our provisional decision, but that is simply a part of the tragedy of being a paleontologist.

While it has to be admitted that within any species or genus the individuals that make it up will inevitably show a considerable range of variation, nevertheless, experience shows that it is wise not to regard any single first find as representing a freak. Instead, we must expect that the first new fossil of any kind that is found lies well within the normal range of its species.

We have to remember that the skull of Rhodesian man was considered by many authorities to represent a "freak" and not a normal member of its kind; so was Neanderthal man and Java man, but subsequent discoveries, in each case, showed that the first discovery was, in all essential respects, normal and not abnormal.

On the other hand, we must remember also, in our paleontological classifications, that until we have a long range of specimens, our conclusions can only be regarded as tentative and that they may have to be modified as further material comes to light. There can be no finality for a very long time, if ever.

HOMINOIDEA CLASSIFICATION AT THE FAMILY LEVEL

The earlier classification of the super-family into only two families Pongidae and Hominidae seems to be no longer tenable, unless we insist upon carrying the process of "lumping" to an extreme degree.

It can be argued (indeed it has been argued) that the differences between the extremes which are, at present, grouped in the Pongidae, are no greater than those which are placed in the single family Equidae. This is probably correct, but if we are to use this particular argument, then I submit that we should either use the term Pongidae for *all* the Hominoidea, or else the term Hominidae to include all the great apes (living and fossil) together with all "men." The name Hominidae has precedence and would therefore have to be chosen.

To do so would be carrying the lumping idea to a degree which some of our colleagues may not like but which I shall develop presently. For the moment I submit that if we feel we must separate the group which includes "man" and the australopithecines into a family distinct from that which includes the gorilla and chimpanzee, then, logically, we must look very closely at those fossils which we, at present, include in the Pongidae and take some of the genera out and place them in new and distinct families.

This is what I proposed some two years ago and I believe that such a step can be fully justified by the available fossil evidence. Even though most zoologists have regarded the living and extinct gibbons as falling within the Pongidae, I believe that they differ every bit as much from the group which is usually regarded as including gorilla, chimpanzee, orang and *Dryopithecus*, as that group does from the Hominidae. Consequently, I proposed, some years ago, the acceptance of a full family, Hylobatidae, to include the living gibbons and their fossil relatives.

Ever since we came to know more about the swamp dwelling Lower Pliocene genus, *Oreopithecus* (whose fossil remains were found in lignites), through the work of Huerzler, this creature has been another great stumbling block to our earlier classification.

Clearly it exhibits characters which make it hard for us to retain it in the Pongidae. On the other hand, it is equally out of place in the Hominidae as at present defined. It differs from both these families every bit as much as they do from each other.

Either we can have one single family in the Hominoidea, or else logically we must give full family status to *Oreopithecus* since it is wholly out of place in both Pongidae and Hominidae.

This is what I proposed in the past for I believe that *Oreopithecus* represents a family fully as distinct from the Hominidae and the Pongidae as the Tapiridae are separated from both Equidae and Rhinocerotidae.

If we follow this line of thought we would then have at least four families, Pongidae, Hylobatidae, Hominidae and Oreopithecidae.

To these four I suggest that we have to add one more family, namely, Proconsulidae, which would include the genus *Proconsul* as well as possibly a part of what is now included in *Sivapithecus* and also a part of what is, at present, treated as *Dryopithecus*.

I believe there are ample differences in respect of the basic structure of the canine teeth, the face and (where known) of the skull to warrant this separation at family level. Clearly new finds might necessitate some modification of our ideas as to which genera to include in Proconsulidae, but I believe we can regard the special nature of the canine teeth, the shape of the mandibular arch, the nature of the glenoid fossa, the total absence of a simian shelf, as among the characters which will be found to separate this family from the Pongidae, should we decide to accept it.

If we were to decide to accept these five families instead of the present two, we might still find ourselves in some difficulty over one or two known primitive members of the Hominoidea.

Where, for example, do *Kenyapithecus* and *Ramapithecus* fit in? I suspect that both may be primitive members of the family Hominidae but I would hesitate to say so, categorically, at present. Simons has recently revived the idea in respect of *Ramapithecus*.

CLASSIFICATIONS AT GENERIC LEVEL

Hominidae. Clearly the genera at present known as *Homo, Pithecanthropus, Australopithecus, Paranthropus, Zinjanthropus, Atlanthropus,* (as well as other less acceptably defined members of the family such as *Africanthropus*—which is, I think, a *nomen vanum—Telanthropus, Palaeanthropus, Meganthropus*) fall within the Hominidae and we would have to consider them further at generic level.

The criteria at present accepted for membership of the family Hominidae include certain dental characters (such as thè nature of the first deciduous lower molar) certain cranial characters (such as the position of the *foramen magnum* and the relative height of the brain case) the structure of the limb bones and pelvic girdle, the nature of the lower third pre-molars and the morphology of the canines.

We can probably separate all the other members of the super-family Hominoidea from the Hominidae by means of these and other characters. It seems, therefore, that the family is well defined and well established, but it seems likely that some of the earlier and more truly primitive members will be less easy to fit in and may lack some of the characteristics.

I believe that we can either carry our "lumping" process to the extreme of having only two genera, *Homo* and *Australopithecus*, or else retain, for the time being, the first five mentioned above.[7]

Hylobatidae. The living species of the gibbon group differ in so many ways from the other living great apes—gorilla, chimpanzee and orang—that many of us feel justified in putting the genera *Hylobates* and *Symphalanges* into a full family. Their teeth, their jaws, their skulls, their limb bones, their way of living, all proclaim them distinct and different. We may, for the time being, reasonably add the fossil genera *Propliopithecus, Pliopithecus* and *Limnopithecus* and *Prohylobates*, while recognizing that they may not all wholly agree with our ·concept of "gibbon" which is based upon the living genera. The validity of having two genera for the living gibbons is a matter for discussion.

Oreopithecidae. The strange creature to which the generic label *Oreopithecus* was given is clearly a problem. In some of its dental characters it even seems to

7. This is what, in fact, we agreed to do at the end of the conference. See the additional note at the end of this article for comment.

stand closer to the Cercopithecoidea than to the Hominoidea. In other characters, however, it may be closer to the Hominoidea and even perhaps in some respects to the family Hominidae.

I feel sure that if we were forced to limit ourselves to only the two families Hominidae and Pongidae, most of us who have seen the original material would prefer to treat *Oreopithecus* as a very aberrant member of the Hominidae, but the rational solution would seem to be the setting up of a full family with (at present) only one known genus. Such a procedure has several precedents in mammalian taxonomy. We thus will have a family with, at present, only one genus.

Pongidae. This family has, as its living representatives, *Gorilla, Pan* and *Pongo*, which are three seemingly distinct genera differing from each other not only in teeth, skull structure and limb bones, but also in many other characters such as mode of life, soft part anatomy, behavior pattern, etc.

All three are characterized by certain features of dentition and skull, such as supra-orbital torus, simian shelf, structure of the canine teeth and limb proportions, which would appear to be characteristic for the family as a whole, including its fossil members, insofar as we know them, which is indeed not very far.

I suspect that *Pongo* may even be sufficiently distinct to warrant sub-family level separation from *Pan* and *Gorilla*. Mayr and Simpson, however, would prefer to treat *Gorilla* as only a specie of the genus *Pan*.

The problem of deciding which of the extinct genera of the super-family Hominoidea should be included in the Pongidae is, in fact, very difficult because we have relatively little material to go on, in most cases. It is far easier to exclude certain genera such as *Proconsul*, than it is to decide what to include.[8]

On the one hand, the classical representatives of the genus *Dryopithecus*, seem to fall clearly within the Pongidae because of their simian shelves and the structure of their canines, as well as the arrangement of their teeth, but I feel that only close revision of all the material at present placed within the genera *Dryopithecus, Sivapithecus, Ramapithecus, Bramapithecus, Surgrivapithecus*, etc., will make it possible to sort out this problem. (This is a matter for the *ad hoc* committee which I have suggested and any conclusion which its members reached would be subject to modification in due course as more and better material became available.)

Proconsulidae. I have suggested the setting up of this family for certain ape-like fossil creatures which differ from the true Pongidae in lacking a simian shelf, in the structure of their canine teeth, in the position of the root of its molar-process maxillary, in the absence of a torus, in the nature of their mandibular condyles and the way their teeth are arranged in the mandible, etc.

Admittedly, we need much more data, but, for the moment, this seems to be a reasonable basis for the group.

I would, for the moment, suggest placing in this family the genus *Proconsul*,

8. See my additional notes.

and some of the fossils at present placed in the genera *Sivapithecus*, *Bramapithecus*, (but not *Ramapithecus*) and possibly *Surgrivapithecus*.

DISCUSSION OF SUB-FAMILIES

It will have been noticed that I have omitted all discussion at sub-family level. I have done so deliberately for I believe that if we accept five full families, such as I have outlined, we shall do better to do without sub-families such as' Australopithecinae, Homininae, Hylobatinae, Ponginae, etc.

I would, however, stress, as indicated before, that if we want to limit our terminology and bring it in line with that of the Equidae and Bovidae THEN WE MUST agree to accept only *one single family* instead of the super-family Hominoidea, calling it either Pongidae or Hominidae. We would then have this one family with a long list of sub-families and perhaps some sub-divisions at tribe level following the Equidae and Bovidae pattern. The sub-families would be Homininae, Australopithecinae, Ponginae, Hylobatinae, Oreopithecinae, Proconsulinae and perhaps some others, all with equal status within the single family.

SPECIAL CONSIDERATION OF CERTAIN EAST AFRICAN HOMINOIDEA

1. *The Genus Proconsul*. When Le Gros Clark and I prepared the 1952 monograph on the Rusinga and other fossil Hominoidea of the Lower Miocene of East Africa, we treated them as members of the family Pongidae because, at that time, the suggestion that there might be more than two families of Hominoidea was not seriously considered, and the genus *Proconsul* was clearly closer to the Pongidae than to the Hominidae. It must be noted, however, that the genus *Proconsul* exhibits certain characters which do not fit in, at all well, with what we normally regard as typical of the Pongidae.

In 1952 we thought that these differences might be due to the fact that since the genus *Proconsul* was of Lower Miocene age, some of the later specialized characters of the Pongidae, such as the simian shelf, the highly sectorial third lower pre-molars, the parallel molar pre-molar series which typify more recent Pongidae, had not yet been evolved. Now, however, we know that a small representative of the Pongidae with a well defined simian shelf and other characters, was contemporary with *Proconsul*, so that our original line of argument is no longer valid. I consider, therefore, that the whole matter is one which needs further study.

2. *"Sivapithecus" africanus*. In 1952 Le Gros Clark and I gave this name to a fragment of maxilla and a few teeth from Rusinga which clearly did not fall within our definition of *Proconsul*, but which were slightly reminiscent of some of the Indian fossils which had been placed in the genus *Sivapithecus*. We recog-

nized and stated that this East African specimen might perhaps have to be taken out of the genus *Sivapithecus*, when more data was available. It now seems possible that *Sivapithecus africanus* should be treated as a species within the new genus *Kenyapithecus*, but one which is more primitive and perhaps ancestral to *wickeri* of the Lower Pliocene. Unfortunately, we still have very little material representing "*Sivapithecus*" *africanus* and until we obtain more we cannot be quite certain.

3. *Kenyapithecus wickeri*. This new fossil is either of uppermost Miocene or Lower Pliocene date. The associated fauna suggests this, and so does the potassium argon dating.

I have published a note on this fossil in a preliminary report, without saying what family it belonged to, but I believe that it may represent a really early member of Hominidae. Only the finding of further material can settle this point.

4. *Limnopithecus*. The genus *Limnopithecus* is now known from many specimens and there can be almost no doubt as to its affinities with the gibbons. The two species at present recognized appear to be quite distinct, but not very far removed from each other. It is possible that they are as different as *Hylobates* is from *Symphalanges* and if that is so, and if the two living genera are really significantly different at generic level, it may be necessary one day to set up a separate genus for *Limnopithecus macinnesi*.

5. *Zinjanthropus boisei*. In my preliminary report, I laid stress upon the fact that this new genus was clearly an Australopithecine, in the present sense in which we use that sub-family term, and that it exhibited characters, some of which recalled *Australopithecus* and others *Paranthropus*. I stressed that it also showed many features in which it differed from both these South African genera, and seemed to stand closer to the Homininae. In particular this was true of the mastoid processes, the position of the *foramen magnum* and of the occipital condyles, together with the morphology of the malar maxillary region of the face. Further study has confirmed this view.

The early geological age of *Zinjanthropus*, together with the many differences between it and the South African Australopithecines, suggests that it may be ancestral to both of them, or else represents a third and distinct branch emanating from a still earlier common stem.

6. *The "pre-Zinjanthropus" juvenile*. It is possible that the most significant differences between the Australopithecines, including *Zinjanthropus*, *Paranthropus* and *Australopithecus*, and the new human type which is represented by juvenile fossil remains from beneath the *Zinjanthropus* level, can be found in the morphology of the teeth and in the size and shape of the brain case.

In one of my notes to "Nature" I have already stressed the differences between the lower dentition of the pre-*Zinjanthropus* child and that of the Australopithecines, but a difference of similar magnitude can also be seen between the upper molars of the two types.

If we take these dental differences, together with those to be seen in the parietals and in the clavicle, I do not think any doubt can remain, that in the "pre-*Zinjanthropus*" juvenile we have a member of the Hominoidea which is closer to the Homininae than to the Australopithecinae, although geologically contemporary.[9] We also have parts of an adult of this type.

This fact means that our whole concept of the earlier stages of the family Hominidae must be drastically altered.

7. *The new skull found with the Chellean stage 3 at Olduvai.* There appears to be a tendency today, in some circles, to regard it as an established fact that hominid evolution went through three stages—an Australopithecine stage, a Pithecanthropine stage and one characterized by the genus *Homo.*

Whatever may be the merits of this *as a theory*, it is certainly not yet an established fact, and the contemporary presence of an Australopithecine and a Hominine in Bed I at Olduvai (see No. 6 above) is a factor against this view.

The new skull found at the level of Chellean stage 3 at Olduvai has, as stated in my preliminary report, certain superficial characters such as a very large brow ridge and relatively low vault, which recall the Pithecanthropines. Some scientists have consequently seen in the new Chellean skull from Olduvai evidence to support their view that a Pithecanthropine stage was widespread in Middle Pleistocene times, and they consequently believe that the new skull will throw further light on the supposed and wholly theoretical evolution of *Homo* from *Pithecanthropus.*

Pending the publication of our full report I must again stress that the similarities between the new skull and the *Pithecanthropus* genus seem to me to be more superficial than real and, while I believe it is possible that the new Chellean skull may perhaps be ancestral to the genus *Homo,* I do not believe it is a Pithecanthropine, in the accepted sense.

In a similar way some scientists have argued that because the mandibles of Ternifine somewhat resemble one or two of the mandibles from Choukoutien, therefore, the Ternifine fossil human remains—called *Atlanthropus*—really represent an African member of the genus *Pithecanthropus.* I would remind my colleagues that the really diagnostic parts of the skull of the genus *Pithecanthropus* are the frontal, the occipital and the temporal bones, none of which are so far known for *Atlanthropus.*

It is at least as likely that the Far Eastern *Pithecanthropus* genus represents the descendents of an early stock which shared a common ancestor with *Atlanthropus* and with the new Chellean skull from Olduvai in the remote past, and that *Pithecanthropus* developed into an extinct branch, while the other continued to evolve towards *Homo.*[10]

Caution is needed and categorical statements about the evolution of the Hominidae are extremely dangerous at the present time.

8. *The Kanam Mandible.* The Kanam mandible was placed by many scientists

9. A further report is now in preparation.
10. After discussion we all agreed to treat the Far Eastern genus as falling within *Homo* as revised. This still does not justify regarding *H. erectus* as ancestral to *H. sapiens* in my view.

in the "suspense account" when I found it and described it in 1931-32 because it did not agree with their preconceived ideas about the evolution of the human mandible, and, particularly, because of false conclusions which had been based upon the now exposed Piltdown mandible.

I have never wavered from my statement that the Kanam mandible was *in situ* in the Kanam West deposits and contemporary with them. I think that it grows increasingly probable that this fragment represents a fragment of the lower jaw of something akin to our Olduvai Bed I fossils.

Returning to the taxonomy of the Hominoidea, if we decide that it is necessary to retain the super-family name then it may be necessary to have two genera within the Hominidae: these would be *Pithecanthropus* and *Homo*.[11] At the moment, I feel that *Atlanthropus* must rank as *incertae sedis* so far as generic affinity is concerned, and the same is true of *Telanthropus*, although I suspect that both will, in time, deserve the generic name of *Homo*.

So far as the Australopithecinae are concerned, in my own mind, I am satisfied that *Australopithecus* and *Paranthropus* are generically distinct, although they may stand closer to each other than either does to *Zinjanthropus*. All three, however, clearly form a closely knit branch of the Hominidae.

Like *Telanthropus* and *Atlanthropus*, the new pre-*Zinjanthropus* juvenile, the Kanam mandible and the new very weathered facial fragment from the Lake Chad region, found by Monsieur Yves Coppens, must also all rank, for the time being, as *incertae sedis* so far as generic rank is concerned.

In all cases additional material is urgently needed.

Concluding this section, let me repeat that I believe that the most logical way to deal with the classification of living and extinct Hominoidea, at the moment, is to have only one family, Hominidae, and then a large number of sub-families— Homininae, Ponginae, Hylobatinae, etc. If we were to do this, we would only have two genera within the Homininae, viz., *Homo* and *Australopithecus*.

THE STAGES OF EVOLUTION OF THE HOMINOIDEA

The available evidence seems to me to suggest:

(1) that the super-family together with the Cercopithecoidea probably evolved in Africa;

(2) that by the end of Oligocene times at least three distinct trends had probably already made their appearance—one towards the Hylobatidae, one towards the Pongidae and one towards the Proconsulidae. It is not impossible that a fourth trend towards the Hominidae was also already developing;

(3) that in Lower Miocene times we already probably had in East Africa representatives of all four trends as shown by (a) the huge range of material of the *Proconsul* group (b) true pongids which may have been forest dwellers (their remains are found with monkeys and galagos) and of which, therefore, we have very few specimens

11. But see our decisions and my additional notes.

because they seldom died under conditions leading to fossilization (c) at least two divergent branches of the Hylobatidae as shown by the different *Limnopithecus* types and (d) what was probably the stock leading to the Hominidae represented by the creature which Le Gros Clark and I originally called *"Sivapithecus" africanus*, but which should be regarded rather as an ancestral type of *Kenyapithecus;* (e) monkeys, galagos and pottos were also present;

(4) that in Lower Pliocene times we have in East Africa (a) a creature, *Kenyapithecus wickeri*, which seems to be headed directly towards the Hominidae and which must, indeed, be regarded as an ancestral member of that family (b) representatives of the Proconsulidae still persisting (c) monkeys;

(5) that by Lower Pleistocene times the Hominid stock in East Africa had already divided into two distinct branches (a) Australopithecine in character represented by *Zinjanthropus* and (b) Hominines represented by the pre-*Zinjanthropus* juvenile type. The Kanam mandible probably belongs with the latter;

(6) that by the end of Lower Pleistocene times and in the Middle Pleistocene, we had in Africa (a) two variable branches of the Australopithecine stock continuing, as shown by *Paranthropus* and *Australopithecus* (b) representatives of the Hominine branch such as the specimen once called *Telanthropus* and which I believe should be treated as *Homo* (c) the new Chellean skull from Olduvai which has some *Pithecanthropus* characters but many which are more truly like *Homo;*

(7) that by the end of the Middle Pleistocene *Homo sapiens* was already evolved as shown by the Kanjera fossils, but that *Atlanthropus* type also survived;

(8) that in Upper Pleistocene times the genus *Homo* was represented by two species, one which is known as *Homo rhodesiensis* and one *Homo sapiens*. Eyasi belongs with the former.

ADDITIONAL NOTES

There is a twofold objective in naming and classifying specimens. On the one hand is the need for a descriptive label so that we can make it clear to others what specimen (or group of similar and associated specimens), we are referring to. On the other hand classification labels must attempt to indicate the degree of relationship that we believe exists between two groups. This is always difficult with fossils, since to a large extent, such relationship is not simply factual but is dependent upon the state of our knowledge at any given time, as well as our personal interpretation of that knowledge.

There are a variety of the reasons, therefore, which may lead to a valid change of name in taxonomic nomenclature, but it must always be remembered that the changing of a label in no way alters the nature of the specimen concerned, but only indicates an altered interpretation of the facts in the mind of the person changing the label.

When, in 1934, I put forward the suggestion that, from the point of view of the zoologist, the fossil Hominids from Choukoutien could not be validly separated, at the generic level, from the fossil skull found in Java, a number of anatomists attacked my point of view. *Sinanthropus*, they said, differed most emphatically at generic level from *Pithecanthropus*.

My view, which subsequently became accepted by nearly all specialists, did not alter any facts about the specimens from China. It simply meant that the relationship difference between the two groups appeared to me to have been overemphasized.

Those of us who attended the 1962 conference of Burg Wartenstein felt strongly that a revision of the taxonomy of the higher Primates was overdue and we each, therefore, made our own suggestions. The simplification which was suggested in my own contribution was not accepted in full but in the end we reached a remarkable degree of overall agreement—more particularly in respect of the members of the family Hominidae. We felt that the generic name *Pithecanthropus* should be reduced to, at most, subgeneric rank and we agreed to regard the genus *Homo* as sufficient to embrace all the forms of Hominidae previously variously classified as *Sinanthropus, Atlanthropus, Pithecanthropus, Palaeanthropus* and many others. We felt that the specimens called *Telanthropus,* from South Africa, were better treated as of uncertain position, owing to its fragmentary nature.

By our action we tried to underline the basic oneness of this hominid branch of the primate stock, as well as its essential differences from the other hominid group variously called the "near men" or Australopithecines. Similarly we agreed to place the genera *Australopithecus, Paranthropus* and *Zinjanthropus,* (as well as the now abandoned *Plesianthropus*) into the single genus *Australopithecus.* We thus reduced the other names to subgeneric rank. We did this in order to show the essentially close morphological relationship of the members of this group one to the other as well as their distinctiveness from the group represented by the genus *Homo.*

When I named the East African representative of the Australopithecinae *Zinjanthropus* in 1959, I made it clear that it was morphologically a member of what we then called the sub-family Australopithecinae but that it differed in an important and significant number of characters from the South African representatives of this group.

These facts remain unaltered, but in order to simplify the taxonomy of the higher primates we agreed that the differences should be regarded as of specific, rather than of generic, significance.

It must be emphasized here that what we have done in no way precludes the possibility that hominid fossil remains may turn up, from time to time, which cannot reasonably be placed in either the genus *Homo* or the genus *Australopithecus,* as at present defined. In other words, we might find a specimen, or several specimens, which require the setting up of a third genus. Equally, it is possible that we may feel that some new find is insufficiently known to place it clearly in one or other of the two genera which we have now accepted, although it requires a name by which it can be provisionally designated.

In such a case it would be necessary either to assign it new generic rank, as a temporary measure, so as to validate its specific name, or else to place it *with an interrogation mark* into one or other of the accepted groups. Then when it became more fully known from better material an exact decision could be arrived at.

In view of the decisions which we took at the conference, and because our views will be read and studied by many whose taxonomic field lies in groups that differ greatly from the higher primates, we feel it is essential to make it clear that what we have done at this conference is to attempt to set our own house in order; that is to say the Hominidae. In so doing we have departed from the current methods of many of our colleagues in other branches of mammalian taxonomy. It must therefore be made clear that when we say, for example, that *Pithecanthropus erectus* and *Homo sapiens* are now regarded by us merely as different species of the single genus *Homo*, it does *not* mean that we believe they stand closer to each other than, let us say, the genera called *Synceros* (the African Buffalo), *Bos* (domestic cattle) and *Bison* (the American Bison), all of which are given generic rank within the family Bovidae. To take another example, when we now place *Zinjanthropus* with *Australopithecus* and *Paranthropus* into a single genus we do not suggest that the very real differences between these three types are any less than those which are regarded by many of our colleagues as of generic value in other mammalian groups; for example when they separate the European and Asiatic wild boar as *Sus* from the African bush-pig which they call *Potamochoerus*.

We hope rather that our colleagues, in other fields of mammalian taxonomy, will follow our example and will review the taxonomic status of many animal groups, with a consequent simplification which is more in keeping with the facts.

Fortunately for the members of our conference there is now adequate material of the Hominidae, as a whole, to reach the conclusions to which we came with a considerable degree of unanimity. The position, however, proved to be somewhat different when we came to discuss the other members of the Hominoidea super-family, and more particularly the fossil members.

There was little disagreement as to the need to treat the fossil and living members of the gibbon group as being distinct from all the other pongids. But while many of those at the conference felt that they deserved full family rank, this view was not unanimous and Simpson has now withdrawn from this point of view. There was also fairly wide agreement that the genera *Pongo*, *Pan* and *Gorilla* belonged to a single family which should be called the Pongidae.

Here agreement virtually ended. Chiefly because the fossil record covering most of the other higher primates is so sketchy.

The exceptions are in respect of *Oreopithecus* and *Proconsul*. There was unanimous acceptance for the view that *Oreopithecus* was a member of the Hominoidea and near unanimity for regarding it as a very aberrant side branch, which required separation at full family level—the Oreopithecidae.

I maintained (and I continue to do so) that the genus *Proconsul* is so different from the true Pongidae (as well as from the Hylobatidae) in so many characters that it should be given full family rank, if we accept full family status for the Hominidae, Oreopithecidae and Pongidae.

It seems to me that there is ample material representing the genus *Proconsul* in the form of skulls, jaws, teeth, as well as post-cranial material, to warrant such a conclusion and I believe that I am better acquainted with this *Proconsul* material than most of my colleagues, and in a better position to judge this issue.

The correct status of the balance of the many fossil higher primates which Simpson lumps together as the "Dryopithecinae" is a much more difficult problem.

Clearly there is a need for drastic revision. The status of many of the genera and species must remain in doubt until more material has been found and some will perhaps eventually need to be placed either in the Proconsulidae or in the Pongidae. I cannot, however, accept Simpson's suggestion that we should continue to use the sub-family Dryopithecinae as a sort of a dust bin in which to place all the doubtful specimens and to include in it *Dryopithecus, Proconsul, Ramapithecus, Kenyapithecus* and a whole number of other fossil genera and species. It would seem to be much more logical to accept at least as a full sub-family (or as I would prefer a full family) the Proconsulinae (or Proconsulidae). This would mean giving recognition to a very well-known group and not placing it with other much less known specimens into a somewhat meaningless category— Dryopithecinae. We would then, for the time being, treat *Dryopithecus, Ramapithecus, Sivapithecus*, as well for the moment, as *Kenyapithecus*, as being *incertae sedis*. Some, like *Ramapithecus* and *Kenyapithecus*, may very well have to be placed either in the Proconsulidae or Hominidae in due course, while others, and in particular a large part of what is known as *Dryopithecus*, will probably go into the Pongidae. My own revised classification, modified as a result of the conference, would therefore be as follows:

Super family	*Hominoidea*
Family	Pongidae
	Pan, Pongo, Gorilla
Family	Hylobatidae
	Pliopithecus, Limnopithecus,
	Hylobates
Family	Proconsulidae
	Proconsul
Family	Hominidae
	Australopithecus
	Homo
Family	Oreopithecidae
	Apidium
	Oreopithecus

Incertae sedis: Kenyapithecus, Ramapithacus, Sivapithecus, Dryopithecus, and many other little known higher primates.

I would like to add that for the moment I consider that the second type of hominid found at Olduvai Gorge and referred to from time to time as the "pre-*Zinjanthropus* type" will need to be described as a member of the genus *Homo*, *with a question mark*, followed by a specific name. I believe that is the correct procedure in order to distinguish it from *Australopithecus*, from which it seems to me to differ very greatly. By putting in a question mark it is left clear that it may be necessary to give it separate generic rank. There will be those who would prefer to put it with *Australopithecus* and to extend the definition of that genus accordingly. Personally, I am not prepared to do this.

QUANTITATIVE TAXONOMY AND HUMAN EVOLUTION

BERNARD CAMPBELL

> The application of mathematics to natural phenomena is the aim of all science, because phenomenal law should always be mathematically expressed. To this end, data used in calculations should be results of well-analysed facts, so that we may be sure that we fully know the conditions of the phenomena between which we wish to establish an equation.
>
> Claude Bernard: *An Introduction to the Study of Experimental Medicine*. 1865.

IT HAS been simple enough in the past to apply the Linnaean system of nomenclature and classification to the single fossils or small groups of fossils which have been discovered during the last 100 years. The system was designed by Linnaeus for the description of real and discrete units, as species were then considered to be, and was therefore found equally suitable for the classification of the sparsely represented fossil remains of animals.

Today, however, these "conveniently" discrete groups are not only theoretically known to be part of a continuous series in time, but as a result of their number are in fact proving to be increasingly difficult to differentiate from their ancestors and descendants. Not only is the Linnaean system therefore theoretically inappropriate, but it is also becoming difficult to apply to a succession of fossil forms.[1] It is the purpose of this chapter to discuss whether it is possible to recognize any characteristic of the evolutionary process which might be made the basis of distinct taxa such as taxonomists are used to handling, and if so, what means can be used to identify it.

In describing a lineage of evolving organisms in taxonomic language according to the accepted Linnaean method, the boundaries that must be placed between taxa may be either real or artificial. Since the position and significance of such boundaries form the framework of any taxonomic scheme, and since the recognition or creation of these boundaries is one of the main sources of controversy in taxonomy, it would appear desirable to examine what evidence must be assimilated by what method in order to arrive at a taxonomy which is least liable to con-

---

1. The current controversy about the status of *Meganthropus palaeojavanicus*, now commonly regarded as an early form of *Homo erectus* but claimed to be an Australopithecine (see Robinson, 1954), illustrates this point. Since genera are continuous, the morphological distance between the most closely related members of two sequent genera will be the same as that between the members of a single subspecies. Do we have evidence that the line between *Homo* and *Australopithecus* has been drawn in the *wrong* place?

troversy, and most nearly represents the reality of the continuous process that it describes.

The *vertical* taxonomic boundaries which delineate contemporary but morphologically dissimilar forms will be considered first.

1. SPECIFIC. All boundaries between separate lineages (between which breeding does not occur by definition) are obviously real, and the groups which they separate are obviously at least specifically distinct in the biological sense, though of course they may also be different genera or families. They can be of two kinds:

a) *Sympatric.*—Here the existence of the biological boundary can be assumed if the groups are distinct morphologically (i.e., do not form a single statistical population in at least some of their morphological characters). Two such groups living together and not interbreeding are two separate evolutionary lineages.

b) *Allopatric.*—In this situation we have no certain evidence that the two groups are different species, though it is reasonably assumed for groups morphologically very different. When the morphological difference is not much greater than that between known sympatric species, the taxonomic boundary should not be assumed to be of higher than subspecific level.

2. SUBSPECIFIC. The reality of clear subspecific vertical "boundaries" cannot be maintained either in neo- or palaeo-zoology. But the existence of different taxonomic populations, subspecies or races, can be demonstrated by a statistical analysis of the appropriate data. The existence of such populations is not in doubt, but they can only be recognized in palaeontological material if the samples are of some size. Taxonomic hypotheses involving subspecific taxa based on the evidence of a single individual are therefore bound to be very tenuous.

The recognition of vertical taxonomic boundaries at both specific and subspecific levels is discussed fully in this volume by Dobzhansky and Mayr. Detailed consideration follows therefore of the taxonomic evaluation of a single evolving lineage and its horizontally separated taxa. The problems which arise in the taxonomy of a simple lineage are universal in evolution, but particularly obvious in a group such as the Hominidae of which the evolution has probably been primarily anagenetic.

HORIZONTAL BOUNDARIES

The reality of horizontal taxonomic boundaries, which separate morphologically continuous, but chronologically distinct forms, cannot be maintained, but it should be considered whether artificial boundaries could be related to distinctive evolutionary phases. Simpson (1943, p. 176–7) has written:

In studying choroclines,[2] some are found to have almost even slope (in graphic terms) from one end to the other while some have distinct plateaus bounded by

2. Simpson's word for a cline (Huxley, 1938) used to distinguish a cline in space from a cline in time, or chronocline.

shorter steep slopes or narrow transition zones. Ideally, it is the latter phenomenon that permits the definition of well-defined and homogenous subspecies. There is considerable evidence that a similar phenomenon occurs in chronoclines on a much larger scale. The chronocline analogue of the steep transition zone in a chorocline is a relatively brief period of relatively rapid evolution. Even within an essentially continuous sequence, an acceleration of this sort provides a definite and natural boundary zone (although not a line or point). For many reasons too complex to list here, it is clear that those zones of acceleration are least likely to be represented by fossils, and are almost sure to be more poorly represented than are the periods of more even and slower evolution. Thus it happens that the inevitable and at first sight merely accidental division of vertical units by gaps in the record does probably tend to approximate a real and important sort of division in phylogeny.

This gives some reason to believe that artificial horizontal boundaries might be founded on some sort of real evolutionary event: that is, a period of fast morphological change and small population. Conversely, it would appear that a taxon would be most reasonably created to include fossil data which exists as a relatively large sample showing slow morphological change in time (see also Rensch, 1959).

From evidence published by Ford (1955) and others, it does appear that a similar situation can be recognized at the level of the evolution of populations (see also Simpson, 1953, p. 124).

From these considerations therefore we not only find some justification, but a real biological basis, for the utilization of gaps in the fossil record as horizontal taxonomic boundaries. The evidence suggests that such boundaries will in fact serve to delineate real and distinct stages in the evolution of a group represented by the successful development and stabilization of different functional complexes of characters.

Artificial boundaries. It is clear that these brief remarks on the form of evolutionary progress depend for their validity on the existence of an environment that not only changes, but changes in jumps. This situation is certainly characteristic of the Pleistocene. However, it should be pointed out that a slow steady change of environment may have occurred in earlier periods, and almost certainly has occurred in the sea, and this will not necessarily result in the sort of evolutionary jumps envisaged, but may result in steady evolutionary change. It is in instances of this sort that we are finally forced to fall back on the conscious creation of artificial boundaries such as are suggested by the 75 per cent rule for the range of the subspecies and the four standard deviations rule for the species (Amadon, 1949 and Haldane, 1949 respectively). This is, of course, an attempt to put into mathematical terms the intention that palaeo taxa should be of similar variability as taxa in neozoology. It would appear that the use of these methods is perhaps neither necessary nor justified unless no natural boundaries can be shown to exist at convenient stages in the evolution of the group. Taxonomy must carry the maximum information content about the groups which it classifies.

Recognition of the horizontal boundaries separating subspecies and all higher categories may therefore be made in the following circumstances:

1. Periods of high selection pressure resulting in apparently fast evolution, and very low variability.
2. Periods of low population density.
3. Periods closely following, or during, environmental (especially climatic) change.

Conversely the taxa may be characterized by:

1. Relatively slow apparent evolution in time, but increasing morphological diversity with evolutionary potential.
2. Increasing population density.
3. Relatively constant environmental conditions.

The following factors therefore would seem to require the attention of the taxonomist:

1. Population density.
2. Morphological variability.
3. Morphological divergence between populations.
4. Rate of evolution between populations.
5. Environmental changes.

It is my purpose in this chapter to examine the methods used in the evaluation of the third and fourth factors listed above: i.e., morphological divergence, and rate of evolution. As has been said, consideration of these two subjects is made primarily in relation to the examination of an evolving lineage (which the second of these factors implies) rather than to the morphological relationships between two such lineages.

THE SELECTION OF TAXONOMIC CHARACTERS [3]

Quantitative assessment of morphological change in populations involves the selection and isolation of taxonomic characters which can be treated in a quantitative form. Such selection involves an understanding of the relationship between structure and function, and between function and behavior.

From the point of view of the study of evolution with which we are concerned, behavior is a connecting link between the environment and the genotype. It plays a dominant role in the invasion of new ecological niches; at the same time the concept of the biological species is, in part, ethological, since it is based, not on the absence of interfertility, but on the absence of interbreeding. Environment therefore acts on the phenotype, not only directly, but through behavior,

3. This section was written as a result of discussion at the Conference.

and so patterns of behavior operate in conjunction with patterns of morphological features.

A single functional morphological complex together with its accompanying behavioral complex should therefore ideally be the object of study for students of human evolution, and it is the nature of these dual complexes that we should aim to evaluate. The significance in evolution of individual characters can only be assessed in relation to such evaluation of a double pattern.

However, all characters interact, and no character can change without correlated changes in other characters. This makes it difficult to delineate functional complexes. Change in five important complexes are recognizable in Hominid evolution: (1) the brain; (2) the size and form of the masticatory apparatus; (3) the balance of the skull on the vertebral column in relation to posture, skull proportions and head movement; (4) bipedalism; (5) manipulative functions of the hand.[4] Subdivision of the skeleton into smaller functional units than these will give increasingly distorted interpretation until the absurd point is reached where distinct functional significance is attributed to every single character.

Individual characters, however, are still essential data in the study of the taxonomy of a monophyletic line:

1. Where the functional relationships of a character are not known or not fully analyzed.

2. As representatives of functional complexes where only fragmentary remains are present: i.e., where one character is taken to indicate the existence of a particular complex.

3. As significant components of functional complexes, consciously selected to facilitate classification, or to serve as data for quantitative study.

It is with these considerations in mind that the selection of characters for quantitative evaluation must be made.

THE MEASUREMENT OF MORPHOLOGICAL DIVERGENCE

From what has just been said it is clear that any method of measuring morphological divergence based on one character can hardly be of real taxonomic value. Univariate methods (Klauber, 1940) are convenient to arrive at an arrangement of populations, but the results are entirely subjective, in so far as the choice of character is subjective, and such an analysis will not give results which reflect biological reality.

Multivariate methods which involve a consideration of a number of characters simultaneously can be divided into three kinds:

a. Coefficients of similarity.
b. Simple "distance" functions.
c. Generalized "distance" functions.

4. Examples of some of these complexes are discussed by Washburn.

a. Sneath (1957), in his studies of bacteria, has developed a simple coefficient of similarity between taxa, which was based on 105 characters (in this example) that either occur in two states (i.e. present/absent, plus/minus) or can be treated as such. Sneath estimated the affinity between his different strains of bacteria by calculating the extent to which different characters in each group occur together. Clearly, if two groups have a lot in common they are likely to be more closely related than two groups with few characters in common. Treatment of continuously variable characters by this method, however, is mathematically unsound and subjective and the method as it stands cannot therefore be satisfactorily applied to the continuous characters of Metazoans.

Another example of the calculation of the correlation of characters between species has been published by Mitchener and Sokal (1957) who recorded 122 characters of 92 species of Megachilid bees. The characters are treated as discontinuous, being divided into a small number of subjectively defined character states. Most of the characters of fossils, however (and the exceptions are few), concern one organ—the skeleton. They are therefore not only continuous, but often mathematically and functionally correlated. This method does not take into account the varied origin of this correlation, and it depends on the data from a large number of characters to give anything approaching a true picture of affinity. The central problem of taxonomy, moreover, has not been overcome, for the results are still subjective, both as a result of character selection, and of character coding (i.e., continuous characters treated as discontinuous). These methods have recently been reviewed by Sneath and Sokal (1962), under the title "Numerical Taxonomy."

Another method more suited to our problems—similar in broad principle, but different in technique—has been proposed by Kurtén (1958). In a study of the Pleistocene Hyena, Kurtén has measured and calculated 24 different "allometry coefficients" (equivalent to shape indices) of five subspecifically distinct taxa. He has then quantified the morphological relationships between them by recording those coefficients significantly different as a percentage of the total to give what he calls a "differentiation index." This appears to be a useful measure of similarity because it is based on continuously variable shape characters (far more significant than size in this context), and the function is quantified as a percentage, so that it is relative, and thus independent of absolute dimensions or dispersion parameters of an absolute nature. Though this method is based (like others, inevitably) on subjective character selection, and though the correlation between characters may somewhat distort the results, this method has much to commend it and deserves further examination in the future.

b. Simple distance measurements have been in use for many years, and were developed by Heinke (1898), Pearson (1926), Zarapkin (1934), and more recently by Clark (1952), Wanke (1953), Penrose (1954), Cain and Harrison (1958), and Sokal (1961). These measures can be divided into two types: the early type (Heinke, Pearson, Zarapkin, Penrose) and the recent type (Clark, Cain and Harrison, Sokal). The early type relates the unit of morphological divergence to the variability within one of the groups and assumes a common dispersion matrix for all groups. The recent type relates the unit of divergence to the variability between the set of groups.

The significance of this distinction is considerable. Between-group variability is not a constant but will change with the addition of a new member group to the investigation: the whole calculation must be redone when new data is added for analysis. Within-group variability, once properly established, can reasonably be assumed a constant, and data of further groups may be added using the original parameters. It appears more logical to measure the divergence between groups in relation to the divergence within groups, rather than to relate it to the divergence between the very groups that are being investigated.[5] In fact the latter method appears to be begging the question of intergroup variability but the results will be meaningful in relation to each other within the investigation, though not in any wider context.

A second point which distinguishes one of these coefficients is that none except that of Penrose take into account the correlation between characters, be it mathematical and necessary, or functional, and so the coefficients will give somewhat distorted results when used to analyze the characteristics of fossils. Penrose has introduced a correction factor based on the mean correlation coefficient between every pair of characters. The remainder are of some value in comparing the divergence between taxa, especially if not closely related, but none of them takes the problem of taxonomic affinity right out of the sphere of subjective assessment. Some of these formulae are listed in Table 1. By far the best assessment of this approach is that by Penrose (1954) who presents the most sophisticated version of them (but see Talbot and Mulhall, 1962, p. 122).

c. The Generalized Distance Statistic (D^2) was first introduced by Mahalonobis in 1930. This multivariate function, closely related to Fisher's discriminant function (Fisher, 1938) makes possible analysis of the relationship between continuous characters, discounting the correlation between them. It will give a true function of morphological distance on the basis of the characters which can be quantified between the groups under consideration. Like many of these measures of divergence already mentioned, it assumes a common dispersion matrix in the different groups—an assumption which need not, it would appear, lead to serious inaccuracies. A very similar function, A^2, has been introduced by Defrise-Gussenhovan (1955).

A further great advantage of a technique based on multivariate analysis is that after a certain point the addition of further data does not affect the accuracy

5. Cain and Harrison (1958) counter this objection by using the "maximum and minimum likely values" of each character for the calculation of a maximum and minimum function of the Mean Character Difference. Surely the maximum likely value should be replaced by some parameter of dispersion. The method has been employed by Sneath (1961); for an interesting comment see Huizinga (1962).

TABLE 1
FORMULAE OF SOME SIMPLE FUNCTIONS OF MORPHOLOGICAL DIVERGENCE

Pearson (1926)
(= Heinke 1898)

$$C.R.L. = \sum_1^m d^2/m$$

Clark (1952)

$$C.D. = \sqrt{\sum_1^m \delta^2/m}$$

Penrose (1954)

$$G.D. = \frac{\sum_1^m d^2 - R\left[\sum_1^m d\right]^2 \Big/ (1 - R + Rm)}{1 - R}$$

Cain & Harrison (1958)

$$M.C.D. = \sum_1^m \Delta/m$$

Sokal (1961)

$$C^2. = \sum_1^m \Delta^2/m$$

m = no. of characters.

R = mean correlation coefficient between characters.

d = Difference between the mean values of a character in two populations expressed in units of the standard deviation of one of those populations.

Δ = difference between the mean values in two populations expressed as a function of the variability of the character amongst the different populations.

δ = difference between the mean values in two populations expressed as a fraction of the mean value of the character amongst the different populations.

of the result, and therefore, so long as sufficient characters are used, the subjective element involved in their choice is lost. Under proper conditions, this technique is an answer to many problems: its great disadvantage is that it involves a tremendous amount of computation, but with modern electronic computers, this is no longer an overriding disadvantage. Owing to its complexity D^2 has not been used as much as it deserves to be. Examples of the use of this function include Trevor (1947), Jones and Mulhall (1949), Mahalanobis et al. (1949), Mukherjee et al. (1955), Blackith (1958, 1959, 1960), Giles (1960), and Talbot and Mulhall (1962).

This function has also been used recently to assess the morphological status of the Swanscombe skull bones in relation to modern man, Neanderthal man, and various intermediate fossil skulls (Weiner and Campbell, 1964). The Swanscombe Committee concluded (1938) that apart from the great thickness of the bones and the unusual biasterionic breadth, the Swanscombe skull bones were indistinguishable from those of modern man. This point of view has been expounded by Vallois (1958). However some palaeoanthropologists—Weidenreich (1943), Gieseler (1957), Breitinger (1955), Stewart (1959)—have believed the skull to be intermediate in form between modern man and Neanderthal man, and upon these two interpretations, two hypotheses of human evolution have been offered. To obtain an objective assessment of the morphological affinities of the skull, every measurement on the skull bones was taken that could be satisfactorily defined for comparative purposes. Eight out of twenty-five were found to be 100 per cent mathematically correlated with one or more of the others, so seventeen measurements were used, which described, quantitatively, so far as it was possible, the bones in question. The measurements concerned, among other things, the nuchal area of the occipital, and the reduction of this area has been a feature in evolution which is functionally correlated with the growth of the brain and reduction of the face and jaws, including the supra-orbital torus.

For the purpose of calcuating the distance function from modern man, 500 skulls from a natural Bronze age population excavated at Lachish in Palestine were used, and the morphological distance of the fossil skull from this *sapiens* population was calculated. Using similar data, the "distances" of certain Neanderthal and intermediate forms were also calculated. The results are shown in Table 2. The position of the Swanscombe skull varies according to the sex

TABLE 2

Skull	Sex	D^2_{10}	D^2_{17}
Skhul V	M	16.71	47.87
Swanscombe	M ⎫	39.67	51.72
Swanscombe	F ⎬	40.93	58.52
Djebel Kafzeh	M	43.56	......
Broken Hill	M	44.67	......
Steinheim	F	45.03	. ...
Skhul IX	M	45.93	......
Tabun I	F	47.91	......
La Chapelle	M	54.29	79.78
Gibaltar	F	58.47	......

D^2 values of fossil skulls from the Lachish mean, based on 10 and 17 measurements of the occipital and parietal bones.

attributed to it, which is as would be expected, since a heavy male skull is more Neanderthaloid in character than a fine-boned female skull—or to put it another way, Neanderthal man and modern man are relatively gerontomorphic, whereas *Homo sapiens* and modern woman are paedomorphic. But whatever the sex of the skull, it falls roughly between the modern group and the Neanderthal group. To Figure 1 have been added the distance figures for a control series of eighteen skulls selected at random from all the subspecies of living *H. sapiens*. These have a considerable morphological distance from the Lachish population but their mean is well clear from the Neanderthal skull (La Chapelle). The *direction* of morphological divergence is of course likely to be different amongst these skulls, and a two-dimensional representation hardly does justice to the facts. (For a discussion of this interpretation, see Weiner and Campbell, 1964.)

Similar data was used to find the comparable values of Penrose's statistic. From

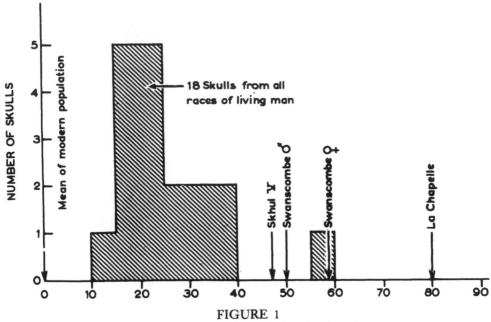

FIGURE 1

D² measurements of three fossil skulls and control series, based on seventeen measurements of the parietal and occipital bones.

Figure 2 it will be seen that the results of the two methods approximate quite closely.

A closely related technique for analyzing morphological divergence has been developed by Ashton, Healy and Lipton (1957) from Fisher's discriminant function. In an investigation into the status of *Australopithecus* these authors derived two orthogonal canonical variates from the multivariate matrix, and were able thereby to reduce a multidimensional function to a two-dimensional one which could be plotted on a simple chart. While this simplification obscures some of the information content of the multivariate function, it does make possible a very satisfying comparison of the main features of morphological divergence between the pairs of groups of which canonical variates are calculated. The technique has more recently been used by Jolicoeur (1959) on the wolf, and by Blackith (1960) on insects, and deserves further use and development.

While there is certainly room for development in techniques of this kind, it would appear that an approach along these lines is the most valuable quantitative aid that exists to evaluate the morphological relationships of evolving organisms. It is recognized that, used to give estimates of morphological divergence between contemporary groups, these techniques will mask the existence of convergent evolution, and the results must be interpreted with great care. However, this does not appear to be a problem involved in Hominid taxonomy since, as has been

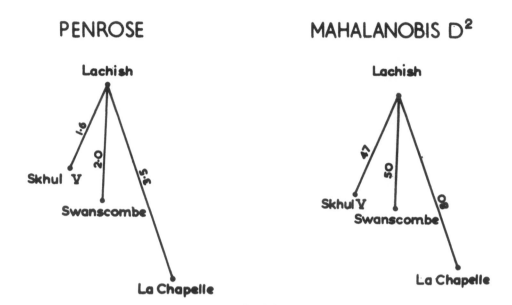

FIGURE 2

Morphological distance of three fossil skulls from the mean of a modern population based on the metrical characters of the occipital and parietal bones according to the methods of Penrose (1954) and Mahalanobis (1930).

pointed out, the evidence suggests a monophyletic descent, and convergence within a species lineage is not a factor which can seriously upset our broad taxonomic conclusions.

Blackith (1961) has recently reviewed the broad field to which multivariate techniques can be applied, and it is likely that we shall witness a great increase in the use of multivariate methods where computers are available for taxonomic research. It is not, however, suggested that quantitative techniques alone should be used for taxonomic evaluation, but that they should be used in conjunction with, and as an aid to, the established qualitative approach of classical taxonomy.

THE MEASUREMENT OF THE RATE OF EVOLUTION OF A LINEAGE

The study of the rate of evolution of animals was discussed in some detail by Simpson (1944 and 1949). I am concerned here with what he has termed "temporal morphologic rate" which is the rate of morphological change with time. Calculation of such a rate is becoming increasingly possible for Pleistocene animals in view of the new and relatively accurate dating techniques available—in particular the Potassium/Argon method.

It is necessary to define, in the first place, what is meant by the rate of evolution of a phyletic lineage. Since evolution is frequently mosaic, and is clearly so amongst the Hominidae, its measurement by means of assessing the amount of change of any single character (e.g., brain or molar size) for a given period is of doubtful value since the character may belong to a slow or fast evolving character complex.

From what has been said above (page 54), it is clear that the development of a functional complex is the most meaningful level at which to consider evolution, and it follows that this is also the most meaningful level at which to consider evolution rates. Therefore, in order to arrive at the rate of evolution of a given lineage, it is necessary first to consider the rate of evolution of different functional complexes which change most during the period under consideration.

Secondly, it has not been essential, in considerations of morphological divergence alone, to distinguish *very precisely* between divergence in time and in space. It may, however, be assumed that some creature *like* Swanscombe man was ancestral to modern Europeans, if not to Mongolians and Australians. When, however, we come to consider evolution *rates* it is necessary to be quite clear that the two samples between which the rate is calculated are assumed to be phyletically related (i.e., as ancestor-descendant). As a result of this necessary assumption, evolution rates over short periods of time cannot at present satisfactorily be measured from fossil material since it cannot be assumed that one sample is a direct descendant of another even if it comes from the same site. Population movements and hybridization between populations could both result in very high apparent rates. It would appear that for reasons of this kind, the data presented by Kurtén (1959) of very high rates of evolution of characters of quaternary and recent mammals may not be taken at its face value. In view of this, measures of the rate of evolution are unlikely at present to be of much help in distinguishing vertically consecutive *subspecies* of a group such as the Hominidae, though theoretically there is no difficulty in the analysis of subspecies in space or time if the fossil record is fairly complete.

On a larger time scale, however, the errors introduced by sampling are going to be relatively smaller and the variation of a group in space will be relatively less important. In the genus *Homo* it is generally supposed that speciation has not occurred, a view which for these purposes may be accepted (see page 65). It is therefore not unreasonable to conclude that fossils from lower levels are certainly conspecific with and not very different from those which gave rise to later forms, especially if they occur in the same geographical region. We can therefore assume, I believe, that Hominid fossils well separated in time were (with some exceptions) phyletically related in a single line of descent.

THE CALCULATION OF EVOLUTION RATES

1) *Methods based on many characters.*—In the calculation of the rate of evolution of a character complex, the most obvious way of approaching the problem

is to take a number of characters representative of the complex and derive a single measure of morphological change during a known period of time. D^2 could serve in this way. Evolution rates could then be obtained, from the example of a D^2 calculation given above (p. 58), as follows:

TABLE 3

	Approx. Time in Years	Change Factor (D^2)	Rate
Skhul V → modern man	40,000	48	12.0×10^{-4}
Swanscombe → modern man	200,000	52	2.6×10^{-4}

However, it is clear that measurements of divergence based on the dispersion matrix of one particular group, or a number of groups, will not give results that can be used outside the investigation for which they were calculated. This means that none of the measures of divergence already considered (type b and c pp. 55–56) is suitable for the measurement of evolution rates as they are calculated on the basis of an absolute parameter—within or between–group variability. For the most useful measurement of evolution rates a factor is required which is based on the relative amount of change, i.e. on percentage increase. Such a factor could be obtained from the functions of similarity (type a p. 55).

In fact, methods of this kind using more than one character have been used by Olson (1944), Westoll (1949) and Kurtén (1958). The first two plotted evolution rates on the basis of a system of scoring the development of a character complex. This is a thoroughly subjective method and will not give comparable results between different taxa, but it is of great interest within its limits (Simpson 1953, p. 22). Kurtén, however, has developed from his "differentiation index" a measure of the rate of evolution based on percentage change per thousand years.

This method of Kurtén is the best available for the assessment of evolution rates on the basis of a group of characters. The same disadvantages, however, are carried by it in this context, as were observed in the previous discussion. The selection of characters is subjective, and their inter-correlations will distort the factor of morphological change. (This disadvantage applies of course to all the other functions of similarity.) The approach based on a number of characters meets, therefore, with difficulties implicit in the calculation of factors of morphological change which cannot be overcome.

2) *Methods based on single characters.*—Most of the calculations of rates of evolution published to date (listed by Boné, 1962) have been based on the rate of change of a single character. In no instances has this character been selected explicitly as a significant abstraction from a character complex. More often characters have been selected either because figures were conveniently available

for them, or because they were evolving fast. Since in most lineages there will be found characters which do not change over long periods, though evolution has occurred, the selection of characters will affect the resultant rate which is obtained. In fact, by the appropriate use of characters, rates from the maximum down to zero could be obtained. The most significant result will be the maximum, and this can of course be calculated on the basis of one character which is found to show the maximum rate of evolution. It would therefore appear to be reasonable to calculate evolution rates on the basis of *one* character selected for its high rate of change. We have no reason to suppose that any one character per se is more significant in evolution than any other, for the significance we attach to characters is assessed *a posteriori*. Characters which change most are of greatest evolutionary significance, because evolution is characterized by change.

The idea of basing a function of the rate of evolution on percentage change in a single character was perhaps first made by Haldane (1949) who suggested the unit *darwin* to indicate an increase or decrease in a character of 1/1,000 part in 1,000 years.[6] As an example, Haldane used a single shape index to calculate the rate of evolution between Pekin man and modern man. He used the length/height ratio of the skull $\dfrac{\text{maximum length}}{\text{opisthion height}}$ which, he wrote, increased from 0.541 to 0.736 in 500,000 years. Using the formula $\dfrac{\log_e x_2 - \log_e x_1}{t}$ (where x_1 and x_2 are the two indices) Haldane obtained a rate increase of 6.2×10^{-7} which is 0.62 darwins, or 620 millidarwins. With a revised age of 400,000 years (Oakley, 1962b) the rate would be 775 md. For comparison, Haldane calculated the rates of evolution of tertiary horses (on the basis of the shape of M3) as averaging 42 md. with a maximum of 78 md. (figures revised by Simpson, 1953, to read 31 and 35 md. respectively). Dinosaurs, on the basis of the size character of body length alone, gave rates of 26 md. mean and 60 md. maximum. More recently Kurtén (1959) calculated the rate of human evolution on the basis of $\sqrt{3}$ brain volume to be as follows (but see footnote 7):

TABLE 4

	Size		Diff.	Rate
Java → Pekin	860	1075	215	1,160 md.
Pekin → Swanscombe/ Steinheim	1075	1200	125	560 md.
S/S → recent	1200	1300	100	290 md.

6. Simpson (1949, p. 210) also discusses these possibilities and with Haldane suggests that the rate might be calculated per generation. The period of generation however does not appear to be correlated with the rate of evolution of a group.

Though not fully discussed by Haldane (1949) this does raise the important question of whether size or shape characters should be used in these calculations. It is generally accepted that size characters are useful in distinguishing races and subspecies, but that shape characters are of greater value in the investigation of higher taxa. Changes in size are in fact reversible in evolution, and changes of shape are rarely so. This is well illustrated by Kurtén's (1959) Table 2, where the size of *Ursus'* LM2 is reversed six times during the Pleistocene and shows evolution rates up to 13.8 darwins! The rate over the whole period (not calculated by Kurtén) was only 0.013 darwins. Not only must we therefore discount size as a character for these purposes, but in this instance, we must as suggested, surely recognize spatial as well as temporal variation.

In conclusion, it seems that although the multi-character method proposed by Kurtén is of great interest, it is in fact a more subjective method than Haldane's in so far as the selection of characters enters into it. In using Haldane's method one index of shape alone is used. This can be selected in a relatively objective manner by the calculation of a number of rates and by the retention of the highest for the two groups under consideration.

EVOLUTION RATE OF THE HOMINIDAE

A few authors have calculated rates of evolution for the Hominidae. The size of the brain has been recognized to be the most meaningful and convenient character in this connection, though it is clearly not the only character which could be selected. As has been stated, Haldane, on the basis of a length/height index of the skull, calculated a rate of 620 md. from "Sinanthropus" to modern man, and Kurtén, a rate decreasing in time from 1,160 to 290 md. From a recalculation of Kurtén's data, however, it is clear that he has not used the formula proposed by Haldane. Whether or not this was intentional, his results cannot be termed *darwins*, nor can they be compared with results outside his own study. These comments apply with even greater force to the recent publication by Boné (1962), who has not apparently used Haldane's formula or any related to it.[7] Neither of these publications therefore can be used for comparative data in the study of hominid evolution rates.

My own calculations of the rate of increase in the $\sqrt[3]{}$ of the volume of the brain based on the formula of Haldane (p. 23) have given the following results:

7. Kurtén appears to have used the formula Rate $= \dfrac{\log_e (x_2 - x_1)}{t}$. This is not mathematically equivalent to Haldane's formula and is without mathematical meaning in this context. Data and results published by Boné (1962, pp. 85–6), do not appear to be based on either formula unless recurrent mistakes or misprints have been overlooked, though he states that he follows the methods of these two authors.

TABLE 5

Fossils	Cranial Capacity	Rate in Millidarwins
Australopithecus (1.2 m. yrs)	500	
to		
H. erectus erectus (0.5 m. yrs)	900	280
H. erectus erectus (0.5)	900	
to		
H. erectus pekinensis (0.4)	1000	351
H. erectus pekinensis (0.4)	1000	
to		
H. sapiens steinheimensis (0.15)	1325	375
H. sapiens steinheimensis (0.15)	1325	
to		
H. sapiens sapiens (0.0)	say 1375	6

For the meaning of the Latin names see p. 66. The "genealogy" implied by this scheme is no more than a rough approximation to a possible course of human evolution.

These results suggest that the development of the brain reached a maximum rate of about 375 md., and after the appearance of Steinheim man (with Swanscombe) dropped to practically zero. Clearly the development of that particular character complex, and all its related changes, was coming to an end. This does not mean of course that Hominid evolution slowed down after the full development of the brain: other character complexes were still changing, and are still evolving, in particular the reduction of the masticatory apparatus.

The discovery of relatively large brained fossil Hominids announced by Leakey (1962) from the lower Pleistocene at Olduvai, will of course alter these figures, but their effect will be to lower the calculated *rate* of evolution of the brain in the early stages, whereas the *shape* of the curve for the evolution of the brain will not be affected.

THE EVOLUTION OF THE HOMINIDAE

The purpose of this chapter has been to suggest certain techniques which might be used for the elucidation of hominid taxonomy. In view of the conclusions reached about the recognition of taxonomic boundaries, it seems worth applying them briefly to the fossil Hominidae as we know them.

Consideration of the *vertical* boundaries that can be recognized in Hominid evolution by Dobzhansky and Mayr appears later in this volume, and it seems unnecessary to add much more here. Clearly there is no justification to postulate more than one species existing at any one time in the evolution of the genus

Homo.[8] Not only is there no *direct* evidence for this (of two contemporary sympatric groups), but there is a considerable body of *indirect* evidence (see Dobzhansky, this book) that the course of evolution was anagenetic, and most workers concur with this view. The classification that follows recognizes this state of affairs.

A number of subspecies can now be recognized at different chronological levels of the evolution of *Homo.* Three quite distinct racial or subspecific groups of modern man today occupy the old world (excluding Australia)—or four according to Garn (1961). With the decreasing motility of man as we pass backwards in time, it does not therefore seem unreasonable to be prepared to recognize at least three subspecies at any one time.

More than one subspecies has been recognied at two broad time levels of the genus *Homo:*

a.   30–50,000 years BP.

Homo sapiens sapiens L. 1758. modern man extended throughout the old world during this period. (Combe Capelle, Kanjera, Niah).

H. sapiens neanderthalensis (King, 1864). Europe, N. Africa, Central Asia. (Neanderthal, Haua Fteah, Shanidar.)

H. sapiens rhodesiensis (Woodward, 1921). Central & S. Africa. (Rhodesia, Saldanha.)

H. sapiens soloensis (Oppenoorth, 1932). S.E. Asia. (Ngandong).

b.   200–300,000 years BP.

Homo sapiens steinheimensis (Berckhemer, 1936). Only one subspecific group can be recognised to include the important but scarce remains from this period. Swanscombe and Steinheim man at present represent the first group of fossils in time which can be classified as *Homo sapiens,* and they do not appear to fall into any of the subspecies which succeeded them, combining as they do, characters from more than one of these later forms. The origin of this group remains obscure, and elucidation of their position in human evolution awaits further discoveries. (See Weiner and Campbell, 1964.)

c.   400–500,000 years BP.

Homo erectus erectus (Dubois, 1892). S.E. Asia. (Java).

H. erectus pekinensis (Black & Zdansky, 1927). China. (Choukoutien).

H. erectus mauritanicus (Arambourg, 1954). N. Africa. (Ternifine, Sidi Abderrahman, Rabat).

It is not felt that the single jaw from Heidelberg, in spite of its unique and distinct morphology, can alone justify the creation of the appropriate subspecies *H. erectus heidelbergensis.* At the same time it should be pointed out that although the Ternifine remains have not been described, it seems clear that, together with other Middle Pleistocene N. African fossils, they may represent a variety of the species *Homo erectus* which can reasonably be associated as a single subspecies (Tobias, 1962). The recently discovered skull from Olduvai Bed II has not been considered in this connection since it still awaits full publication.

8. By general agreement amongst many interested anthropologists and taxonomists, the genus *Pithecanthropus* is now considered sunk and the species *P. erectus* is transferred to the genus *Homo.*

d. 0.6–2.0 million years BP.

As we pass backward into the *Australopithecus* stage of Hominid evolution, gene flow is restricted and the picture changes to what is now generally accepted to be a cladogenetic situation. Whether we accept the more recent (Kurtén, 1962) or earlier dating (Oakley, 1962) for the later Australopithecines in Africa (middle or lower Pleistocene), it seems probable that they existed contemporaneously with the early forms of *Homo erectus* in Java. Not only does it seem *a priori* probable that the extreme "Paranthropus" type of *Australopithecus* should have continued to exist in Africa while the more advanced *Homo erectus* appeared in Java, but the dates confirm this, and morphological evidence suggests that the two groups were likely to be specifically distinct, though allopatric. Whether they existed *sympatrically* as well, as Robinson has claimed, would appear to be as yet not proven, though there is some evidence of this in S. Africa ("Telanthropus" and "Paranthropus" at Swartkrans), in East Africa (*Zinjanthropus* and "Co-Zinjanthropus" at Olduvai, and in Java ("Meganthropus" and "Pithecanthropus" at Sangiran).

Amongst the genus *Australopithecus* itself we have no watertight evidence of two sympatric and contemporary groups in S. Africa. Morphological evidence suggests two species, and in fact there seems no reason at present to question a classification on this basis, though such a classification may have to be revised in the face of more detailed evidence. Though the recognition of four subspecies is accepted by many (see especially Robinson, 1954), the arrangement listed below is merely put forward as a record of the correct designation of the fossils in question should such a classification be acceptable.

It must be pointed out that the following classification does not take into account Leakey's recent discoveries in East Africa, the publication of which is awaited.

Australopithecus africanus africanus Dart, 1925. Taung.

A. africanus transvaalensis (Broom, 1936). Sterkfontein, Makapan.

A. robustus robustus (Broom, 1938). Kromdraai.

A. robustus crassidens (Broom, 1949). Swartkrans.

It cannot be too strongly emphasized that this classification is subject to revision both in so far as evidence is still lacking to justify the recognition of two different species, and in so far as the quantity of the material is not at present considered to be sufficient to justify the recognition of all these four subspecies from South Africa.

Consideration of the *horizontal* taxonomic boundaries only serves to confirm the situation which has been implied in this discussion. The fossil evidence (Fig. 3) points to four main groups:

1. Present to 100,000 years.
2. About 400,000 years.
3. About 600,000 years.
4. About 1 million years, or more.

These can be described as

1. *Homo sapiens.*
2. *Homo erectus pekinensis.*
3. *Australopithecus* later types, and *H. erectus erectus.*
4. *Australopithecus* early types.

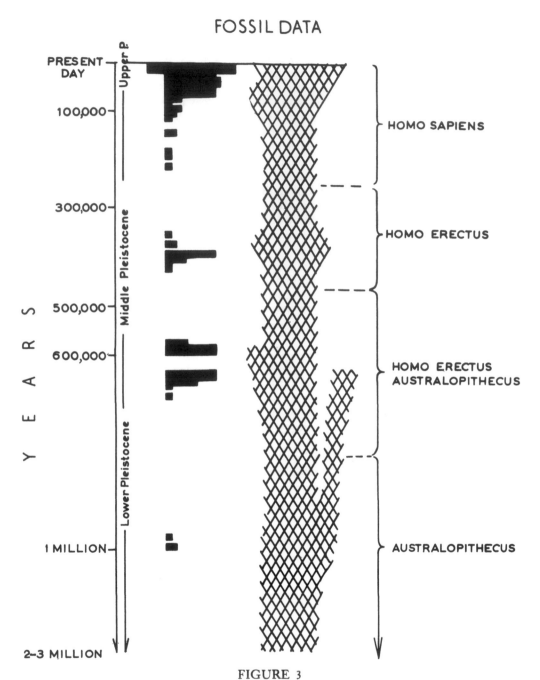

FIGURE 3

Diagram showing the frequency of fossil material during the Pleistocene. One small square corresponds to one fragmentary skull or its approximate equivalent. All dates are approximations, and those of the discoveries opposite 1 million years are quite uncertain. The "tree" is a purely diagrammatic representation of the evolution of the Hominidae as a minimal hypothesis.

The fourth group has not been mentioned in this paper and is based on the discoveries by Leakey in East Africa and Coppens at Lake Chad. The date of these finds is most uncertain, but is certainly much earlier than the S. African group of *Australopithecus*. The taxonomic status of these finds cannot be considered before their publication.

The ranks to be accorded to these chronologically separated groups are bound to be a matter of opinion, but the arrangement presented in this chapter accords well with the views of most anthropologists and taxonomists.

CLASSIFICATION OF THE HOMINIDAE [9]

The following classifi ation summarizes the conclusions reached in this discussion, and represents a statement of the situation which is broadly accepted by many workers at the present time.

Family: Hominidae Gray, 1825. Worldwide.
Genus: *Australopithecus* Dart, 1925. Africa.
Species: *A. africanus* Dart, 1925. (includes *Plesianthropus* Broom, 1936.) S. Africa.
               A. robustus (Broom, 1938). (includes *Paranthropus* Broom, 1938, *P. crassidens* Broom, 1949.) S. Africa.
Genus: *Homo*. L. 1758. Worldwide.
Species: *H. erectus* (Dubois, 1892). (includes *Pithecanthropus* Dubois, 1894, *Sinanthropus* Black & Zdansky, 1927.)
Subspecies: *H. erectus erectus* (Dubois, 1892). S.E. Asia.
                   H. erectus pekinensis (Black & Zdansky, 1927). China.
Species: *H. sapiens* L. 1758. (includes *Javanthropus* Oppenoorth, 1932, *Cyphanthropus* Woodward, 1921, *Protanthropus* Bonarelli, 1909, and many species of *Homo*.) Worldwide.
Subspecies: *H. sapiens steinheimensis* (Berckhemer, 1936). Europe.
                   H. sapiens neanderthalensis (King, 1864). Europe, Asia.
                   H. sapiens soloensis (Oppenoorth, 1932). S.E. Asia.
                   H. sapiens rhodesiensis (Woodward, 1921). Africa.
                   H. sapiens sapiens L. 1758, and other living subspecies.

The following taxa and the fossils upon which they are based have not been considered in the preparation of the above classification, since they have not yet been fully published:

Atlanthropus mauritanicus Arambourg, 1954.
Zinjanthropus boisei Leakey, 1959.

Apart from these two taxa, at least 105 different species have been created in this family, of which 63 are invalidated by the rules of nomenclature. The crea-

9. Prepared as a result of discussion at Burg Wartenstein.

tion of the majority of the remainder (38 out of 42) is considered unjustified.[10] Besides the genera mentioned in the classification as being sunk, the following well-known taxa fall into this group:

Africanthropus helmei Dreyer, 1935.
Homo heidelbergensis Schoetensack, 1908.
Meganthropus africanus Weinert, 1950.
Meganthropus palaeojavanicus von Koenigswald, 1945.
Telanthropus capensis Broom & Robinson, 1949.

The material designated by these binomials is included in one or other of the taxa listed above, but the names are available for use should this become justified on the discovery of further data.

REFERENCES TO AUTHORS OF LATIN NAMES

Arambourg, 1954. *C. R. Acad. Sci. Paris*, 239, 893–5.

Berckhemer, 1936. *Forsch. Fortsch.* 12, 349–50.

Black & Zdansky, 1927. *Paleont. sin.* Ser D. 7, 1–24.

Bonarelli, 1909. *Boll. Soc. Geogr. Ital.* 8, 827–51; 9, 953–79.

Broom, 1936. *Nature, Lond.* 138, 486–8.

———1938. *Nature, Lond.* 142, 377–9.

———1949. *Nature, Lond.* 163, 57.

Broom & Robinson, 1949. *Nature, Lond.* 164, 322–3.

Dart, 1925. *Nature, Lond.* 115, 195–9.

Dreyer, 1935. *K. Acad. Sci. Amsterdam*, 38, 1, 119–28.

Dubois, 1892, *Verslag. Mijnwesen*, 3, Batavia.

Dubois, 1894. *Pithecanthropus erectus, eine menschenähnliche Uebergangsform aus Java*. Batavia Landes-Druckerei.

Gray, 1825, *Ann. Philos.* 10, 337–44.

King, 1864. *Quart J. Sci.* 1, 96.

von Koenigswald, 1945. in Weidenreich F. 1945. *Anthrop. Pap. Amer. Mus.* 40, 1–134.

Leakey, 1959. *Nature, Lond.* 184, 491–3.

Linnaeus, 1758. *Systema Naturae*, 10th Edn. Stockholm.

Oppenoorth, 1932. *Wet. Meded. Dienst. Mijnb. Ned-Ind.* 20, 49–63.

Schoetensack, 1908. *Der Unterkiefer des Homo heidelbergensis*. Leipsig.

Weinert, 1950. *Z. Morph. Anthrop.* 42, 113–44.

Woodward, 1921. *Nature, Lond.* 108, 371–2.

10. This opinion is that of the author.

BIBLIOGRAPHY

ASHTON, E. H. *et al.*
1957. "The descriptive use of discriminant functions in physical anthropology." *Proc. Roy. Soc. B.*, 146:552–72.

BLACKITH, R. E.
1958. "An analysis of polymorphism in social wasps." *Insectes Sociaux*, 5:263–72.
1959. "Morphometric differences between the eye-stripe polymorphs of the Red Locust." *Sci. J. Roy. Coll. Sci.*, 27:13–27.
1960. "A synthesis of multivariate techniques to distinguish patterns of growth in grasshoppers." *Biometrics*, 16:28–40.
1961. Multivariate statistical methods in human biology. *Medical Documentation*, 5:26–28.

BONÉ, E. L.
1962. "Rhythmes Evolutifs Comparés des Hominidés et des Mammifères Pléistocènes." *Bibl. primat.*, 1:71–92.

BREITINGER, E.
1955. "Das Schadelfragment von Swanscombe und das 'Praesapiensproblem.'" *Mitt. Anthrop. Ges. Wien*, 84–5:1–45.

CAIN, A. J. and G. A. HARRISON
1958. "An analysis of the taxonomist's judgement of affinity." *Proc. Zool. Soc. Lond.*, 131:85–98.
1960. "Phyletic weighting." *Proc. Zool. Soc. Lond.*, 135:1–31.

CLARK, P. J.
1952. "An extension of the coefficient of divergence for use with multiple characters." *Copeia*, 2:61–4.

DEFRISE-GUSSENHOVEN, E.
1955. "Mesure de divergence A^2." *Inst. Roy. Sci. Nat. Belgique*, 31:56.

FISHER, R. A.
1938. *Statistical Methods for Research Workers.* 7th Ed. Edinburgh: Edinburgh Univ. Press.

FORD, E. B.
1955. "Rapid evolution and the conditions which make it possible." *Cold Spring Harbor Symposia on Quantitative Biology*, 20:230–8.

GARN, S. M.
1961. *Human Races.* Springfield, Ill.: Thomas.

GIESELER, W.
1957. "Die Fossilgeschichte des Menschen." In Heberer, G., *Die Evolution der Organismen.* 2nd Ed. Stuttgart.

GILES, E.
1960. "Multivariate analysis of Pleistocene and recent Coyotes from California." *Univ. California Publ. Geol. Sci.*, 36:369–90.

HALDANE, J. B. S.
1949. "Suggestions as to quantitative measurement of rates of evolution." *Evolution*, 3:51–6.

HEINKE, F.
1898. *Naturgeschichte des Herings*. Berlin: Salle.

HUIZINGA, J.
1962. "From DD to D² and back: the quantitative expression of resemblance." *Proc. Kon. Ned. Akad. Wet. Amsterdam*, Ser. C, 65: 380–91.

HUXLEY, J.
1938. Clines: an auxiliary taxonomic principle. *Nature, Lond.*, 142:219–20.

JAHN, T. L.
1961. "Man versus machine: a future problem in Protozoan taxonomy." *Syst. Zool.*, 10:179–92.

JOLICOEUR, P.
1959. "Multivariate geographical variation in the Wolf *Canis lupus* L." *Evolution*, 13:283–99.

JONES, G. I. and H. MULHALL
1949. "The physical type of certain peoples in S.E. Nigeria." *J. Roy. Anthrop. Inst.*, 79:11–19.

KLAUBER, L. M.
1940. "Two new subspecies of Rhyllorhynchus." *San Diego Soc. Nat. Hist.*, 9:195–214.

KURTEN, B.
1958. "A differentiation index and a new measure of evolution rates." *Evolution*, 12:146–57.
1959. "Rates of evolution in fossil Mammals." *Cold Spring Harbor Symposia on Quantitative Biology*, 24:205–15.
1962. "The relative ages of the Australopithecines of Transvaal and the Pithecanthropines of Java." In Kurth, G. 1962, pp. 74–80.

KURTH, G.
1962. *Evolution und Hominisation*. Stuttgart: Fisher.

LEAKEY, L. S. B.
1961. "New finds at Olduvai gorge." *Nature, Lond.*, 189:649–50.

MAHALANOBIS, P.
1930. "On tests and measures of group divergence." *J. Asiatic Soc. Bengal*, 26:541–88.

MAHALANOBIS, P., et al.
1949. "Anthropometric Survey of the United Provinces." *Sankhya*, 9:89–324.

MITCHENER, C. D. and R. R. SOKAL
1957. "A quantitative approach to a problem in classification." *Evolution*, 11:130–62.

MUKHERJEE, R. et al.
1955. *The ancient inhabitants of Jebel Moya*. Cambridge: Cambridge Univ. Press.

OAKLEY, K. P.
1962a. "The earliest toolmakers." In Kurth, G. 1962, pp. 157–69.
1962b. "The emergence of man." *Adv. Sci.*, 18:415–26.

OLSON, E. C.
1944. Origin of mammals based upon cranial morphology of the Therapsid sub-orders. *Geol. Soc. Amer. Special paper*, 55.

PEARSON, K.
1926. "On the coefficient of racial likeness." *Biometrika*, 18:105.

PENROSE, L. S.
1954. "Distance, size and shape." *Ann. Eug.*, 18:337–43.

RENSCH, B.
1959. *Evolution above the species level.* London: Methuen.

ROBINSON, J. T.
1954. "Taxonomy of the Australopithecinae." *Amer. J. Phys. Anthrop.*, 12:181–200.

SIMPSON, G. G.
1943. "Criteria for genera, species and subspecies in Zoology and Palaeozoology." *Ann. New York Acad. Sci.*, 44:145–78.
1944. *Tempo and mode in evolution.* New York: Columbia Univ. Press.
1949. "Rates of evolution in animals." in Jepsen, G. L. *et al.* 1949. *Genetics, Palae-ontology and Evolution.* Princeton: Princeton Univ. Press. 205–28.
1953. *The major features of evolution.* New York: Columbia Univ. Press.

SNEATH, P. H. A.
1957. "Some thoughts on bacterial classification" and "The application of computers to taxonomy." *J. Gen. Microbiol.*, 17:184–226.
1961. "Recent developments in theoretical and quantitative taxonomy." *Syst. Zool.*, 10:118–39.

SNEATH, P. H. A. and R. R. SOKAL
"Numerical taxonomy." *Nature, Lond.*, 193:855–60.

SOKAL, R. R.
"Distance as a measure of taxonomic similarity." *Syst. Zool.*, 10:70–9.

STEWART, T. D.
1959. "Indirect evidence of the primitiveness of the Swanscombe skull." *Amer. J. Phys. Anthrop.*, 18:363.

SWANSCOMBE COMMITTEE.
1938. "Report on the Swanscombe skull." *J. Roy. Anthrop. Inst.*, 68:17–98.

TALBOT, P. A. and H. MULHALL
1962. *The physical anthropology of Southern Nigeria.* Cambridge: Cambridge Univ. Press.

TOBIAS, P.
1962. "Early members of the genus *Homo* in Africa." In Kurth, G. 1962, Pp. 191–204.

TREVOR, J. C.
1947. "The physical characters of the Sandawe." *J. Roy. Anthrop. Inst.* 77:61–78.

VALLOIS, H. V.
1958. "La Grotte de Fontéchevade." *Arch. Inst. Paléont. Hum. Mém.*, 29.

WANKE, A.
1953. "Metoda badan czestosci wystepowania zespolow cech." *Przeglad Antrop.*, 19:106–47.

WEIDENREICH, F.
1943. "The skull of *Sinanthropus pekinensis*." *Palaeont. Sin.*, 127:(N.S. D.10).

WEINER, J. S. and B. G. CAMPBELL
   1964. "The taxonomic status of the Swanscombe skull." In Ovey, C. D. 1964. *Swans-combe—A Survey of Research on a Unique Pleistocene Site*. London: Royal Anthro-pological Institute.

WESTOLL, T. S.
   1949. On the evolution of the Dipnoi. In Jepsen, G. L. *et al*. 1949. *Genetics, Palaeontology and Evolution*. Princeton: Princeton Univ. Press. Pp. 121–84.

ZARAPKIN, S. R.
   1934. "Zur phaenoanalyse von geographischer Rassen und Arten." *Arch. Naturgesch.*, 3:161–86.

I should like to record my thanks for the stimulating and valuable discussions of these topics which we enjoyed at Burg Wartenstein. In particular I am most grateful to Professor Dobzhansky and Dr. Ernst Mayr who have given me much help and encouragement.

SOME CONSIDERATIONS IN THE FORMULATION

OF THEORIES OF HUMAN PHYLOGENY

G. A. HARRISON and J. S. WEINER

INTRODUCTION

THERE ARE SIGNS that closer attention is being paid by physical anthropologists to the principles of evolutionary systematics. In retrospect it would be appropriate to date this increased awareness from the appearance of the notable essays by Colbert (1949) and G. G. Simpson (1953) in which were adumbrated a number of fundamental phylogenetic principles for the specific guidance of the physical anthropologist. A further step was taken by Le Gros Clark (1955) in his discussion of "morphological and phylogenetic problems of taxonomy in relation to hominid evolution."

Two events in particular have had a special effect in fostering a more critical attitude. The dispute over the hominid versus pongid status of the australopithecines raised acutely (and still does) many taxonomic issues, and has had one beneficial outcome at least—the publicizing of the uses of multivariate methods in the analysis of affinity (Bronowski and Long, 1952). The other event was the Piltdown exposure (Weiner, Oakley and Clark, 1953) which has forced human evolutionists (not all!) to take a rather more cautious view of the weight of interpretation and speculation that can be placed on fossil remains of a fragmentary nature or of disputed chronology. The removal of so extreme and peculiar a form as Piltdown also made anthropologists realize that the phylogeny of the higher hominids involves relatively small and perhaps subtle morphological differences and therefore that a more rigorous and less impressionistic approach to their interpretation is necessary.

It seems opportune to bring together, even in somewhat cursory fashion, those principles and propositions which offer guidance in human taxonomy and phylogeny and which the paleoanthropologist ignores at his peril.

THE NATURE OF PHYLETIC AFFINITY

In the first place it is necessary to be quite clear about what is meant by the term phyletic affinity (or evolutionary relationship). In the past, there has been much confusion because it has not been realized explicitly that the term is used

75

for two quite distinct concepts. Firstly, there is relationship based upon that similarity between forms which can be attributed to a common ancestor. The more similar two forms are in their ancestral characters the more closely they are related. Alternatively, relationship is expressed in the genealogical sense of the number of generations which separate forms from their common ancestor: the more recently two forms had a common ancestor, the more closely they are related. The first type of relationship has been called "patristic affinity," the second "cladistic affinity" (Cain and Harrison, 1960). If evolution solely involved divergence of forms at a constant rate, then both types of evolutionary relationship would be the same; but it has long been recognized that rates of evolutionary change vary enormously, and two very different forms may well have had a recent common ancestor, whilst the common ancestor of two very similar forms may be remote, even in the absence of convergence.

Much of the controversy that has raged over the evolutionary affinity of the australopithecines, for example, would seem to have arisen from failures to distinguish between patristic and cladistic affinity. Let it be assumed first that *Australopithecus* represents an early stage on the phylal lineage (or clade) leading to *Homo* and secondly that the hominid lineage has evolved more rapidly than the pongid lineage since their division. If the latter not unreasonable condition is satisfied, it is more than likely that the patristic affinity of *Australopithecus* is nearer the present day Great Apes than *Homo*, although "by assumption" its cladistic affinity is with the latter genus.

It is important to appreciate here that while paleoanthropologists are mainly concerned with ascertaining cladistic affinity, most of the morphological evidence presented by a fossil is only relevant to establishing patristic affinity. Clades can only be certainly recognized when they are well represented by a sequence of forms showing, from one to another, a high degree of patristic affinity. If the evolutionary change along a clade is very rapid, then it is obviously necessary to have a much better fossil record for recognizing the lineage than if the rate of change is slow.

ESTABLISHING PHYLOGENIES

The basic problem confronting the phylogenist is one of evaluating the evolutionary changes that are most likely to have occurred. In other words he attempts to establish which of a variety of possible relationships between a group of forms is most consistent with the known pattern of evolution. The nature of this pattern is evidenced by present day laboratory and field studies, but is most conclusively demonstrated by groups of organisms whose fossil record is so good that there can be no reasonable doubt about their phylogenetic relationships.

The most fundamental quality about evolution that emerges is that change is gradual. If this were not so, it would be quite impossible to determine either patristic or cladistic affinities. As it is, even with a poor fossil record certain of the theoretically possible schemes of relationship are excluded since they would involve evolutionary "jumps." Not even the most bizarre anthropologist considers that the gorilla was transformed into man overnight! However, after one has excluded such absurdities, invariably there remain a number of possible alternatives and the following propositions should help in deciding which is the most likely.

1. THERE SHOULD ALWAYS BE AN ECONOMY OF HYPOTHESIS

One of the basic guiding principles in formulating a phyletic hypothesis is that "plurality is never to be posited without need." In other words, all the reliable taxonomic material available, and its phenetic affinities, should, if possible, be embraced by a single coherent scheme. Ancillary hypotheses, as far as possible, are to be avoided and on the whole, the fewer the assumptions made the less likely is error to creep in. In fact, Occam's principle applies as much to phyletic reasoning as it does to other scientific methodologies. The fraudulent nature of the Piltdown remains was first suggested by the fact that, however hard one tried, they could not be fitted into any unitary evolutionary scheme which would fit the rest of the hominids.

It may be mentioned here that the economy principle should be extended to the activity of naming new fossil finds. Economy would be justified solely on the basis that, since current taxonomic practice deals in discrete categories, it must tend to misrepresent the continuum of phylal forms that evolution produces. But the activity is more insidious than this, since it can generate the concept of a complicated phylogeny which itself then necessitates the further naming of new discoveries, and so on. Rather than filling gaps which exist, naming tends to produce gaps that do not exist. It seems that some human paleontologists regard the binomial system as a means of giving every hominid a Christian and surname and for creating phylogenetic schemes like family trees: an activity which starts with the nonsensical premise that the individual and not the population is the unit of evolutionary change.

2. INCOMPLETE FOSSIL MATERIAL MUST NOT BE USED AS PRIMARY OR SELF-SUFFICIENT EVIDENCE

Any phyletic hypothesis must be based primarily on well authenticated, reliably dated and fairly complete fossils alone. The probable or possible status of disputed fossils should then be assessed in the light of such an hypothesis. If the specimen (or our evaluation of it) is consistent with the hypothesis, the interpretation which best fits the hypothesis would appear the most reasonable to adopt. If the specimen does not conform to the hypothesis (and the evidence it presents is reliable) it may be possible to modify the hypothesis; but if it does not conform, and it is not possible to revise the hypothesis, the hypothesis would then seem to emphasize still more the uncertainty of the specimen in question (Weiner, 1958).

3. A Character Whose Function Is Known Can Give More Reliable Phyletic Information than One Whose Function Is Not Known

Since adaptation to an environment is achieved by changes in the way an organism functions, the phenotypic units on which natural selection directly acts are obviously units of function. It seems reasonable therefore that one will get more phyletic information from such units than from morphological differences which have been abstracted arbitrarily. In mammals, at least, the functional significance of most structures is appreciated to some extent, and workers do on the whole make their comparisons with reference to function. Thus jaws, brain, orbits, nasal chamber, etc. are typically recognized either explicitly or implicitly as units for comparison. However, it is very easy to isolate some varying attribute in a series of forms, which because it does not contain the maximum functional information leads one badly awry in determining phyletic relationships. To take a particular example, the presence of a sagittal crest is a very striking morphological character and at first sight it would seem reasonable to conclude that, so far as this character is concerned, forms with a crest were more closely related patristically than forms without one. However in this case the units on which selection is operating are the size of the temporalis musculature and the size of the braincase which is available for its attachment. Two forms without a crest may, therefore, be phyletically more remote from one another than one of them from a slightly crested form, since in one case the supra-temporal lines may almost but not quite meet, and in the other the temporalis may be small and have a very low origin on the braincase.

A knowledge of function also enters into the making of phyletic judgments, by indicating which characters are necessarily correlated (Cain and Harrison 1960). Such necessary correlation may exist between quite discrete structures. Thus, for instance, the occlusal surfaces of teeth which function together just cannot vary independently. The shape of the anterior and posterior surfaces of the ape upper canine tooth would seem to be completely specified by the shape of the posterior surface of the lower canine and the anterior surface of the first lower premolar respectively. Likewise much if not all of the upper molar morphology is determined by the lower molar morphology and vice versa. It is evident that such necessary correlates can give no independent phyletic information, since if one character changes the other must also change in a specified way to maintain the functional integrity of the system.

As will be indicated later, function enters into many other phyletic judgments but it does not follow that because variation in two or more characters is correlated, that this correlation is functionally necessary. Changes in many different systems, produced by a change of ecology for example, will appear to be correlated, but these are not necessary in the sense that the organism can only survive if they come together. Hominid evolution is particularly characterized by changes in the limbs, jaws and brain, but there is no direct functional determination of one by the other, and to some extent, at least, the systems can and do evolve separately. Non-necessarily correlated characters can be used independently for giving phyletic information.

It may be mentioned here that variation in a character due to environmental modification of development must obviously not be used in determining any form of evolutionary relationship. In the past it has usually been automatically assumed that environmental lability is so unimportant in determining at least the skeletal morphology of mammals that it could be neglected. Recent experimental work, however, has suggested that such an assumption is probably very wrong (Washburn, 1951; Harrison, 1960).

4. Judgments of Phyletic Affinity Should Not Be Based on Single Characters But on Total Morphological Pattern

Whether the function of particular structures is known or not, the phyletic status of fossil remains must, as Le Gros Clark has cogently argued, be based not on the comparison of individual characters considered in isolation, but on a consideration of the total pattern which they represent. Admittedly the precision gained by introducing into phyletic consideration all of the non-necessary variations between forms is usually not worth the trouble involved, but since characters can and do vary independently, single character analyses can lead one into absurdity. One of the most significant recent advances in taxonomic practice has been the devising of statistical methods to express variation in the total morphological pattern in some single objective meaningful way (Bronowski and Long, 1953; Ashton, Healey and Lipton, 1957; Sneath, 1957; Michener and Sokal 1957; Cain and Harrison, 1958).

5. The Recognition of Functional and Morphological Trends Facilitates the Recognition of Phyletic Lineages

One of the most striking features about the evolution of groups with good fossil records is that in them marked morphological and functional trends can be discerned. Such trends are the consequence of interspecific competition ever limiting the ecological niche available, and intraspecific competition demanding ever more efficient exploitation of that niche.

Within the hominids it is possible to arrange forms so that they show progressive trends in perfection of the upright posture, reduction in jaw size and enlargement of the brain. Other trends (such as reduction of the supra-orbital torus and occipital crest) though probably dependent upon these primary ones, can also be clearly discerned. In general, it seems probable that a trend will represent a single lineage, or, at least, a group of closely related lineages, and phyletic schemes which cut across trends are likely to be incorrect. It often happens that the apparent position of a fossil in a trend does not fit with its geological age. In this case it is possible that the form represents some unknown lineage showing the same trend but evolving at a different rate. On the other hand, if but a single specimen, it could equally well represent some extreme variant of the expected population. The molar teeth of Heidelberg man differ little in size from those of modern Australian aborigines although there is a marked trend to reduce molar size in hominids. It also looks as though Swanscombe man is more advanced on the morphological trend than one would expect from its dating. However, the position of neither of these fossils is really anomalous when one considers the magnitude of variation that exists in single present day human populations.

The existence of trends, as Le Gros Clark has recognized, also helps in connecting up phylal lineages which are severely interrupted by gaps in the fossil record. An early group of forms is probably related to a late group if it shows the beginning of a trend amplified in the latter. If it is confirmed that *Oreopithecus* was incipiently bipedal it would seem very reasonable to consider it an early ancestor of the hominids.

6. PHYLETIC SCHEMES SHOULD INVOLVE THE MINIMUM OF "GAPS"

Even when no clear cut trends occur, the fact that evolution is a gradual process means that morphological continuity exists between all forms in a lineage. It follows from the economy of hypothesis proposition that only absolutely necessary gaps should enter into judgments of phyletic relationship. In actual fact there seem to be few gaps in the broad outline of human phylogeny. The so-called "Meganthropus" acts as a link between *Australopithecus* and *Pithecanthropus* [1] (Robinson, 1953). The latter genus, as represented by Peking man, is clearly connected morphologically with classical Neanderthal man through Solo man, and there can be little doubt that the progressive variety of Neanderthal man gave rise to present day Homo sapiens. Admittedly the ancestries of progressive Neanderthal man and of Rhodesian man are obscure, but it would be absurd to postulate some completely unknown and independent ancestry for these forms. One can feel certain that their relationship to already known forms will ultimately be established.

7. PROPOSED PHYLETIC SCHEMES WHICH INVOLVE REVERSIBLE CHANGE SHOULD BE PARTICULARLY CRITICALLY EXAMINED

It is probably true that if the appropriate environmental demand is present any evolutionary change, whether it involves reversibility or not, is possible. Never-

1. The view of the Congress was that the forms included in *Pithecanthropus* should now be referred to as *Homo erectus* and we are prepared to accord with this judgment.

theless it is, to say the least, rare for characters once lost to reappear in exactly the same form in an evolutionary lineage. The reason usually advanced for this so-called irreversibility of evolution is the improbability of building up the same gene complex on two separate occasions. Actually this argument would seem to be irrelevant since the same phenotype can have many different genetic bases, and as has been repeatedly emphasized, it is on the phenotype that selection directly operates. It is also to be remembered that even when some character remains unchanged in evolution, its genetic basis is almost certainly changing. The evolution which is occurring in other characters practically necessitates this.

The reason why evolutionary change is typically irreversible is that it is highly unlikely that exactly the same selective forces will operate on two or more separate occasions. The very fact that evolution has occurred in other organisms between these occasions means that the overall environmental demand must be different. Further, even if some specific selecting force does reappear from time to time, it can rarely produce the same character on each occasion, unless there is no other change in the phenotype. The form of the variation offered for selection by a character is at least in part determined by the variation in some other characters. Nevertheless, it is to be expected that characters which are related to a specific physical environmental factor, and are not profoundly dependent on other characters, can display reversible change, e.g., pigmentation.

Although in general terms one cannot exclude the possibility of evolutionary reversibility, there would seem to be one set of circumstances in which it is quite impossible. This is the case where an adaptation to some particular demand turns out to be a general improvement, i.e., a character which facilitates survival in many different environments. Such characters cannot be lost in any of these environments. This proposition has particular relevance to the phylogenetic position of classical Neanderthal man. It is possible that the differences between this form and modern man represent quite particular adaptations to different environments, but it seems much more likely that the distinctive characters of modern man are truly general improvements. It is, therefore, very significant that the so-called progressive Neanderthalers show many of these features, for if they are general improvements it would mean that a progressive Neanderthal ancestry for the classical Neanderthalers could not be entertained, despite the apparent paleontological sequential relationship between the two groups (Weiner, 1960).

It may be mentioned here that the concept of "specialization" has frequently been invoked to exclude some fossil hominid from the ancestry of later forms. Although the term is often used merely to indicate some rather striking distinctive feature, in the above sense a "specialization" is a character, or group of characters, which cannot be lost and, therefore, restricts the evolutionary opportunities of its possessor. Unfortunately "specializations" can only be certainly recognized *post hoc*. If a lineage becomes extinct then it may fairly be said that it was too specialized to meet the changing environment; but excepting the case of the general improvement, if one makes some subjective judgment about whether or not a character can be lost, one is almost certainly bound to be wrong. Even the so-called general improvement may not be advantageous in every environment, and the evolutionary record itself offers innumerable examples of how forms have escaped from apparent specializations. As Le Gros Clark has argued it is nonsense to exclude *Australopithecus robustus* from the main hominid lineage on the grounds of its sagittal crest, or to exclude *Proconsul* from the position of common ancestor to the later pongids and hominids because it has large canine teeth. It is doubtful if under any circumstances these characters could be regarded as specializations.

8. If a Character Difference between Two Forms Can Be Shown To Represent Different Ways of Meeting the Same Environmental Demand, the Two Forms Concerned Are Less Likely To Be Cladistically Related than if the Character Difference Is Due to Different Environmental Demands

Whilst this proposition would seem to give very reliable phyletic information, in practice it is almost impossible to determine whether a particular structural difference does represent alternative adaptive responses. If both linearity of physique, and small body size represent more or less equivalent ways of adapting to high temperatures and are concerned solely in temperature regulation, it does seem unlikely that a population which has become adapted to the tropics by becoming tall and linear will suddenly change its mode of adaptation by becoming a population of pygmies. But if other factors start to favor small body size then certainly this will happen. The only circumstance in which the proposition is unequivocal is when gradual change from one form to another would necessitate passing through less adaptive states than the existing one, i.e., if evolution from the tall linear physique to the pygmy condition necessarily involved, through a loss of heat tolerance, an overall loss of fitness.

9. Forms Which Occur in Similar Ecological Niches May Well Be Convergent

Since the environment of organisms operating through natural selection determines the phenotype, one would expect forms with the same ecology to become convergent or evolve in parallel without respect to their ancestry. If the ancestors of the convergent forms were not dissimilar or judgments are made on fragmentry material, it is quite possible for such convergence to go unrecognized. It seems just as likely that the peoples of Africa and Melanesia are convergent as that they had a negriform common ancestor, and it is possible that Solo and classical Neanderthal man evolved in parallel. However, as Simpson makes quite explicit "closeness of parellelism tends to be proportioned to closeness of affinity."

Forms are only likely to evolve in parallel if they can seize the same evolutionary opportunities, and this ability they typically inherit from a common ancestor. One, therefore, probably need not be seriously concerned with the possibilities of convergence, at least within such groups as the hominids, but it is worth noting that taking out from affinity tables all necessary correlates will greatly reduce the chance of error from convergence.

10. Forms Which Occur in Different Ecological Niches May Well Be Closely Related Cladistically although They Differ Greatly in Morphology

If all the differences between a group of forms can be seen to be specific adaptations to the different environments inhabited, then it is to be expected that they arose rapidly. The power of natural selection is greatest when forms change their ecology and the early differentiation of the hominids as terrestial bipeds must have been extremely rapid. One important consequence of this is that the number of individuals that existed in the main transformation phase must be small and it is therefore not surprising that one of the most critical phases of human evolution is so poorly represented in the fossil record.

One surprising fact, emerging from these ten propositions, is the relative unimportance of genetic information. This is not merely because such information is inevitably unavailable, but rather because the speed and direction of evolution is determined far more by the nature of the environment than by the supply of new variation. Nevertheless the genetic system imposes some inertia on rates of evolutionary change and whatever the force of natural selection, *Australopithecus* could not have evolved into *Homo sapiens* in a hundred generations.

BIBLIOGRAPHY

Ashton, E. H., M. J. R. Healy and L. Lipton
  1957. "The descriptive uses of discriminant functions in physical anthropology"
  Proc. Roy. Soc., B 146:552.
Bronowski, J. and W. M. Long
  1952. "Statistics of discrimination in anthropology." *Amer. J. Phys. Anth.*, 10:385.
Cain, A. J. and G. A. Harrison
  1958. "An analysis of the taxonomist's judgment of affinity." *Proc. Zool. Soc. Lond.*, 131:85.
  1960. "Phyletic weighting." *Proc. Zool. Soc. London.*, 135:1.

84 CLASSIFICATION AND HUMAN EVOLUTION

COLBERT, E. H.
1949. "Some palaeontological principles significant in human evolution" in *Early man in the Far East*. Amer. Assoc. of Phys. Anth.

HARRISON, G. A.
1960. "Environmental modification of mammalian morphology." *Man*, 60:3.

LE GROS CLARK, W. E.
1955. *The fossil evidence for human evolution*. Chicago: Univ. of Chicago Press.

MICHENER, C. D. and R. R. SOKAL
1957. "A quantitative approach to a problem in classification." *Evolution*, 11:130.

ROBINSON, J. T.
1953. "Meganthropus, Australopithecus and Hominids." *Amer. J. Phys. Anth.*, 11:1.

SIMPSON, G. G.
1953. "Some principles of historical biology bearing on human origins." *Cold Spring Harbor Symp.*, 15:55.

SNEATH, P. H. A.
"Some thoughts on bacterial classification." *J. gen. Microbiol.* 17:184.

WASHBURN, S. L.
1951. "The new physical anthropology." *N. Y. Acad. Sci.*, 13:298.

WEINER, J. S.
1958. "The pattern of evolutionary development of the genus *Homo*." *S. Afri. J. med. Sci.*, 23:111.
1960. "The evolutionary taxonomy of the Hominidae in the light of the Piltdown investigation." *Proc. Vth Int. Anth. Congress, Philadelphia*, 1956:741.

WEINER, J. S., K. P. OAKLEY and W. E. LE GROS CLARK
1953. "The solution of the Piltdown problem." *Bull. Brit. Mus. (ant. Hist.)* 2:139.

AGE CHANGES, SEX DIFFERENCES, AND VARIABILITY AS FACTORS IN THE CLASSIFICATION OF PRIMATES

ADOLPH H. SCHULTZ

INTRODUCTION

THE CLASSIFICATION OF PRIMATES with all of its implications for phylogenetic conclusions still rests chiefly on dental and cranial characters and in general not nearly as close attention has been paid to other features. This is most evident in the classification of fossil material of which skulls and teeth have been much more frequently collected, or at any rate described, than other parts. Fossil finds usually represent at best very inadequate samples of a population so that it is impossible to determine whether a given specimen stands near the average of its species or happens to be an extreme variation. The age and sex of fossil fragments are further factors of frequent uncertainty in interpretations. The degrees of variability, of sex differentiation and of the intensity and speed of age changes can differ very widely among primates and certainly need not at all resemble these conditions in recent man, as is tacitly assumed by many anthropologists. It is particularly in the important selection of such features as are most useful for classifying recent and extinct primates that we must constantly bear in mind their variability, ontogenetic change and sexual dimorphism to avoid reaching untenable conclusions.

All systematic and phylogenetic studies can be merely tentative until they have considered a great many different characters and these with due regard for age, sex and variability which calls for extensive comparative observations. Inadequate analysis and rash overrating of localized morphological differences have produced an amazing amount of quite unnecessary, new taxonomic names, unjustified classifications and unconvincing evolutionary speculations particularly for the higher primates. For instance, according to Simonetta's (1957) depressingly long lists of hominoid synonyms, chimpanzees have been given twenty-one different generic names and at least seventy-three specific names! Frechkop's (1954) proposal of separate families of Simiidae and Gorillidae for the recent great apes may serve as only one of many examples of the gross overrating of differences in

85

merely a few variable morphological features. Similarly, such well-known theories of human evolution, as those by Weinert (1932, 1944) or by Wood Jones (1948), are wide open to criticism on account of their insufficient consideration of the real conditions of variability in the characters selected for support, as had been shown by the writer in previous publications (1936, 1950).

The stimulating fetalization theory of Bolk (1926) has also failed to explain man's specializations as the result of general retardation in development because of its great limitations in the characters considered and in the material used for comparisons (see also Starck, 1962). The same can be said about many other claims in regard to human evolution and particularly concerning man's distinctive features, which have been published without an adequate scale of evaluation based upon comparative data on more than a few other primates. It has been stated, e.g., that man's hairlessness is a symptom of his "self-domestication," ignoring the fact that a far greater difference in density and amount of hair exists between gibbons and chimpanzees than between the latter and at least some recent races of man (Schultz, 1931). Another example of such assertions, unsupported by facts, is the frequently seen statement that the great toe of man has become uniquely lengthened, even though it can readily be shown to be proportionately shorter than in gibbons and not significantly longer than in chimpanzees and some gorillas.

Many more such rash and erroneous claims could be cited and some will be briefly discussed below to show the need for more comparative primatological work with due regard to age, sex and variability. Only with the latter can we assemble a reliable list of man's distinctive characters as the basis for determining man's place in the pedigree of primates.

Some authors have distinguished sharply between qualitative and quantitative characters, regarding only the former as having diagnostic value in classification. Such a distinction, however, can rarely be maintained if all conditions of age, sex and variability are taken into account, when it is usually found that so-called qualitative differences are really quantitative ones. For instance, we should no longer simply state that one form of primate has diastemata and some other one has none, since actually there merely exist quantitative differences in the distribution of variations of this feature within series of the same race, sex and age.

Diastemata are primarily connected with the often highly variable rates of growth of the *os incisivum*, rather than with the size of the canines, as can here be merely indicated by the examples in Figure 1. The famous occurrence of a small diastema in one specimen of *"Pithecanthropus"* has very limited significance in view of the facts that just as large diastemata have been found in occasional recent men, while none exist in many pongids, especially chimpanzees and female gorillas (Schultz, 1948 and new observations).

In a similar way the misnomed "simian shelf," commonly regarded as a constant and hence reliable diagnostic character of all adult pongids, is found to be so extremely variable in large series that its many transitional formations, including total absence, differ quantitatively even among the great apes them-

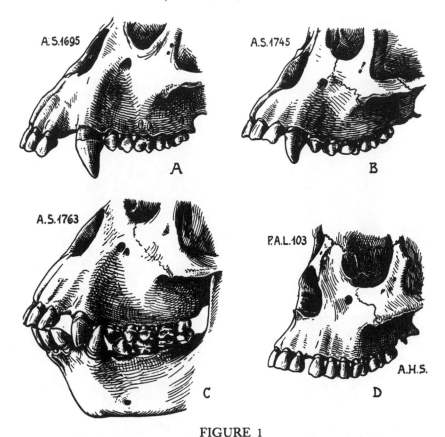

FIGURE 1

A = adult male chimpanzee with large diastema and large premaxilla; *B* = adult female chimpanzee with very small diastema and small premaxilla; *C* = adult female chimpanzee with diastema, congenital lack of lower second premolar and lower canine occluding behind upper one; *D* = adult female Negro with diastema
(from Schultz, 1948).

selves. This has recently been demonstrated in careful detail by Vogel (1960) and can here be shown by the chimpanzee mandibles in Figure 2, of which only the largest one possesses what can be called a real *shelf*. Incidentally, not infrequently a "shelf" merely appears to exist on account of an exceptionally large *fossa genioglossi*, which can deeply excavate the lingual side of the symphysis, thereby causing the lowest symphyseal part to resemble a thin plate or a rounded torus according to the exact location of the fossa above.

With this brief introduction the fact has already been emphasized that there are no "short-cuts" in the interpretation and proper evaluation of characters used for classifying fossil or recent primates because the factors of age, sex and variability must always be fully considered. This will be further shown by the following examples, selected for their special interest in the study of human evolution.

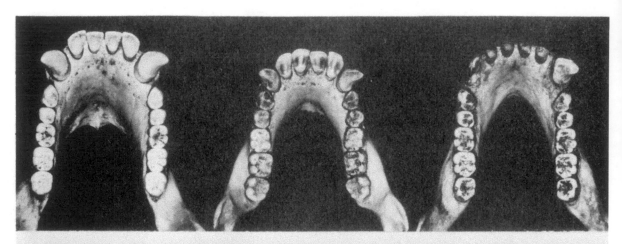

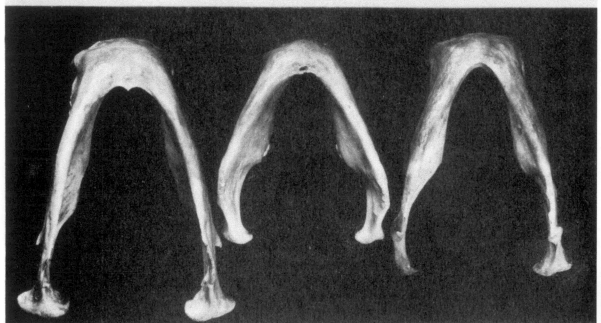

FIGURE 2

Mandibles of the same three West-African chimpanzees from above and from below, showing a very large "simian" shelf in the male on the left, a mere trace of one in the female in the center, and the lack of a shelf in the male on the right. Incidentally, the total lengths of the postcanine teeth happen to be exactly alike in these specimens, but the total mandibular lengths are very different.

AGE CHANGES[1]

It is a matter of course that in comparisons for purposes of species diagnosis great care must be taken to use specimens of really comparable age. This, however, is an ideal which is not always as easily attained as it may seem. It is not yet generally realized, e.g., that in many primates growth and development can continue long after the completion of the dentition, which forms the customary definition for the beginning of "adulthood." At this stage of dental development nearly all epiphyses have become fused with their diaphyses in recent man, fewer in the African apes, whereas only a small minority of them in monkeys. The physiological ages of newborns also are not strictly comparable since the stages of maturation at birth differ widely among primates, being much less advanced in man and apes than in all monkeys studied so far. Ontogenetic differences, such as these, complicate exact comparisons between primates, but are in themselves highly significant conditions for phylogenetic investigations.

It is well known that the main sutures of the neurocranium become closed much later in man than in the apes, because—it has been claimed—the brain grows more intensively and over a longer period in man. In most platyrrhines, however, the sutures remain open to even later relative ages than in man (Chopra, 1957) and there is evidence of continued cranial growth well into the middle of adult life (Schultz, 1960). The very early disappearance of the facial part of the intermaxillary sutures has commonly been regarded as one of the fundamentally distinctive characters of man and as representing a unique ontogenetic occurrence. This local acceleration in human development becomes far less significant when adequately compared with similar conditions in other primates. Thus, the same sutures begin to close during early infancy in most chimpanzees and some orangutans, whereas in many monkeys they are among the very last sutures to close, often being still open even in senile specimens. Furthermore, the internasal sutures become fused before birth in macaques and in chimpanzees, whereas in man this takes place only in rare individuals of extremely old age.

Many similar changes in the *relative* ages of localized developmental processes have become known and, if marked and stabilized, do represent good taxonomic characters. In no single species, however, do either retardations (as implied by the "fetalization theory") or accelerations occur exclusively, since every single feature can independently shift its place in the sequence of ontogenetic processes in either direction. Without the knowledge of such ontogenetic changes it is frequently impossible to evaluate correctly apparent species differences in specimens not exactly of comparable age. For instance, mastoid processes have been claimed to be practically absent in apes, small in fossil men and extremely large in recent

---

1. Detailed accounts of the age changes referred to here can be found in the writer's publications of 1956 and 1960, which also contain further relevant literature.

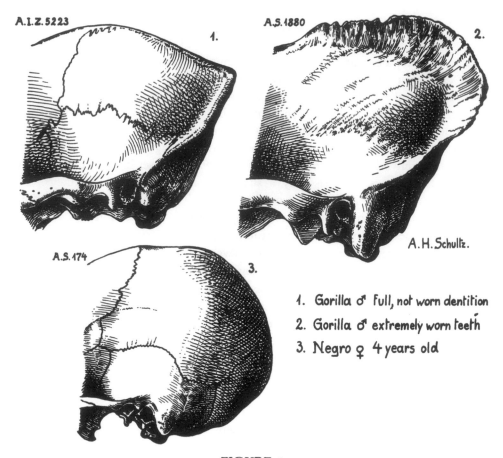

1. Gorilla ♂ full, not worn dentition
2. Gorilla ♂ extremely worn teeth
3. Negro ♀ 4 years old

FIGURE 3
Posterior cranial parts of two adult gorillas and one infantile negro, showing relative sizes of mastoid processes (from Schultz, 1957).

men. Actually, the essential differences seem to be ontogenetic ones which also produce the marked variability of these structures in apes.

In most recent races of man the mastoid processes begin their development soon after birth and have reached their full size in young adults, whereas in the great apes they appear late, but continue to grow throughout adult life to become finally as large as in man in most old male gorillas and some old chimpanzees (Schultz, 1952). These facts can here merely be indicated by the examples in Figure 3. In fossil man the mastoid development may simply not yet have become as much accelerated as in recent man, since the final size of these processes in really old individuals of fossil man is still unknown.

Many such changes in the relative ages of ontogenetic processes can produce at least temporary morphological distinctions among the types of primates affected thereby. For instance, the various bony elements of the sternum, which normally remain separate throughout life in monkeys, tend to become fused in all Hominoidea, but this process has become most accelerated in recent man whose *corpus sterni* solidifies into a single bone before the completion of the dentition, when the *corpora sterni* of apes are still composed of multiple pieces which fuse only during more advanced ages to become eventually single bones as in young adult men.

The topographic relations between the bones in the temporal region, in the medial orbital wall and between the sphenoid and ethmoid in the cranial cavity have been claimed to differ significantly in apes and man. Wood Jones (1948), e.g., had stressed the latter two features, though without adequate analysis, in support of his theory that man must have started on his separate evolutionary development at an extremely early time. In the light of the ontogeny of these conditions during cranial ossification, however, it appears that such differences are merely the result of slight variations in the rates of early growth in the respective bones and represent only quantitatively distinct characters.

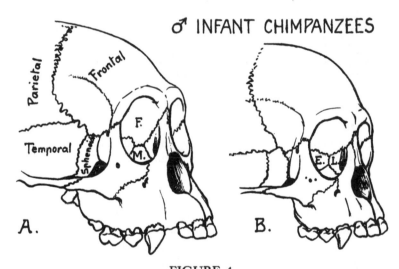

FIGURE 4

Infantile chimpanzee skulls from Cameroon on left and from south
of Congo river on right, showing different relations between
bones at temple and on medial orbital wall.

In many species such local topographic relations of cranial bones are remarkably uniform and hence have taxonomic value, but in other and at times quite closely allied species the same relations can be extremely variable, indicating that the relative rates of growth in the bones concerned have become very unstable. For instance, among many hundreds of skulls of monkeys and Asiatic apes the author (1952) found the lacrimal and ethmoid bones to meet and form a common suture

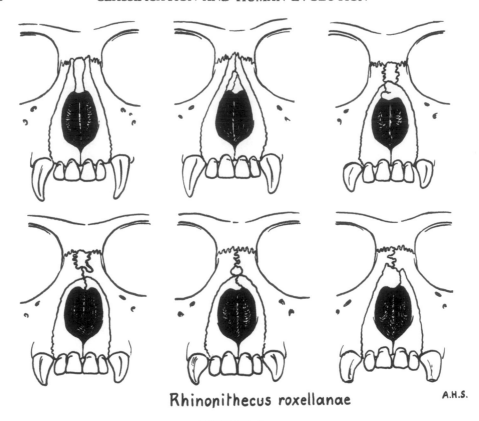

Rhinopithecus roxellanae   A.H.S.

FIGURE 5

Different variations in the intermaxillary and nasal bones of some adult *Rhinopithecus* (from Schultz, 1957).

without any exception, but this formation was encountered in only 43 per cent of gorillas, 50 per cent of ordinary chimpanzees, 86 per cent of pygmy chimpanzees, and 97 per cent of men. In the remaining cases of the latter primates the lacrimal and ethmoid had become separated through more intensively growing frontal and maxillary bones, which met each other behind the lacrimal, as in one of the skulls in Figure 4. Another example of exceptionally irregular growth of some cranial elements the writer found in an isolated northern Chinese species of *Rhinopithecus*, which is distinguished by an incredible variability in the development of its intermaxillary and nasal bones, with not even two of the fourteen skulls examined being alike (see Figure 5).

Fossil and recent men alone retain throughout life the forward position of the occipital joint which exists in all simian primates in fetal and infantile stages, only to change later to much more aboral positions in monkeys and apes (Schultz, 1955). Recent man has lost also the postnatal migration of his orbits from underneath the braincase to a more oral position, which is so extremely marked in pongids. This latter lack of change with age is not unique in man, but exists also in various platyrrhines.

Alongside such retentions of fetal conditions, or retardations, however, man has also acquired many accelerated age changes, such as his prenatal descent of the testes, the embryonic fusion of the *centrale* with the *naviculare* and the examples already mentioned. Furthermore, man is certainly not "fetalized" in the sense of Bolk's theory, in regard to the intense postnatal growth of his legs or the early and profound changes in his feet. On the other hand, man closely follows ape fetuses in the early development of hair, which begins in all of them with the eyelashes, eyebrows and hair on the scalp, but, while little more is added postnatally in man, apes quickly acquire hair on most of their remaining body surface. With these final examples it is pointed out once more that accelerations, retardations or even suppressions in the development of single characters can and do occur, thereby producing at times very marked and possibly rapid changes in the adult form.

SEX DIFFERENCES

Paleoanthropologists with limited experience in primatology usually take it for granted that large and robust specimens must be male and the smaller and delicate variations female. This popular rule is probably more or less valid for a majority of recent catarrhines, but certainly not for Hylobatidae and for many platyrrhines and prosimians. Sex differences in average body weight vary to an amazing degree among primates and this even in closely related forms, such as gorillas and chimpanzees or different species of macaques. While in some species the males can attain weights of more than twice those of females, in other species the females are on an average somewhat heavier. Among the Hominoidea the intraspecific variability in body size seems to be particularly great.

In gibbons, siamangs and chimpanzees one frequently finds some males smaller than some females of the same local populations. Among fully grown *Hylobates lar* from northern Thailand, e.g., we obtained females weighing as much as 6.12 kg. and males as little as 4.08 kg. (Schultz, 1944). Even among adult orangutans with their great *average* sex difference in size, the writer has seen evident runts and giants in both sexes. It appears, therefore, that "sexing" fossil finds must remain a very hazardous undertaking[2] and one not always helped by the canines, since they may or may not have become sexually differentiated without overlapping of their ranges of variations.

Contrary to widely held opinions, sagittal crests are not necessarily indicative of the male sex since they can also occur in females of apes and monkeys, provided they happen to possess large jaws in combination with a small braincase. These crests never develop before the dentition is completed, but thereafter they can keep on growing until old age, especially in males.

2. Genovés (1954) had reached the same conclusion in regard to certain fossil hominids.

From an extensive metrical analysis of the correlation between the degrees of sexual differentiation and of change during postinfantile growth in primate skulls the author (1962) could demonstrate that on an average sex differences become most marked in adults in such characters and such species as change the most during growth and vice versa. For instance, sex differences become more pronounced as a rule in the splanchnocranium (particularly the jaws) than in the neurocranium, which changes postnatally much less in size and shape, and such species as gorillas, orangutans and baboons develop far more marked sex differences than man or capuchin monkeys whose skulls change with growth comparatively not nearly as much.

Consistent secondary sex differences appear in adult primates in the pelvic inlet, which is proportionately larger in females than in males of the same species and this even in the great apes which are the only primates with birth canals amply large for the modest size of their full-term fetuses. In all other primates studied so far there exists generally such an incredibly close correlation between the size of the maternal pelvic ring and that of the newborn's head that any unfavorable change of only a slight degree could quickly lead to the extinction of a species (Schultz, 1949, 1961). This is illustrated by the example in Figure 6. While selection undoubtedly tends to favor large diameters of the female pelvis, it must also act against any prolongation of gestation or at least against unduly large newborns which, incidentally, can attain up to 10 per cent of the maternal weight in some monkeys (Schultz, 1961).

VARIABILITY

The degrees of intraspecific variability in morphological characters can differ widely among recent primates (Schultz, 1947) and most likely have done so among those of the past. Quite generally speaking it appears that recent man, most Old World monkeys, and some of the New World, tend to be decidedly less variable than are the great apes, gibbons and certain prosimians. Strictly speaking one cannot assess the variability of a species or a population, but merely that of single features, such as skull shape, vertebral formula or hair color, and it is only after one has established the degrees of variability of many different features that one becomes justified to generalize in regard to the uniformity of a given species and the probability that an available sample permits a valid description of typical conditions. At least among recent higher primates it is usually found that the face-part of the skull is even more variable than the brainpart and that such structures as the sternum, the scapula and the sacrum are more variable in size and shape than the long bones, hip bones, or the lumbar vertebrae.

The following few random examples of the ranges of variations within populations of recent primates, quoted chiefly from the writer's own studies, will suffice to show that even some very impressive differences in size and shape among fossil finds may well represent merely individual variations and not

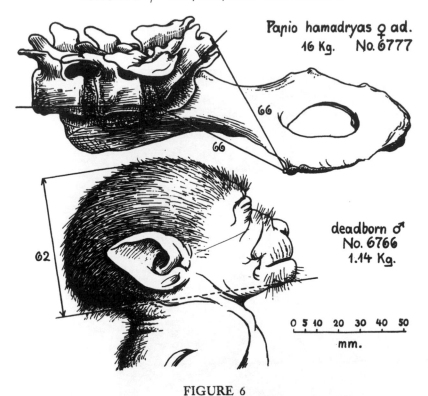

Papio hamadryas ♀ ad.
16 Kg.   No. 6777

deadborn ♂
No. 6766
1.14 Kg.

0 5 10   20   30   40   50
mm.

FIGURE 6

The pelvis of an adult baboon and the head of its full-term fetus.
drawn to same scale. Mother and young had died immediately
after the difficult birth (from Schultz, 1961).

necessarily specific distinctions nor evolutionary changes. One of the foremost
authorities on primate teeth, Remane (1961), has expressed his similar conclusions
even more emphatically as follows: "Wir wissen nicht, wieviele und welche von
den zahlreichen aufgestellten Arten und Gattungen [der fossilen Pongiden]
wirkliche Gattungen und Arten sind. Wenn man mit gleichen Methoden die
Zähne und Kiefer der rezenten Genera bearbeitet, würde jede Art in fünf bis
zehn Genera und 20–30 Arten zerfallen." In a preceding paper Remane (1954)
had already stated that "Mindestens seit dem unteren Miozän besassen die
Stämme der Pongidae und Hylobatidae eine ungewöhnliche Breite individueller
Vielgestaltigkeit."

The much discussed length-breadth proportion of the braincase varies in a
series of 248 adult wild *Hylobates lar* between 68 and 86 and among 105 adult
orangutans even between 70 and 92. In 80 fully grown lowland gorillas this range
extends from 57 to 79 and the breadth-height proportion of the facepart of the
skull varies even between 61 and 88. Since anthropologists dwell on the fact
that nearly all skulls of early men were dolichocran, they should also realize that

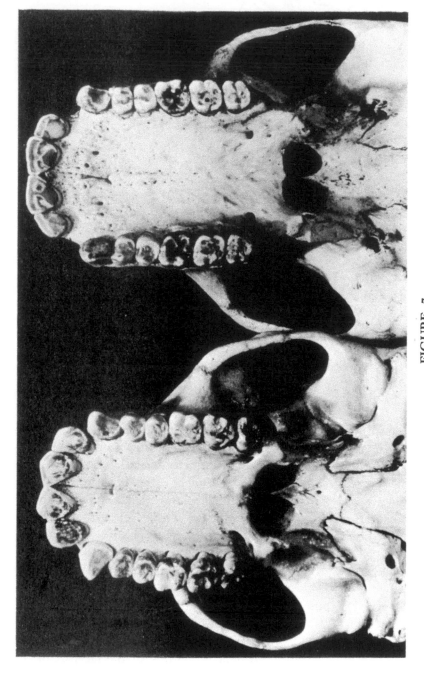

FIGURE 7

Palates of two adult female chimpanzees from Cameroon, showing many marked differences.

this particular character was far less variable than it is in recent apes with their ranges from dolicho- to brachycrany.

The cranial capacity is another feature of great interest in the study of human evolution, but one of very limited significance in isolated cases. In a series of only 58 full-grown male gorillas the capacity was found to vary between 423 and 752 cc. with a variation coefficient of over 10. Since the capacities of fossil skulls usually can be only estimated, it may also be pointed out that actually measured capacities show very little correlation with the outer dimensions of the same skulls in recent gorillas and men (Schultz, 1962), so that mere estimates can rarely claim to be more than rough approximations.

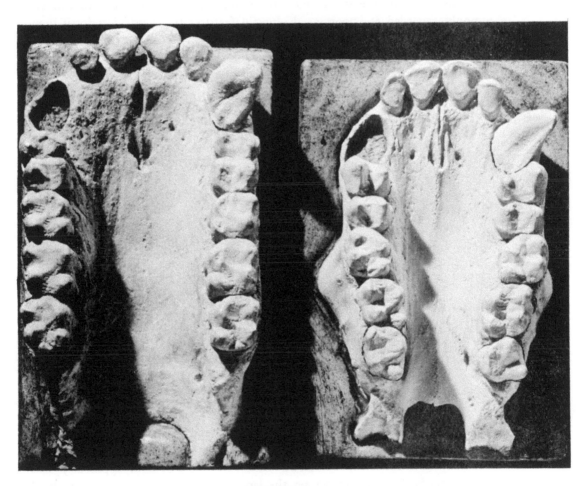

FIGURE 8

Palates of two adult male West-African gorillas, showing many marked differences in size and shape. The long palate on the left is also very much deeper than the short and narrow one on the right. Photographed from exact casts because one skull is perfectly white and the other one is stained deep-brown.

The shape and relative size of the adult palate vary intraspecifically to a surprising degree among all recent apes, and this quite independent of the dentition. The total length of the teeth behind the canines appears to be remarkably constant when contrasted with the greatest length of the palate, which can extend little or much behind the last teeth and which can project in its intermaxillary part in front of the canines in widely varying degrees. The latter variations determine directly the development of diastemata which range from total absence to very wide gaps without any close dependence on the size of the lower canines, as has already been mentioned. If the examples of palates of adult wild chimpanzees and gorillas, shown in the Figures 7 and 8, had been found as fossils at different places, they would quite likely have been assigned to at least different species.

The formation of the supraorbital torus and the shape of the forehead are commonly regarded as important diagnostic features for classifying fossil men, since these structures have demonstrably changed with evolution. It seems, however, that the significance of detailed individual differences in these characters has been much overrated at times and that such differences can readily be ex-

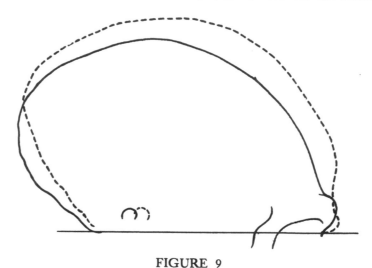

FIGURE 9

Superimposed sagittal sections of the two adult Neanderthal skulls from Spy, oriented according to level of nasion and of porion. Note difference in formation of frontal curvature and of glabellar region.

plained as natural variations, occurring with differing frequencies in local populations. Very low foreheads combined with heavy browridges apparently had become remarkably common among the later Neanderthal men, but undoubted exceptions have already been found, such as in the two skeletons from Spy which certainly were contemporary (Figure 9), or among the Mt. Carmel material and that from Krapina.

In some other primates there occur even more striking individual variations in

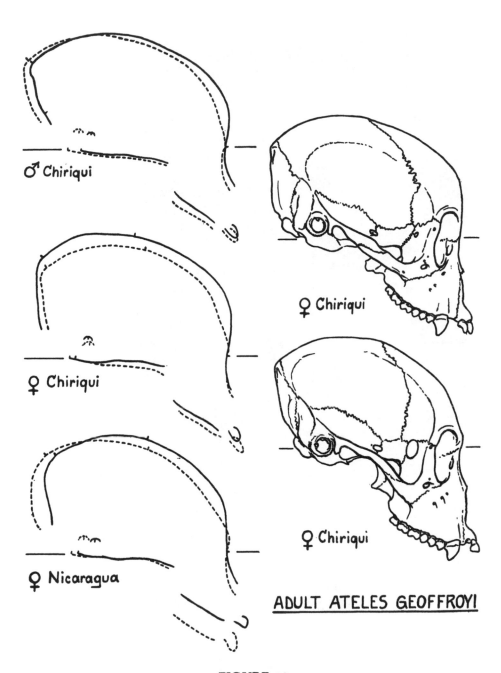

FIGURE 10

Midsagittal sections and dioptrographic side views of skulls of wild spider monkeys from corresponding local populations, oriented in the nasion—basion horizon. Note differences in formation of frontal curvature and glabellar region (from Schultz, 1960).

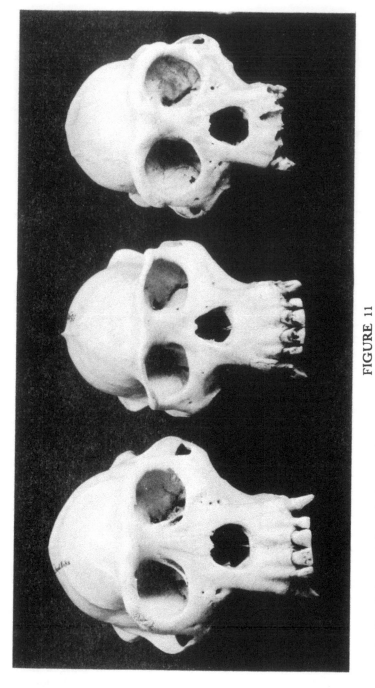

FIGURE 11

Skulls of three adult male West African chimpanzees showing many marked differences in size and shape.

precisely the same features. For instance, among spider monkeys some show a flat forehead and clear browridge, while others from the same locality have a bulging frontal region and practically no supraorbital torus (Figure 10). In chimpanzees the supraorbital region can be so differently developed that it might seem incredible that they belong to one and the same local race, especially since their skulls can also be so widely different in many other respects, as is evident from the examples in Figure 11. Among adult male gorillas it is particularly the formation of the huge supraorbital torus which varies individually to an amazing degree, as can be seen from skulls, illustrated by the author in a previous publication of this series (1961, Figure 17).

So much has become known about the great intraspecific variability of the teeth of at least the recent Hominoidea that our faith in the reliability of detailed dental characters for classificatory purposes has been considerably shaken. One merely has to recall here the incidence of *Hesperopithecus*, the premature diagnosis of *Gigantopithecus* as a hominid instead of a now undoubted pongid, or the earlier views on *Oreopithecus* to indicate some of the errors resulting from an inadequate appreciation of dental variability. Even the entire dental formula is subject to intergeneric as well as intraspecific variations which can lead to unjustified classifications. Thus *Callimico* had been assigned to the Cebidae on the basis of its having three molars, instead of only two as the Callithricidae, with which it agrees perfectly in all other features (Hill, 1959).

Congenital suppression of third molars occurs also in many other recent primates, but so far in never more than a varying minority of the cases. This is best known for recent man in whom there exist marked racial differences in the frequency of missing third molars.[3] While this condition is extremely rare in the great apes, I have found it in 4 per cent of gibbons, in 3.5 per cent of Panamanian spider monkeys and in 15 per cent of Nicaraguan spider monkeys of the same species, among which this molar agenesis happens to have become unusually frequent.

Characters which have changed, to become specializations of one or several related primates only, show frequently an exceptionally high variability, thereby indicating that the particular innovation is still incomplete and has not yet changed to one and the same new stage in the entire population. For instance, the number of cervical vertebrae deviates (by only one segment) very rarely from the ancient and nearly constant number of seven. In sharp contrast to this, the number of caudal vertebrae, which has become extremely reduced in all the "tailless" Hominoidea, varies in the latter between 0 and 6, with as many as 9 segments still occurring temporarily in early development.

As further examples of this sort can be mentioned the highly variable middle phalanges of the degenerated fifth toe in man, which have become even completely lost in about one fourth the persons of some European populations and in a majority of Japanese. The corresponding phalanges of non-human primates

3. Third molar agenesis has been found in as much as 50 per cent of some series of modern Europeans (Hellman, 1936) and had affected 27 per cent of old Hawaiians (Snow, 1962).

are well developed without exception. In a similar way the greatly reduced first toe, typical of orangutans exclusively, lacks its distal phalanx unilaterally or bilaterally in over 60 per cent of these apes and in the remaining cases the same phalanges are extremely variable in size. Spider monkeys are supposed to have completely lost an outer thumb, but actually this digit is still retained as a useless appendage in considerable percentages of specimens from certain localities and may or may not even bear a nail.

A great many more instances of this kind are known in which structures suddenly have become so highly variable in some species that they have become "specialized" through selection in definite directions. The congenital lack of third molars and of phalanges of the thumbs, index fingers, great toes and lateral toes and retention of the embryonic webbing between the second and third toes etc., have been regarded as teratological, if occurring sporadically in man, but each one of these conditions is found so regularly in one or another species of primates [4] that it has become "the norm," since it exists in at least a majority of the cases with typically high variability.

GENERAL AND DETAILED CLASSIFICATION OF HIGHER PRIMATES ACCORDING TO DIFFERENT CHARACTERS

General descriptions of the taxonomic characters of the higher primates · in particular usually dwell far more on the skull, teeth and brain than on the neck, trunk and limbs and limit themselves to conditions in adults. Ontogeny and variability have rarely been considered according to their importance in comparative surveys for phylogenetic conclusions. That the much neglected conditions of the trunk and limbs, of age changes and of variability can be at least as decisive as those of the head in classifying primates, will be shown by the following brief discussion of some examples, chosen at random from a great many findings that have become available.

The number of segments in the combined thoracic and lumbar regions of the spinal column fluctuates among primates between fifteen and twenty-four, with nineteen of these vertebrae representing most likely the original condition, and still forming the modal number in the great majority of the recent genera, according to the data in Table 1. Only in the entire family of Lorisidae has this number become greatly increased and exclusively in the superfamily of Hominoidea has it clearly decreased. The latter reduction in the number of segments has affected chiefly the lumbar region, so that the pelvis has migrated much more closely toward the thorax than in any lower primates (Schultz, 1961). This unique and profound phylogenetic change, characterizing all Hominoidea, is convincing proof of their close interrelationship through common descent, even though this trend has progressed less far in gibbons and man than in siamangs

4. For further relevant details see Schultz, 1956.

TABLE 1

PERCENTAGE DISTRIBUTION OF VARIATIONS IN THE NUMBER OF THORACIC + LUMBAR VERTEBRAE AND AVERAGE NUMBERS OF THESE VERTEBRAE IN THE DIFFERENT GENERA OF PRIMATES (AFTER SCHULTZ, 1961)

Genus	Specimens	Number of thoracic + lumbar vertebrae										Average
		15	16	17	18	19	20	21	22	23	24	
Tupaia	16	–	–	–	6	81	13	–	–	–	–	19.1
Lemur	31	–	–	–	10	80	10	–	–	–	–	19.0
Hapalemur	17	–	–	–	12	82	6	–	–	–	–	18.9
Lepilemur	12	–	–	–	–	–	17	75	8	–	–	20.9
Cheirogaleus	5	–	–	–	–	–	100	–	–	–	–	20.0
Microcebus	19	–	–	–	–	11	89	–	–	–	–	19.9
Avahi	11	–	–	–	–	9	91	–	–	–	–	19.9
Propithecus	20	–	–	–	–	10	85	5	–	–	–	19.9
Indri	4	–	–	–	–	–	50	50	–	–	–	20.5
Daubentonia	11	–	–	–	9	91	–	–	–	–	–	18.9
Loris	12	–	–	–	–	–	–	8	17	59	8 8	22.8
Nycticebus	26	–	–	–	–	–	–	–	15	4 46	4 31	23.2
Arctocebus	2	–	–	–	–	–	–	50	50	–	–	21.5
Perodicticus	31	–	–	–	–	–	6	23	58	13	–	21.8
Galago	21	–	–	–	–	67	33	–	–	–	–	19.3
Tarsius	18	–	–	–	22	78	–	–	–	–	–	18.8
Callithrix	14	–	–	–	14	79	7	–	–	–	–	18.9
Leontocebus	75	–	–	–	–	1 96	3	–	–	–	–	19.0
Callimico	1	–	–	–	–	100	–	–	–	–	–	19.0
Aotes	15	–	–	–	–	–	13	87	–	–	–	20.9
Callicebus	3	–	–	–	–	33	67	–	–	–	–	19.7
Pithecia	3	–	–	–	–	100	–	–	–	–	–	19.0
Cacajao	4	–	–	–	–	100	–	–	–	–	–	19.0
Alouatta	32	–	–	–	–	69	28	3	–	–	–	19.4
Saimiri	16	–	–	–	–	–	100	–	–	–	–	20.0
Cebus	32	–	–	–	3	34	3 60	–	–	–	–	19.6
Lagothrix	8	–	–	13	50	37	–	–	–	–	–	18.2
Ateles	29	–	–	7	83	10	–	–	–	–	–	18.0
Macaca	216	–	–	–	4	1 89	2 4	–	–	–	–	19.0
Cynopithecus	5	–	–	–	20	80	–	–	–	–	–	18.8
Papio	66	–	–	–	3	97	–	–	–	–	–	19.0
Theropithecus	4	–	–	–	–	100	–	–	–	–	–	19.0
Cercocebus	12	–	–	–	17	83	–	–	–	–	–	18.8
Cercopithecus	53	–	–	–	6	90	4	–	–	–	–	19.0
Erythrocebus	3	–	–	–	33	67	–	–	–	–	–	18.7
Presbytis	122	–	–	–	5	1 92	2	–	–	–	–	19.0
Rhinopithecus	13	–	–	–	–	100	–	–	–	–	–	19.0
Nasalis	50	–	–	–	2	98	–	–	–	–	–	19.0
Colobus	8	–	–	–	12	88	–	–	–	–	–	18.9
Hylobates	319	–	–	4	1 70	2 23	–	–	–	–	–	18.2
Symphalangus	29	4	7	3 41	7 38	–	–	–	–	–	–	17.3
Pongo	127	18	1 72	2 7	–	–	–	–	–	–	–	15.9
Pan	162	–	25	4 67	1 2	1	–	–	–	–	–	16.8
Gorilla	81	–	39	4 56	1	–	–	–	–	–	–	16.6
Homo	125	–	7	91	2	–	–	–	–	–	–	17.0

and the great apes which are distinguished, as expected, by an extremely high variability in this extremely altered character.

The number of vertebrae fusing to form the sacrum varies between only two and four in the great majority of primates, three being by far the most frequent variation and undoubtedly the original number. In only two groups of primates has this number of sacral vertebrae become significantly changed and this by very marked increases (see Table 2). Among prosimians it is again the family of Lorisidae in which the increase extends to as many as nine sacral vertebrae in some slow loris and pottos.

The second and independently acquired increases are characteristic of again the entire group of Hominoidea in which as many as seven or even eight sacral vertebrae have been found and among which the great apes have developed higher average numbers than have the Hylobatidae and man. Only among these genera with greatly increased numbers of sacral vertebrae do the individual variations extend over four or even five different numbers in contrast to all the many other primates with much smaller ranges of variations for this character.

Though all species of primates have the same number of seven cervical vertebrae and this with remarkable constancy, as has already been mentioned, the length of the cervical region in relation to the total presacral spine length as well as to the trunk length is greater in all recent Hominoidea than in lower primates with the only exceptions of spider monkeys and tarsiers, which have developed similarly long necks (Schultz, 1961).

The general shape of the trunk differs sharply in the higher and the lower primates, since it has become much stouter in all of the former. The chest girth and the breadth of the shoulders and the hips in their relation to the trunk length show much higher averages in gibbons, great apes and man than in any monkeys even by including the consistent and at times marked sex differences in these proportions (Schultz, 1956).

Another clear-cut distinction of exclusively the higher primates exists in the shape of the thorax in postnatal life, when it always becomes much broader than deep, in contrast to all monkeys which retain permanently the fetal shape of a narrow and deep thorax, hanging underneath the vertebral column. Together with this great widening of the chest in all apes and men the shoulder blades migrate ontogenetically to the broad back and do not retain their lateral position as in monkeys and all other quadrupedal mammals. Furthermore, the vertebral column shifts far toward the center of the thoracic cavity and, hence, toward the center of gravity in erect posture in varying, but always significant degrees in all adult Hominoidea. This is in contrast to the unspecialized, dorsal position of the vertebral column in lower primates. The ventral shift of the column, jointly acquired by all higher primates, is undoubtedly advantageous for an upright position of the body which they can maintain with decidedly greater ease than can monkeys.

The general widening of the trunk, typical of all higher primates, has been accompanied also by the development of much broader sterna (hence the old

TABLE 2

PERCENTAGE DISTRIBUTION OF VARIATIONS IN THE NUMBER OF SACRAL VERTEBRAE AND
AVERAGE NUMBERS OF THESE VERTEBRAE IN THE DIFFERENT GENERA
OF PRIMATES (AFTER SCHULTZ, 1961)

Genus	Specimens	Number of sacral vertebrae								Average
		2	3	4	5	6	7	8	9	
Tupaia	16	–	94	6	–	–	–	–	–	3.1
Lemur	31	7 3	90	–	–	–	–	–	–	2.9
Hapalemur	17	–	100	–	–	–	–	–	–	3.0
Lepilemur	12	8	58	33	–	–	–	–	–	3.2
Cheirogaleus	5	–	100	–	–	–	–	–	–	3.0
Microcebus	19	11	84	5	–	–	–	–	–	2.9
Avahi	11	–	64 9	27	–	–	–	–	–	3.3
Propithecus	20	–	75 5	20	–	–	–	–	–	3.2
Indri	4	–	50	50	–	–	–	–	–	3.5
Daubentonia	11	9	73	18	–	–	–	–	–	3.1
Loris	12	8	67	8	17	–	–	–	–	3.3
Nycticebus	26	–	–	–	15 4	50	23	4	4	6.3
Arctocebus	2	–	–	–	–	50	50	–	–	6.5
Perodicticus	31	–	–	–	13	36 3	29	16	3	6.6
Galago	21	–	81	19	–	–	–	–	–	3.2
Tarsius	18	6	94	–	–	–	–	–	–	2.9
Callithrix	14	14	86	–	–	–	–	–	–	2.9
Leontocebus	74	4 1	95	–	–	–	–	–	–	2.9
Callimico	1	–	100	–	–	–	–	–	–	3.0
Aotes	15	–	87	13	–	–	–	–	–	3.1
Callicebus	3	–	100	–	–	–	–	–	–	3.0
Pithecia	3	–	100	–	–	–	–	–	–	3.0
Cacajao	4	–	–	50	50	–	–	–	–	4.5
Alouatta	32	–	69	31	–	–	–	–	–	3.3
Saimiri	16	–	100	–	–	–	–	–	–	3.0
Cebus	32	–	97 3	–	–	–	–	–	–	3.0
Lagothrix	8	–	63	37	–	–	–	–	–	3.4
Ateles	29	3	93	4	–	–	–	–	–	3.0
Macaca	214	2 1	90 3	4	–	–	–	–	–	3.0
Cynopithecus	5	–	40	60	–	–	–	–	–	3.6
Papio	66	–	80 3	17	–	–	–	–	–	3.2
Theropithecus	4	–	100	–	–	–	–	–	–	3.0
Cercocebus	12	17	75	8	–	–	–	–	–	2.9
Cercopithecus	52	–	96	4	–	–	–	–	–	3.0
Erythrocebus	3	–	100	–	–	–	–	–	–	3.0
Presbytis	119	4 1	94	1	–	–	–	–	–	3.0
Rhinopithecus	13	–	92	8	–	–	–	–	–	3.1
Nasalis	50	–	92	8	–	–	–	–	–	3.1
Colobus	8	–	100	–	–	–	–	–	–	3.0
Hylobates	319	–	2 1	40 2	50 1	4	–	–	–	4.6
Symphalangus	29	–	–	28 10	55 7	–	–	–	–	4.7
Pongo	125	–	–	3	57 2	36	2	–	–	5.4
Pan	161	–	–	1	33 3	53 2	8	1	–	5.7
Gorilla	81	–	–	1	32 4	56	6	1	–	5.7
Homo	116	–	–	3	72	24	1	–	–	5.2

term *Latisternalia* for Hominoidea) as well as pelves. This increase in the width of the hip bones has affected chiefly the parts of the ilia lateral to their attachment with the sacrum, whereby these parts, naturally, had to follow closely the direction of the last pair of ribs which they approach far more than in monkeys. The formation of a promontory at the lumbo-sacral border during postnatal growth is not, strictly speaking, a hominoid distinction in varying degrees, since in baboons the sacrum is also much tilted dorsally as a necessary adaptation for widening the birthcanal throughout its length (Schultz, 1961; see Fig. 6).

It is of special interest that most of these conditions which characterize the trunk of higher primates appear only in the course of individual development and thus are good examples of phylogenetic changes having resulted primarily from ontogenetic modifications.

On account of the obscuring, but ineradicable anthropological custom of expressing limb lengths in relation to that biologically meaningless measure, stature, of which roughly half is a limb length, it is still rarely realized that man shares with apes an arm length-trunk length ratio in excess of this same ratio in at least all Old World monkeys. This increase in the length of the arms, relative to the trunk, has reached unique extremes in the best of all brachiators, the Hylobatidae and orangutans. The relative arm length of man also stands well above the corresponding averages of monkeys, except such few pronounced brachiators as *Ateles* and *Lagothrix*. Many human beings, particularly among negroids, have proportionately longer arms than have some chimpanzees and many gorillas, especially the short-armed eastern variety (Schultz, 1934 and later observations).

There is hardly more need for further support of the conclusion that all recent higher primates have so much in common that they can and must be placed in one and the same superfamily, the Hominoidea. It is in regard to the subdivisions of this natural group and, most of all, to man's place within it, that opinions still diverge according to the various categories of evidence which happen to be preferred and relied upon. Since these problems will be solved with general agreement only when all available facts have been fully considered, some of the newer evidence which often has received too little, if any, attention from students of hominoid phylogeny and taxonomy, may be briefly mentioned here.

Some of the latest textbooks still copy from one another the old lists of characters which have become specialized in different directions or degrees in the various subgroups of the Hominoidea. These lists had been compiled mostly with regard to adults only and without adequate consideration of the corresponding conditions in lower primates as a scale of evaluation of these hominoid specializations. For instance, it is customarily mentioned that ischial callosities exist in all Old World monkeys and in all Hylobatidae, but are lacking in Pongidae and man. With due regard to age changes and variability, however, these supposedly sharp distinctions become less profound since callosities appear very early in the prenatal life of monkeys, whereas not until birth or even later in gibbons and siamangs, when they replace primary lanugo, and among the great apes perfectly good callosities still develop during postnatal life in not a few individuals (Schultz, 1933, and later observations).

Man and the African apes are set apart in the suborder of Simiae through the lack of an *os centrale* in the wrists of adults. Again this distinction changes from a qualitative into a quantitative one, if studied ontogenetically and in large series, since the *centrale* simply disappears through fusion with the *naviculare* in embryonic life of man, around the time of birth in chimpanzees and even later in at least some gorillas. In the Asiatic apes it can become lost through the same fusion in occasional old individuals, just as it still can remain separate in a small percentage of human beings. The possession of a free *centrale*, therefore, differs only in regard to age and frequency of occurrence (Schultz, 1936).

The intensively studied differences in the adult skulls of the Hominoidea become very marked only during growth. Thus, the size and direction of the face, relative to the skull base, are far more alike in apes and men in the young than in the adult. While the great apes develop the comparatively highest and most protruding faces, gibbons retain a proportionately smaller face than does man. This shows that such striking differences among adult Hominoidea are largely due to corresponding differences in relative rates of growth, as could be demonstrated in detail in a recent study of the writer (1962). As mentioned above, the relative position of the occipital condyles is another character which has become specialized in very different ways among the higher primates, but these distinctions also appear only during postnatal growth, when the condyles retain in man their fetal position far orally (typical for all Simiae), whereas they shift aborally in all apes and especially in gorillas and orangutans (Schultz, 1955).

Apparently as aids in carrying their poorly balanced, heavy heads, the great apes have acquired exceptionally long dorsal spines on their third to seventh cervical vertebrae in sharp contrast to the very short corresponding spines of man, gibbons and the majority of all other primates (see Figure 12). Incidentally, the conditions of these spines in Neanderthal man are not more ape-like than in recent man, as has commonly been claimed, but fall well within the relevant ranges of variations in the latter (Toerien, 1957; Stewart, 1962).

No other skeletal part has become more highly specialized in man than his pelvis, and this not only in its size and shape, but also in its exact position within the trunk, where symphysis and sacrum have become the floor and roof of the pelvic cavity rather than the ventral and dorsal walls, as in all non-human primates (Figure 13). The most impressive distinctions between the pelves of man and of all apes are found in the ilium which, though greatly widened in all, is short in the former and extremely long in the latter. In its percentage relation to the trunk length the ilium length equals on an average only 24 in man, whereas it is 33 in gibbons and even 36 to 38 in the great apes. The latter far surpass all Old World monkeys in this respect since their averages range only between 23 and 28 (Schultz, 1950). The human ilium has also become unique in regard to the great size of its sacral surface and the short distance between the latter and the hip

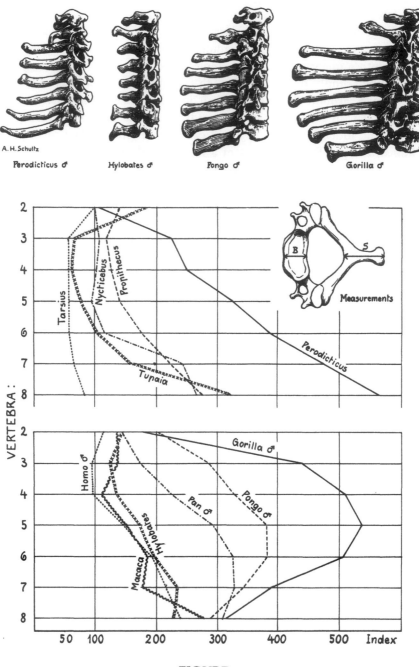

FIGURE 12

The first eight vertebrae in some adult primates, showing the different lengths of the dorsal spines and (below) the lengths of these spines in percentage of the midsagittal diameters of the corresponding vertebral bodies in different adult primates (from Schultz, 1961, where more relevant data can be found).

joint, which itself is of exceptionally great size. All these striking human specializations of the pelvis are undoubtedly direct adaptations to erect posture
and seem to have evolved also in a closely corresponding manner in the
australopithecoids.

The scapulae show far greater generic differences among Hominoidea than
among all the many genera of monkeys and are also much more variable (and
even asymmetrical) in the former than in the latter.

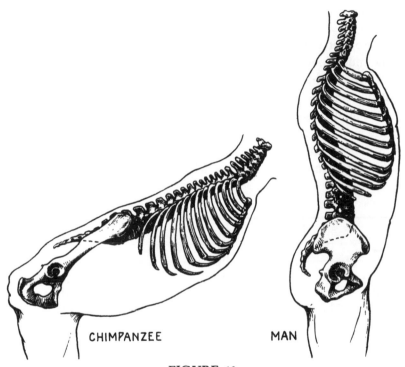

FIGURE 13
The position and relative size of the pelvis in adult ape and man
(from Schultz, 1961).

In the general shape of the thorax some significant differences also exist within
the group of higher primates, since it resembles a barrel in gibbons and man, but
is like a funnel in all great apes, in which the transverse diameters of the chest
cavity increase continually and very markedly from the first to the last ribs
(Schultz, 1961).

The limbs of the recent Hominoidea show many widely differing specializations as direct adaptations to their respective modes of locomotion. Thus, the arms
of the Asiatic apes, the most pronounced brachiators, have become far more
lengthened than have the legs of bipedal man. In the hands we find the most
slender and elongated forms in gibbons and the opposite extreme in gorillas. In

the Hylobatidae the long thumb has become unique by being free from the palm in a large metacarpal part. Only the great apes among all primates walk on their knuckles and have their fingers automatically flexed with dorsoflexion of the metacarpus.

The feet of the Hominoidea show much more diverging differentiations than the hands. In the writer's current investigation of several hundred primate foot skeletons it is very evident that those of all monkeys are remarkably uniform, whereas those of the apes and man very unlike each other with particularly great differences in the relative sizes of the main parts (Figure 14). In man and gorilla the tarsus has become extremely large and the center of the talus joint has shifted far distally, so that the power lever of the foot, in its relation to the load lever, has become much longer than in other primates. The phalangeal portions of the toes II to V are enormously elongated in gibbons and orangutans, whereas they are so extremely reduced in man that they have degenerated to mere vestiges, especially the middle phalanges. The first toe has become comparatively longest in Hylobatidae and, like the thumb, is "free" even in part of its metatarsus. Orangutans are the only primates in which the "great toe" has clearly degenerated, as has already been pointed out. All these widely different morphological specializa-

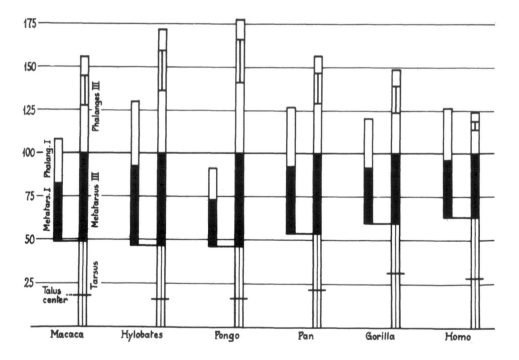

FIGURE 14

Diagrammatic representation of the average relative lengths of the main parts of the foot skeletons of adult higher primates and of macaques, all reduced to the same tarsal + metatarsal length.

tions in the feet of adult hominoids are closely connected with corresponding functional differences. It is significant also that these same differences are as yet merely indicated in early ontogenetic stages.

Besides these few examples of the great many significant morphological differences among recent hominoids, some generic specializations of a very different kind may also be mentioned here. The periodic sex-swelling, well-known in baboons and some macaques, exists also in chimpanzees in an extreme form and can at times even persist during pregnancy. In gorillas sex-swelling is barely indicated, in orangutans it appears only toward the end of pregnancy, when it can become very marked (Schultz, 1938), and in gibbons and man it is totally lacking. The durations of the main periods of life have become greatly lengthened in all Hominoidea, when compared with lower primates, and this least in gibbons and by far the most in man, except the gestation period. The group differences in this respect become steadily the more marked the later in life the period occurs (Schultz, 1961). This sort of difference, naturally, deserves the same careful consideration in classifications and for phylogenetic conclusions as do morphological characters, even though for the former kind we can expect no help from paleontology.

CONCLUSIONS

Each of the many chapters in the still growing handbook of primatology (Hofer, Schultz and Starck, eds.; Basel: Karger, 1956–63), contributed by outstanding experts in their special fields of anatomy, contains a mass of evidence for essentially the same final conclusion, namely the close, general and basic similarity of all recent Hominoidea. Regardless of which anatomical system is being described or which ontogenetic processes, the higher primates invariably appear as one closely interrelated group within which there have evolved innumerable structural specializations, representing either differing degrees of the same trends or else more or less diverging adaptations for new functions.

From the topics presented in this paper, selected largely from the writer's own studies, no different general conclusions can be drawn. The sum total of the accumulated comparative-anatomical data, observations on age changes and findings on measurable, quantitative variations in recent primates can be most satisfactorily explained by the assumption of one single origin of all Hominoidea, no matter when and where that may have been, from which all descendants had become endowed with a multitude of corresponding potentialities for later changes. The enlargement and perfection of the brain, the increased durations of life periods, the immaturity of the newborn, and a thousand other such characters, common to all, permit no other interpretation besides the one just mentioned.

From more detailed results of primatological studies it appears that in general we are also forced to the conclusion that the Hominidae must have started on an independent course nearly as early as the Hylobatidae had begun their mostly more modest specializations and certainly before the splitting of the Pongidae into the three surviving branches. Among the latter the two recent genera of African apes have *in general* more in common than either one has with the Asiatic orangutan, but this difference does not seem to call for more than their generic separation nor for anything beyond the adequate single subfamily of Ponginae. The exceptionally great variability in many characters of all recent great apes supports the assumption that their diverging specializations must have originated at comparatively recent dates, and this particularly in the case of the two African genera.

Such modern discoveries of extremely close similarities between man and chimpanzee and (or) gorilla in regard to their chromosomes or certain biochemical qualities, as reported in this book, cannot as yet justify any radical revision in the classification of these primates. Such evidence, though of greatest interest, is more than counterbalanced by the mass of profound differences found in all sorts of other characters of recognized reliability. Among the latter dermatoglyphics represent a good example, also discussed in this volume. How careful we still have to be in the taxonomic interpretation of the results of modern biochemical studies may be indicated by the following example: Bamann and Gebler (1960) also had found practical identity in the stereochemical specificity of the liver esterase in man and chimpanzee, but marked corresponding differences between gibbon and orangutan and between the latter and chimpanzee. However, they later reported (Bamann *et al.*, 1962) that stereochemical identity exists also in such widely different mammals as domestic goats and chamois.

Since many impressive specializations of recent higher primates are commonly and more or less convincingly blamed on different types of locomotion, it may be pointed out that certain basic anatomical preparations for brachiation as well as simultaneously for erect posture have developed in all Hominoidea and that consequently all can use either bimanual or bipedal locomotion with at least some ease. Many boys still can brachiate fully as well as any old gorilla, but no ape can sustain upright walk nearly as long as can every man.

The development of adaptations for bipedal locomotion must have begun sooner and been more decisive for hominids than the modifications for preferred brachiation which seem not to have become perfected until rather late in the evolution of the few extreme brachiators. That even the earliest known forms of *Homo* had already acquired fully erect posture sooner or later during their infancy can be concluded from all available evidence. Possibly in connection with this precocious and one-sided specialization has man remained more conservative than apes in regard to some phylogenetic changes particularly in the vertebral column. Man's conservative nature is also evident from his lack of any such extreme sexual dimorphisms as have developed in a great variety of apes and monkeys.

BIBLIOGRAPHY

BAMANN, E. and H. GEBLER.
1960. "Die Superfamilie *Hominoidea* vom Gesichtspunkt der Konfigurationsspezifität der Esterase der Leber von Silbergibbon, Orang-Utan, Schimpanse und Mensch." *Hoppe Seyler's Z. f. physiol. Chemie*, 319:180–190.

BAMANN, E., H. GEBLER, B. SCHUB and H. FORSTER.
1962. "Die stereochemische Spezifität der Esterase der Leber, der Lunge und der Niere—ein biologisches Merkmal zur Klärung entwicklungsgeschichtlicher Zusammenhänge." *Naturwissensch.*, 49:281–282.

BOLK, L.
1926. *Das Problem der Menschwerdung*. Jena.

CHOPRA, S. R. K.
1957. "The cranial suture closure in monkeys." *Proc. Zool. Soc. London*, 128:67–112.

FRECHKOP, S.
1954. "Les relations génétiques entre les anthropomorphes et les autres singes." *Vol. Jubilaire Victor van Straelen* (Bruxelles), 2:1027–1062.

GENOVÉS, S.
1954. "The problem of the sex of certain fossil hominids, with special reference to the Neanderthal skeletons from Spy." *J. Roy. Anthropol. Inst.*, 84:131–144.

HELLMAN, M.
1936. "Our third molar teeth; their eruption, presence and absence." *Dental Cosmos*, 78:750–762.

HILL, W. C. O.
1959. "The anatomy of *Callimico goeldii* (Thomas)." *Transact. Amer. Philos. Soc.*, n. s. 49:1–116.

REMANE, A.
1954. "Methodische Probleme der Hominiden-Phylogenie, II." *Z. f. Morphol. u. Anthropol.*, 46:225–268.
1961. "Probleme der Systematik der Primaten." *Z. f. wissensch. Zool.*, 165:1–34.

SCHULTZ, A. H.
1931. "The density of hair in primates." *Human Biol.*, 3:303–321.
1933. "Observations on the growth, classification and evolutionary specialization of gibbons and siamangs." *Human Biol.*, 5:212–255 and 385–428.
1934. "Some distinguishing characters of the mountain gorilla: *J. Mammal.*, 15:51–61.
1936. "Characters common to higher primates and characters specific for man." *Quart. Rev. Biol.*, 11:259–283 and 425–455.
1938. "Genital swelling in the female orang-utan." *J. Mammal.*, 19:363–366.

1944. "Age changes and variability in gibbons." *Amer. J. Phys. Anthropol.*, n. s. 2:1–129.

1947. "Variability in man and other primates." *Amer. J. Phys. Anthropol.*, n. s. 5:1–14.

1948. "The relation in size between premaxilla, diastema and canine." *Amer. J. Phys. Anthropol.*, n. s. 6: 163–179.

1949. "Sex differences in the pelves of primates." *Amer. J. Phys. Anthropol.*, n. s. 7:401-424.

1950. "The specializations of man and his place among the catarrhine primates." *Cold Spring Harbor Symp. on Quant. Biol.*, 15:37–53.

1950. "The physical distinctions of man." *Proc. Amer. Philos. Soc.*, 94:428–449.

1952. "Vergleichende Untersuchungen an einigen menschlichen Spezialisationen." *Bull. Schweiz. Ges. f. Anthropol. u. Ethnol.*, 28:25–37.

1952. "Über das Wachstum der Warzenfortsätze beim Menschen und den Menschenaffen." *Homo*, 3:105–109.

1955. "The position of the occipital condyles and of the face relative to the skull base in primates." *Amer. J. Phys. Anthropol.*, n. s. 13:97–120.

1956. "Postembryonic age changes." *Primatologia (Basel)*, 1:887–964.

1956. "The occurrence and frequency of pathological and teratological conditions and of twinning among non-human primates." *Primatologia (Basel)*, 1:965–1014.

1957. "Past and present views of man's specializations." *Irish J. Med. Science*, 1957:341–356.

1960. "Age changes in primates and their modifications in man." *Human Growth*, Oxford: Pergamon Press. Pp. 1–20.

1960. "Age changes and variability in the skulls and teeth of the Central American monkeys *Alouatta*, *Cebus* and *Ateles*." *Proc. Zool. Soc. London*, 133:337–390.

1961. "Some factors influencing the social life of primates in general and of early man in particular." *Viking Fund Publ. in Anthropol.*, 31:58–90.

1961. "Vertebral column and thorax." *Primatologia (Basel)*, 4, Liefer. 5:1–66.

1962. "Die Schädelkapazität männlicher Gorillas und ihr Höchstwert." *Anthropol. Anz.*, 25:197–203.

1962. "Metric age changes and sex differences in primate skulls." *Z. f. Morphol. u. Anthropol.*, 52:239–255.

SIMONETTA, A.
1957. "Catalogo e sinonimia annotata degli ominoidi fossili ed attuali (1758–1955)." *Atti Soc. Tosc. Scienze Naturali*, B, 64:51–112.

SNOW, C. E.
1962. "An old Hawaiian population on Oahu." *Amer. J. Phys. Anthropol.*, n. s. 20:69–70.

STARCK, D.
1962. "Der heutige Stand des Fetalisationsproblems." *Z. f. Tierzüchtung u. Züchtungsbiol.*, 77:1–27.

STEWART, T. D.
1962. "Neanderthal cervical vertebrae." *Bibliotheca primatol.*, Fasc. 1:130–154.

TOERIEN, M. J.
1957. "Note on the cervical vertebrae of the La Chapelle man." *S. African J. Science*, 53:447–449.

Vogel, C.
  1960. "Variabilität und Formenentwicklung der Unterkiefer rezenter Anthropoiden." *Diss., Univ. Kiel* (not yet published).

Weinert, H.
  1932. "Ursprung der Menschheit." *Ueber den engeren Anschluss des Menschengeschlechtes an die Menschenaffen.* (last edition 1944) Stuttgart.

Wood Jones, F.
  1948. *Hallmarks of mankind.* Baltimore.

THE EVALUATION OF CHARACTERISTICS OF
THE SKULL, HANDS, AND FEET
FOR PRIMATE TAXONOMY

JOSEF BIEGERT

INTRODUCTION

THE MORPHOLOGIST will welcome modern efforts (DeVore and Lee, 1963) to obtain an understanding of the behavior of primates in their natural habitats. For morphology is in many respects an expression of organ function and thereby is the result of an organism's adaptation. A knowledge of primate ecology therefore is an important key to the understanding of the particular status of organs and their structures.

The following example illuminates the type of complex functional-anatomical relationships which are involved and the extent to which causal dependencies can be uncovered. The skull of male baboons differs in many respects from that of females. These differences, however, result from the fact that the masticatory apparatus of males is much larger than that of females. In addition, differences in the size of head organs can often be traced back to differences in body size. In baboons, among which males weigh twice as much as females, the size of the masticatory apparatus is not only absolutely but also relative to brain size substantially larger in males than in females. DeVore and Washburn (in press) have recently demonstrated from field observations that special ecological selective factors are operative in producing this sexual dimorphism. Thus, the sexual dimorphism of the baboon skull is also conditioned by special ecological influences.

Such an approach which utilizes different biological modalities not only satisfies the need to demonstrate causal interrelationships but also has considerable significance for the "evaluation" of characteristics used for taxonomic purposes. To cite an example, the deeply indented incisura ischiadica of the ilium of *Australopithecus* is one of many characters which *Australopithecus* has in common with *Homo*. This character immediately attains higher significance if we understand its meaning. Such a deeply indented incisura ischiadica signifies an approach of the sacral articulation toward the hip joint, as well as a shifting of the glutaeus superficialis muscle behind the transverse axis of the hip joints. From our insights into the relationships between form and function in recent primates

116

we can conclude that *Australopithecus* had developed an erect posture. This manner of locomotion is, however, a distinctive characteristic of the Hominidae. For this reason the character "indented incisura ischiadica" demands the placement of *Australopithecus* within the family Hominidae. This example also shows that for taxonomic problems we should not evaluate characters per se but instead evaluate through an understanding of their functions the actual living condition of an organism.

If we are to use skeletal structures for taxonomic purposes—in fossils the only things available apart from the dentition—it is important that we concurrently learn through experiments and observation on recent primates how these structures are conditioned, i.e., we must understand their biological significance. Reliance only on topography for taxonomic purposes frequently leads to erroneous classification as will be shown by citing examples from the skull. Often an entire series of characters are influenced by one particular causality. Conversely, identical characters can be conditioned in entirely different manners.

Examples taken from the hands and feet demonstrate how different structures of an organ have different adaptabilities. For example, not only the proportions and the structures of bones and joints but also the size of the volar pads and the pattern of flexion creases allow recognition of close adaptive responses to particular manners of locomotion. The embryonically determined dermatoglyphic pattern, however, is predominantly a species character. Dermatoglyphic patterns obviously reflect a polygenic background. Viewed taxonomically it is a system of characters analogous to the cusps and furrows of the teeth. The typical conditions of the teeth reflect species and group characters more clearly than they reflect adaptations to definite functions (compare bilophodonty, *Dryopithecus* pattern). In the same manner, dermatoglyphics are often taxonomically valuable per se.

An understanding of the functional abilities of hands and feet in primates can be obtained only in part from a study of their structure. This results from the fact that the motor and tactile abilities of the chiridia depend also on the organization of definite relevant centers of the brain. A "higher" development of such centers signifies a considerable enhancement of the abilities of the hands and feet. We can decipher only in part from skeletal structure (as in fossil primates) the degree of motor capability and can glean nothing at all about the degree of tactility of the chiridia. The bones of the hands and feet usually are significant taxonomic markers only in the later stages of specialized lines (for example the pongid hand, the human foot).

THE SKULL

In conjunction with the dentition, characters of skull structure have long been used as criteria for taxonomic studies of fossil primates. In the long course of reliance on features of the skull for taxonomic purposes it has become apparent that different characters are of differing taxonomic value. Thus there is little doubt as to the relatively high taxonomic value of the osseus ear region (vascular pattern, position and structure of the tympanicum, composition of the bulla, etc.). Conversely, the taxonomic significance of the topography and shape of the skull is very different, and its value is often questionable.

As can be seen in the literature, cranial characters were given different values by different taxonomists and many features were readily abandoned in the light of later studies. Thus, *Necrolemur* was, on the basis of its *Tarsius*-like skull, included with the Tarsiiformes. A subsequent analysis of the ear region, however, revealed a typical lemuriform condition, so that today, in spite of similarities in skull form, a close relationship between *Necrolemur* and *Tarsius* is no longer acceptable.

What, then, is the explanation for the inadequacies of skull structure in reference to taxonomic applications? A survey of both modern and fossil primates leads to the undeniable observation that similar skull topographies have evolved in widely divergent groups, i.e., there have been numerous occurrences of convergent or parallel evolution in many different lines. Such convergencies in cranial structure resulted from the fact that the head organs themselves have often evolved along similar lines. This is an expression of exposure to similar environments, and therefore to similar selective factors which are active in arboreal environment, the "basic milieu" of the primates. We find in both the prosimians and the simians the tendency towards a reduction of olfactory organs, the development of color- and stereoscopic-vision, the enlargement of the neopallium, the specialization of the masticatory apparatus etc. All of the changes in the skull occurring concomitantly with such phylogenetic changes in the head organs therefore are (without further investigation) not significant qualities for a differential taxonomic diagnosis. Thus, if it were not for more significant characters (of the ear region and dentition), the use of skull topography alone would be disputable to establish an unequivocal placement of the Archaeolemurinae with the simians. If, however, such possibilities for mis-classification are so apparent, then it is important for the taxonomist to understand the influences of head organs on skull structure.

At this point it must be emphasized that presently we are not in the position to understand the entire complex array of effects which are important in altering the structure of the primate skull, but are only in a position where approximate ideas may be obtained. Still, in recent years, much progress was reported—e.g., Biegert (1956, 1957); Hofer (1952–60); Kummer (1952, 1953); Moss and Young (1960); Starck (1953), with special literature—which justifies a discussion of some of the problems involved.

Experimental and morphological analyses of the primate skull in which the head organs and the entire pattern of existence were taken into consideration leave no doubt that the skull is somewhat plastic in the sense that it adapts itself to spacial, mechanical and functional trends which the head organs have undergone. Alterations of skull topography are in this respect results of changes in size, form and function of head organs during the course of evolution. These changes are to be regarded as selectively determined adaptations to a definite environment and behavior. This, however, implies that the organs of the head can evolve independently of one another. Changes in head organs which have occurred during the evolution of primates and which have special "formative" significance for the skull structure are listed as follows:

1. Enlargement of the brain as characterized by (a) a general enlargement corresponding to an enlargement of body size and (b) enlargement of special definite regions of the brain, among which the neopallium is of particular significance in this context.

2. Enlargement of the masticatory apparatus (in the widest sense) whether (a) in conjunction with an increase in body size, or (b) in the sense of specialized enlargements.

3. Functional specializations of the masticatory apparatus over the generalized form (as is found in Tupaiidae) which are (a) similar to those found in rodents (as in *Daubentonia*) and (b) adaptations in the "service" of the development of an intensified "working-over" of the food by the molars through a grinding movement (the "universally specialized type").

4. Changes in the size of the eyes which are (a) dependent upon body size, (b) specialized adaptations, and

5. a tendency to rotate the eyes forward to facilitate stereoscopic vision.

6. A tendency towards reduction of olfactory organs (reductions of nasal chamber, loss of turbinals).

7. Specializations in the sizes of organs in the subbasal space between the cervical vertebrae and the mandibular symphysis (vocal apparatus, laryngeal sacs).

To this sort of changes in head organs during the course of evolution the responses of the skull have been at times more regional, while at other times they have involved the general topography. In the process of evolution, the adaptations of the enveloping and propping framework known as "skull" have occurred in a manner to assure the functional integrity of the head.

In this connection it must be mentioned that the type of adaptation which resulted depended upon when (in historical time) and where (in the phylogenetic tree) the changes in head organs occurred. The sort of adaptations which the skull undergoes in response to new selective factors may differ if organisms of different evolutionary lines are involved. Because the skull—on the basis of earlier alterations in head organs—has already become differently organized, an enlargement of the masticatory apparatus induces in the Hominoidea (e.g., *Gorilla*) quite different topographical results compared with those found in the Lemuriformes (e.g., *Megaladapis*).

At this point in the discussion the manners in which alterations in the size, shape and function of cephalic organs can affect the skull should be demonstrated.

(A) The expansion of the brain during the course of the evolutionary history of the primates and in particular the repeated occurrences of special enlargements of the neopallium with respect to the rest of the brain, have had important influences on primate skull topography. Alterations not only in brain size, but also in brain shape have resulted from these changes. Roughly speaking, the brain shape became rounder compared to its primitively elongate structure. Adaptations in the structure of the cranium accompanied these changes in the size and the shape of the brain. These adaptations are particularly impressive when viewed in sagittal section (see Fig. 1).

In connection with the expansion of the neopallium we find

(a) a super-elevation of the brain cavity over the fore- (pre-sellar) and rear- (post-sellar) base of the skull. The skull cap, so to speak, expanded over the base of the skull and we find the development of a frontal and occipital vaulting. Such changes in the skull cap could also be imitated in dwarf forms which have relatively large brains (see Starck, 1953, example of Pekinese dog). However, in such

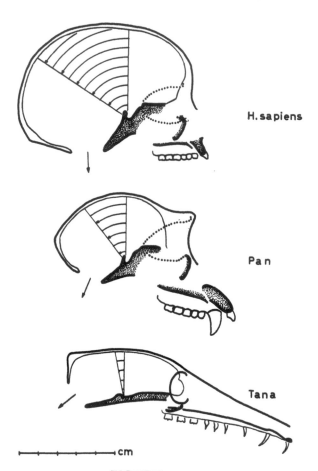

FIGURE 1

The rolling up" of the neurocranium in connection with the enlargement of the neopallium. These and the following sagittal sections of the skulls are oriented on the pre-sellar skull-base (*Tana* 2/1, *Pan* 1/2, *Homo sapiens* 1/2 to the scale).

cases we do not find the second important criterion of primate brain evolution, i.e., an increased kyphosis of the base of the skull.

(b) Not only the skull cap, but also the base of the skull adapts to changes in the form of the brain. Corresponding to a progressive enlargement of the neopallium the primitive elongated base of the skull (as in *Tupaia*) becomes more and more bent (increasing skull-base kyphosis). Generally the more pronounced this kyphosis is, the greater the special enlargement of the neopallium. This is evident, e.g., in the series *Tupaia—Pan—Homo sapiens* (see Fig. 1).

Such a phylogenetically increasing kyphosis of the skull base has great significance for the topography of the entire skull. It can fundamentally alter the sagittal position and orientation of the posterior part of the skull (nuchal plate, foramen magnum, post-sellar skull base and occipital condyles) with respect to the entire anterior part of the skull (frontal part of the brain case, orbitae, pre-sellar base and splanchnocranium). If one orients the skull horizon consistently either on the cerebral side of the post-sellar base (Clivuskoordinate, Kaelin, 1954–56) or on the cerebral side of the pre-sellar base (Biegert, 1957) these relationships become clear.

On didactic grounds we choose here to represent pictorially the orientation with respect to the pre-sellar base of the skull (see Figures 1–5). Such an orientation shows that corresponding to an increased kyphosis of the skull base there follows, so to speak "automatically," an increasingly horizontal orientation of the originally vertical nuchal plate, a shifting of the foramen magnum from the back of the skull to its basal surface, so that it comes to face downward instead of directly backward, and an increasingly more oral position of the occipital condyles. An increased kyphosis of the base of the skull is like a rolling-up of the entire neurocranium (Weidenreich, 1941), as a counterclock-wise rotation of the posterior cranium with regard to the anterior brain case (see Fig. 1). Therefore, the alterations in the topography of the skull of *Australopithecus* as contrasted with that of the Pongidae (Le Gros Clark, 1955) are proof of a distinct enlargement of the neopallium (see Fig. 5).

All of these characters are, contrary to a widely held opinion, not criteria for upright locomotion. Comparative investigations, considering the entire organization of the organism, conclusively show that between the manner of locomotion on the one hand and the occipital condyle position. the orientation of the foramen magnum and the condition of the base of the skull on the other hand, no causal relationship whatever exists. The kind of locomotion of fossil primates can be determined only from the condition of the post-cranial skeleton.

(B) The neurocranial structure can in addition be strongly influenced by the masticatory apparatus. In this context the relative sizes of the brain and the masticatory apparatus have predominant significance. If the masticatory apparatus, e.g., in conjunction with an increase in body size, becomes especially large relative to the brain, then its influence on the structure of the neurocranium is impressively exerted. It causes changes in the development of the superstructures, especially deviations of the tabula externa with regard to the tabula interna of the skull cap. Since such conditions of the superstructures via the masticatory apparatus, are dependent on body size, their taxonomic value has become recognized as being worthless. Therefore, *Australopithecus robustus* with its sagittal crest is not less hominid than *Australopithecus africanus* which lacks such a structure.

The size of the masticatory apparatus has also an influence on the rolling-up of the neurocranium (see above), and therewith an influence on the extent of the kyphosis of the base of the skull. A masticatory apparatus which is large relative to the brain reduces the rolling-up of the neurocranium. It is a consequence of this, that we, by comparing two fossil forms, can only correctly estimate the extent of neopallium enlargement from the topography of the skull if the relative sizes of the masticatory apparatus and the brain are approximately equivalent. Therefore we will not be wrong in saying e.g., that *Saimiri* has an evolutionarily more highly expanded brain than *Aotes* (see Fig. 2). In forms with relatively large masticatory apparatuses, the possibility exists that the extent of neopallium enlargement will be underestimated. Therefore we should be correct in comparing *Australopithecus* with *Pan* and the female *Gorilla*, but not with the adult male *Gorilla*.

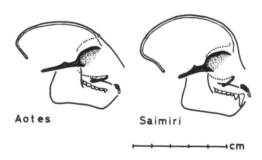

Aotes   Saimiri

├─────┼─────┼─────┤cm

FIGURE 2

The topography of the skull in *Saimiri* and
Aotes (1/1 to the scale).

The relative sizes of the brain and masticatory apparatus are influenced by body-size. Quite generally it can be stated that allometry influences the brain negatively whereas the masticatory apparatus does so positively or isometrically. Compared with the brain the masticatory apparatus is affected to a greater degree by an enlargement in body size. Primates with large bodies, therefore, often have a larger masticatory apparatus relative to brain size than small primates. Differences in skull topography, therefore, can be expected in consequence of differences in body size alone.

This principle, however, can be circumvented if a specialized expansion of the neopallium (compare *Ateles* with *Alouatta*) or a specialized reduction of the masticatory apparatus occurs (compare *Homo* with Pongidae). Since we are here concerned with taxonomically significant changes in the evolution of head organs which affect skull topography we must always refer back to body size in order to recognize the effects brought about by these evolutionary changes in the head organs on the structure of the skull. For example, the differences in the sizes of the brain and also of the masticatory apparatus of *Pan* and *Gorilla* are "conditioned" mainly by different body size. Only in this manner can it be demonstrated that members of the genus *Homo* have undergone a specialized reduction of the masticatory apparatus.

Similarly, from the brain volume (skull capacity) and the topography of the skull of *Australopithecus* we are able to recognize that *Australopithecus* exhibits a slight enlargement of the neopallium, compared with Pongids of the same body size. It follows, therefore, that the brain volume, the size of the masticatory apparatus and with it also skull topography have little taxonomic significance per se. These characters become significant only when their causal relationships are understood. Without reference to body size the evaluation of characters of the skull is often impossible.

Finally, a comparison of the skulls of *Saimiri* and *Homo sapiens* may serve as an instructive example. In both forms the skull caps are roundly vaulted and lack superstructures (sagittal and occipital cristae, torus supraorbitalis), the back of the head protrudes and the occipital condyles, foramen magnum, orbital funnels and splanchnocranium have acquired a subcerebral position. Systems of classification, which have been and are still being used for fossil Hominoidea (compare the discussion on *Australopithecus* and *Oreopithecus*) and which rest only on resemblances, should inevitably lead to the conclusion that there is a specially close relationship between *Saimiri* and *Homo*. Such resemblances, however, prove to be causally unrelated: they are in *Saimiri* a reflection brought about by the small body size of this genus, whose brain is relatively large and the masticatory apparatus small. The man-like form of the skull is a result of the smallness of the body. The condition in *Homo sapiens* has an entirely different basis. Here, the *Saimiri*-like skull configuration is also largely determined by the relative preponderance of the brain over the masticatory apparatus. However, the relative largeness of the brain in *Homo* is a product of an extreme evolutionary development of the neopallium and the small masticatory apparatus is the product of a specialized evolutionary reduction. In this instance topographically similar characters have entirely different taxonomic significance.

(C) However, not only the size but also the functioning of the masticatory apparatus is important for primate skull topography. In this respect the manner in which the nourishment is masticated is important. In primates which have remained primitive the jaw apparatus serves only for the grasping, killing and almost immediate devouring of the food (Tupaiidae) while in the more highly evolved type of simian primates ("universally specialized") the food is intensively "worked over" by a grinding movement in the molar region prior to ingestion.

These different manners of mastication call for different mechanical constructions which in themselves alter skull topography.

In the generalized form (Biegert, 1956) the dental arch and the mandibular joint lie on the same level (see Fig. 3, *Lemur*) while in the "universally specialized" type the mandibular joint has moved higher up than the dental arches. This change produces uniformly a super-elevation of the mandibular condyles with respect to the base of the mandible. Such a construction of the masticatory apparatus shows the principle of the nutcracker, in which the distance between the hinge (mandibular joint) and the crushing surfaces (tooth rows) results in

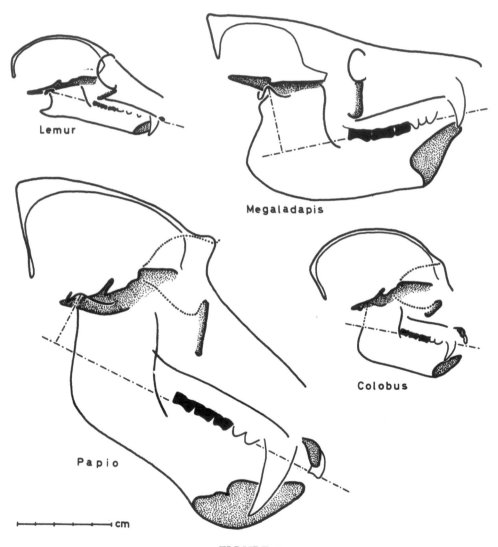

FIGURE 3

The topography of the splanchnocranium in connection with the size and functional type of the masticatory apparatus (*Megaladapis* 1/2, others 1/1 to the scale).

bringing the whole surfaces into immediate contact with each other when the pinchers are closed.

In these primates we find an elevated and steeply inclined mandibular ramus and as a functional adaptation to the more highly differentiated, now three-dimensional, mobility of the mandible we also find a rounded angulus mandibulae in place of the primitively long and slender processus angularis. Therefore, within the simiae the mandible—especially when found as an isolated fossil, as Vogel, 1961, has recently shown—has comparatively little value and is always much less significant than the dentition for taxonomic purposes.

The accompanying changes in the mandibular joints also are very uniform. The taxonomically distinctive character is the new development of an eminentia articularis (tuberculum articulare). The articular tubercle, contrary to reports in the literature, is not a special character of hominids, nor are small post-glenoidal processes and a posterior delimitation of the articular pit by the tympanicum (Biegert, 1956). If *Australopithecus* in its mandibular joint seems to show a hominid tendency, this is due primarily to the angle formed between the tympanicum and the pars glenoidalis, viewed sagitally, is more acute than in the majority of the pongid skulls. This condition, however, is a consequence of the fact that, compared to the Pongids, there is a greater "rolling-up" of the neuro-cranium which has also altered the lateral region of the skull base (see above).

The fact that the required vertical displacement between the articulation and the masticatory surface of the upper tooth row—in contrast to the mandible—was reached in different manners, is very important in interpreting the topography of the skull. The following possibilities, either singly or more often in conjunction with each other, have arisen as conditions which can produce this required vertical displacement: (a) an alteration in the position of the upper jaw with respect to the neurocranium by an upward-tilting of the upper jaw (airorrhynchy after Hofer, 1952; see Fig. 3, compare *Megaladapis* with *Lemur*); (b) by a ventral displacement of the floor of the nasal chamber and of the dental arch with respect to the neurocranium (anterior skull base) and the orbital funnels (compare in Fig. 3 *Papio* and *Colobus* with *Lemur*); and finally by (c) an additional ventral displacement of the dental arch with respect to the floor of the nasal chamber, i.e., through the development of a palate vaulting. Conditions (b) and (c) are found in more highly evolved forms (simians), while condition (a) may be regarded as a specialization (*Megaladapis*) occurring in lower evolutionary stages (Lemuriformes). Corresponding with conditions (b) and (c) one finds in simian primates concomitantly with a reduction in the olfactory organs the tendency to develop a higher and subcerebrally placed upper jaw, i.e., a more vertically oriented face.

The tendency in the "universally specialized" type of masticatory apparatus to move the molars topographically under the neurocranium is evident. Functionally, a subcerebral location of the teeth is advantageous for forms in which the molars have a considerable grinding and tearing capacity. In this manner the very considerable mechanical stresses involved in chewing are transferred, via the gums, nasal chamber and orbital funnels (which are to be regarded as typical frame constructions) directly to the skull-cap (which is comparable to a vault) where they are absorbed. This sub-cerebral positioning of the dental arches represents so to speak the optimal adaptation of skull topography to the "universally specialized" masticatory apparatus and, therefore, is frequently found in the simian primates.

However, a "universally specialized" masticatory apparatus has also developed in some prosimians (e.g., Indriidae with Archaeolemurinae) together with a tendency toward reduction of the olfactory organ and a rotation of the eyes toward the front of the skull. Therefore, it is not surprising that prosimians can be found which exhibit extraordinarily monkey-like skull conditions (for example *Hadropithecus*). The undoubtedly convergent evolutionary tendencies of the masticatory apparatus, eyes and olfactory organs cited above have "forced" an evolutionary alteration of skull structure which is otherwise characteristic of monkeys and apes. This is a particularly instructive example for demonstrating that similarities in skull structure are not in themselves criteria for advocating close evolutionary relationships since in many evolutionary branches of the primates similar trends for specialization of head organs have been manifested.

(D) In simians with a masticatory apparatus not surpassing the brain in size one finds as a rule the "favorable" sub-cerebral location of the jaws. This sub-cerebral location of the splanchnocranium exists particularly in the smaller platyrrhines and catarrhines. For many reasons it seems that small body size and sub-cerebral location of the splanchnocranium are generalized characters of the simians. If the masticatory apparatus has become particularly large compared with the brain in some simians there is always a secondary forward relocation of the dental arch with respect to the brain case (compare *Papio* with *Colobus; Pan* and *Gorilla* with *Hylobates*). As already mentioned the masticatory apparatus increases in size much more than the brain in response to the evolutionary increase in body size. Thus, there can be a forward relocation of jaws on the basis of an evolutionary increase in body size, resulting in greater prognathism. It can be said with certainty that the character "prognathism" is influenced by many factors which must be taken into consideration for an evaluation of the taxonomic significance of this feature. As far as the head organs are concerned, a pronounced prognathic condition can be "caused" by (a) a large olfactory organ, (b) an increase in body size and thereby in the masticatory apparatus and (c) an increase in the size of the organs in the subbasal space (see below). For systematic evaluation the character "prognathy" per se is of doubtful value. It is a primitive condition only in such cases in which it is induced by a larger olfactory organ (nasal chamber with numerous turbinals) as in some prosimians. On the other hand a "vertical face" is by no means a specific character of hominids, as is often assumed.

The location of the dental arch is mainly responsible for the shape of the frontal part of the brain case and the facial skeleton. Among the simians with an enlargement of the masticatory apparatus the dental arch and the upwardly directed masticatory stresses were shifted forward in relation to the brain cavity. In order to maintain the spatial and functional continuity of the skull there occurs a forward relocation of the entrance of the orbital funnel and the anterior root of the jugal arch compared to the brain cavity. A torus supraorbitalis, therefore, represents an anterior "arch-construction" for the reception of the masticatory stresses. The development of the torus supraorbitalis, as a sagittal extension of the orbitae and the postorbital constriction etc. are thus dependent on the sagittal position of the molar row with respect to the brain cavity and are in this limited sense not genetically independent.

However, even for the particular structure of the torus supraorbitalis other influencing factors are not excluded. It is probable that the directions of force vectors of the masticatory musculature play a role. Also selective factors, as for instance a "terrifying countenance," might be involved, for a strongly developed torus has its effect on the exterior of the face. In this connection the bony "cheek swelling" of male mandrills and dogueras can be mentioned. The far more imposing appearance in mandrills cannot be explained on the basis of mechanical requirements alone. The bony cheek swellings in male mandrills can be explained through their contribution to the formation of the "imposing face." Thus selective factors influencing "outer appearance" have obviously been effective in altering skull characters.

(E) Changes in size of the eyes and in their position through a forward rotation to facilitate stereoscopic vision have influenced not only facial skull structure but also the structure of the presellar base of the skull. If the eyes are relatively large compared to the neurocranium (as specializations in *Tarsius* and *Aotes*, or dependent on body size as in *Saimiri* and Callithricidae) and are at the same time placed close together, then the planum sphenoideum approximates in its position and form the niveau of the orbital roofs. The pre-sellar cerebral skull base becomes elevated in its entirety with a predominant elevation of its anterior portion (in accordance with the funnel shape of the orbits). In extreme cases (*Tarsius*) this can lead to an elevation of the brain (airencephaly) as pointed out by Starck, 1953.

As already shown the sagittal position of the orbits is influenced by the position of the dental arch relative to the brain case. One finds no distinct correlation between the size of the eyes and the size of the orbits (Schultz, 1940) for the following reasons: The orbital funnels through a sagittal forward relocation of the tooth row can become distended and thereby enlarged. The eyes (as parts of the brain) are negatively allometrically related to body size and beyond this can undergo specialized enlargements (as in *Tarsius* and *Aotes*).

The degree of approximation between the orbits depends also upon the size of the olfactory organ (nasal chamber with turbinals) in that a relatively large olfactory organ prevents the eyes from coming close together. Therefore the eyes are much closer together in the microsomatic simians than in most of the prosimians.

(F) The reduction in size of the primitively macrosomatic olfactory organ is a characteristic arboreal adaptation of the primates. In the primitive condition (compare Tupaiidae) the length of the snout and therewith the degree of prognatism is determined almost in entirety by the size of the olfactory organs. In the simians, however, the length of the snout and the degree of prognatism are evidently more dependent upon the size of the masticatory apparatus. Therefore, in the simians one can draw no conclusions from the size of the nasal chamber in regard to the development of the olfactory organ. The nasal framework in the simians is used for the reception and transference of the masticatory stresses in the same manner as the orbital funnels. In the simians, therefore, the olfactory organ—unlike the conditions found in many prosimians—is of subordinate significance for the degree of prognatism and the facial profile. Here, specializations in the size of the organs in the subbasal cavity play a role in the determination of such facial characteristics.

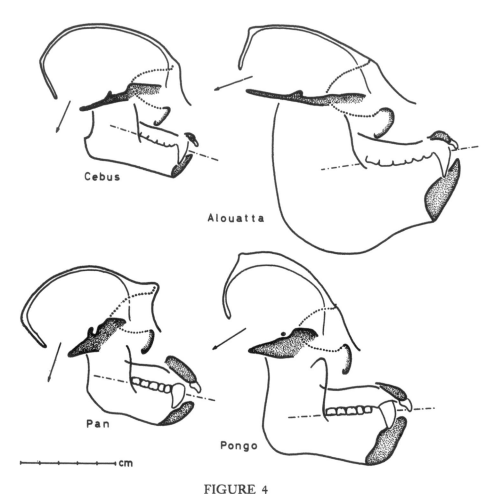

FIGURE 4

The topography of the skull in *Alouatta* and *Pongo* as a consequence of the specialized enlargement of organs in the subbasal space (*Cebus* and *Alouatta* 1/1, *Pan* and *Pongo* 1/2 to the scale).

(G) Skull topography can be influenced by organs which lie subbasally between the cervical vertebral column and the mandibular symphysis. In this context *Alouatta* with its specialized enlargement of the vocal apparatus is very impressive. In order to satisfy the great spatial requirements of these organs (compared to the size of the head) there results; (a) an extreme forward relocation and tilting up of the dental arch (airorrhynchie) through which the symphysis of the mandible is forwardly relocated and more vertically oriented; (b) an extreme backwards orientation of the foramen magnum and a posterior relocation of the occipital condyles through which the cervical column comes to lie far aborally with respect to the head; (c) also the peculiar form of the ramus mandibulae is caused by this specialization of the vocal apparatus in *Alouatta*.

In many respects the extreme alterations in skull topography in this Platyrrhine (compare *Alouatta* with *Cebus* in Fig. 4) are in the first instance doubtlessly an expression of the peculiar evolutionary adaptation of the vocal apparatus. Thus the special enlargement of the vocal apparatus causes the over-emphasized so-called primitive topographical characteristics. Also an extremely extended base of the skull in *Alouatta* is a result of the specialization of the vocal apparatus, for even in *Tupaia* with its very primitive brain there is no similarly extended base of the skull.

By comparing the skull topography of *Pongo* with that of the African apes (see Fig. 4, *Pongo* and *Pan*), one recognizes in the foreward position and tilting up of the dental arch, the aboral position of the occipital condyles and the marked posterior orientation of the foramen magnum that *Pongo* shows divergent tendencies which are in many respects the same as in *Alouatta*. This brings to mind the extreme development of the laryngeal sacs in *Pongo*. It seems highly likely that this divergent skull topography of *Pongo*, compared with *Pan* and *Gorilla*, is determined almost completely by this evolutionary specialization of the laryngeal sacs. Such characters as the extreme forward position and tilting up of the dental arch with respect to the neurocranium, the exteme posterior orientation of the foramen magnum and the elevation of the nuchal plate are in *Pongo* not evidence of a lower evolutionary level than in *Pan* and *Gorilla*, but evidence of high specialization.

Since skull topography changes during evolution and there are convergencies in many evolutionary lines, and also since many differences are merely quantitative ones, then the differences which do arise are not in themselves very reliable keys to use for the determination of systematic relationships in the primates. This is especially so in the evaluation of fossil primates in which the head organs are often quite differently developed than in their recent descendants.

Thus *Proconsul*, on the basis of skull structure, can hardly be placed with certainty in the family of the Pongidae. Decisive is the more generalized structure of the head organs, particularly of the brain (see LeGros Clark and Leakey, 1951). This explains why the skull of *Proconsul africanus* (the only relatively well preserved specimen) is in many regards more similar to the generalized catarrhine model, as represented by types found among recent Cercopithecoidea, than to

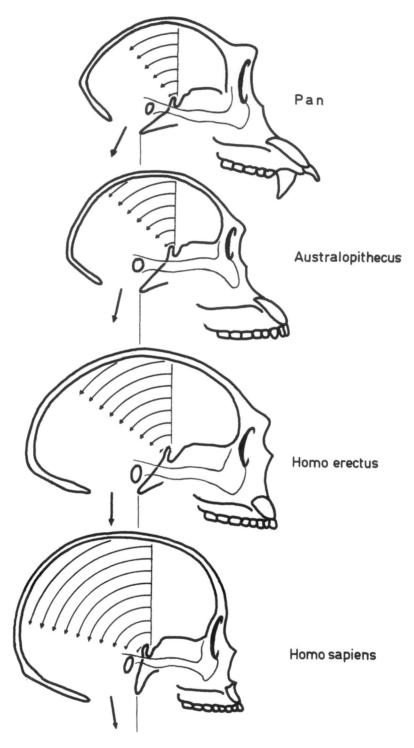

FIGURE 5
Comparison of hominid skulls with *Pan* (all reduced to the
same extent).

recent Pongids. Indeed *Proconsul* represents an even lower evolutionary stage of skull development in which the dentition has already attained pongid features. The systematic relation of *Proconsul* to the Pongidae can be settled with reference to the dentition because in the pongid line the dentition has reached the definite pongid pattern earlier than has the skull. For this diagnosis, therefore, the dentition is far more valuable than skull structure.

The same situation exists in Hominidae in which we find considerable differences in the evolutionary levels of the head organs and therewith also in the topography of the skull. Diagnostic keys, still prevalent in the literature, take the requirement of referring to different evolutionary levels in no way into consideration, since they rely only on recent members of the Hominidae.

Comparisons between presently known pleistocene Hominids already lead us to expect confidently that sooner or later we will find a "genuine" Hominid with pongid-like head organs. We thus expect to find in the early phases of hominid evolution a skull which has a topography of a kind intermediate between *Proconsul africanus* and *Pan*. The differences between the later hominid skull and the pongid rest in the long run upon (a) an enormous enlargement of the neopallium and (b) a specialized reduction of the masticatory apparatus. The accompanying changes of the skull can perhaps be determined by comparing *Australopithecus*, *Homo erectus* and *Homo sapiens* (see Fig. 5). In this listing *Australopithecus* shows the generalized hominid and at the same time is close to a pongid-like type of head organs, while *Homo sapiens* has attained the "highest" evolutionary level. Therefore, the skull topography of *Australopithecus* is in many respects ever so similar to that of the Pongidae: i.e., the skull of *Australopithecus* as a diagnostic tool for the determination of "hominidness" is of much lesser importance than the dentition or the pelvis. Not until *Homo erectus* and particularly *Homo sapiens* do we find a skull which is a "typically" hominid because only in the members of the genus *Homo* have the enlargement of the neopallium and the reduction of the masticatory apparatus become impressively distinct from the conditions found in the Pongidae.

Skull topography acquires diagnostic value for distinguishing Hominidae from Pongidae only in the later, progressive phases of hominid evolution. In this respect there are marked taxonomic differences between *Australopithecus* on the one hand and *Homo* on the other. Since the head organs of *Australopithecus* are of a generalized pattern, *Australopithecus* can be unreservedly regarded as a predecessor to *Homo* on the basis of his skull structure. This is particularly justifiable for *Australopithecus africanus*. On the other hand *Australopithecus robustus*, because of its skull and in particular on account of the specializations of the jaw complex (which are apparently directly dependent on the specialized size relationships between the molars and the front teeth), is less readily interpreted as an ancestral type.

For the taxonomic diagnosis of Hominidae in their entire temporal scope the skull is of much less significance than the pelvis and lower limbs. This, as has become clear in recent years, is due to the fact that upright posture and the concomitant emancipation of the hands are the biological and thereby also taxonomical key characters of the Hominidae (see Rensch, 1959, Washburn and Howell, 1960). For these reasons an entirely new selective plateau has been reached and new selective factors have become effective which in their turn continued the enormous enlargement of the neopallium and the reduction of the masticatory apparatus which are final results.

HANDS AND FEET

Only skeletal elements of the hands and feet of primates are found as fossils and even then there is rarely any degree of completeness of the material. Even from the best fossil finds only a very approximate idea of the functions can be obtained. Our knowledge of the evolution of primate chiridia, therefore, is mainly dependent upon results obtained from comparative studies of modern primates.

The often mentioned fact that the primitive pentadactyl condition of the primates and also other generalized conditions of the skeleton, musculature and the "walking pads" have been more tenaciously preserved than in many other orders of mammals, obscures the fact that the chiridia of primates are in many ways highly evolved organs. This paradox can be resolved through the realization that alterations in the structure of the brain played an equally important role as structural alterations in the chiridia themselves. Undoubtedly the higher evolutionary development of the motor and tactile capabilities of primate chiridia resulted to a large degree, if not decisively, from the special enlargement and differentiation of the precentral motor and postcentral sensory cortical regions of the brain and their mutual interconnections. This, however, implies that the degree of development of the hands and feet of fossil primates can be determined only with a great deal of uncertainty from the skeletal structure. Information which can be more readily extracted from skeletal structure concerns specialized adaptations for different manners of locomotion, which, however, is but one of many functions of primate chiridia.

The chiridia of primates evolved from a primitive organ for claw-climbing and walking—as it is still found largely unaltered in modern Tupaiidae—through the attainment of increased mobility of the digits II to V and particularly of the first digit's ability for grasping. The development of a grasping-climbing foot with a relatively active "big toe" which could be moved in various directions as an early adaptation to arboreal locomotion is one of the characteristics of primates.

Opposability of the thumb is not always present. Among the prosimians it developed as an adaptation to arboreal locomotion in the Lorisidae and in the Indriidae, while in the simians this condition is characteristic only of the Catarrhina. In the latter the opposable thumb exists less as an adaptation to locomotion, than as a result of the higher evolutionary development of the hand for use in numerous other functions such as procurement of food, hygiene, care of offspring, defense, three-dimensional touch, etc. This type of a highly developed hand, dependent upon a special development of the motor and sensory cortex

(v. Bonin and Bailey, 1961; Woolsey *et al.*, 1952), is a biological and taxonomic key-character of the catarrhine primates. Ultimately the attainment of the highest evolutionary perfection of the hand, i.e., its development in the Hominids as an organ for culture, resulted from the total emancipation of the hand from use in locomotion through the development of upright posture.

Primarily this process was achieved by a special expansion of the motor and sensory cortex regions (Penfield and Rasmussen, 1950), and only secondarily through structural alterations of the chiridia themselves. Already in the early phases of catarrhine evolution the biological foundations were created for the development of the hominid hand as an organ for culture. Evolutionary alterations in the position of the body, i.e., erect sitting and subsequently erect posture with a partial to complete "emancipation" of the hands, are found to be causal factors in the development of the catarrhine hand and these have become most important for the evolution of the hominid hand.

The higher development of the hands and feet of primates depends also upon the development and the differentiation of the volar integument with its special mechanical and tactile abilities. The latter development is only to be understood within the wider framework of the organism's entire organization and in particular with reference to the sense organs.

It is generally known that the adaptive plateau of the arboreal habitat resulted in a new orientation of the sense organs. The primitive predominant dependence on olfaction for distant perception declined strongly in favor of visual sensitivities which became highly developed with the acquisition of three-dimensional vision (optical spatial orientation). Much less attention has been given to the fact that among the other sense organs the touch receptors underwent a significantly higher development. One primitive mammalian organ for sensing the immediate environment, the tactile rhinarium on the snout, became lost in *Tarsius* and all simians, while on the volar surfaces of the hands and feet a special integument developed for the reception of tactile stimuli (friction skin, volar skin).

It is evident that the differentiation of touch receptors on organs such as the chiridia which are capable of multidimensional movement and cooperative activities brings about a vastly better sense of touch (three-dimensional touch, equivalent to the sense of spatial extension) than that attained by the rhinarium of the snout. Besides the differentiation of the skin and its tactile receptors, the higher development of the sensory region of the cortex and their inter-connection with the motor and visual areas of the brain is of greatest significance for the higher development of the tactility. In this sense again all the biological needs of a human hand were adaptively acquired long before the appearance of Hominidae and thus may be construed as "pre-adaptation." Primarily the hominid hand differs from that of other primates only through the degree of its differentiation.

The volar integument, like the skeleton, musculature, nervous tissue etc., is anatomically and functionally integral constituent of the primate chiridia. It is distinguished from ordinary integument by the following combination of characters: volar pads, flextion creases and ridge-forming skin. Wherever such skin is found (for instance also on the ventral side of the tails in *Tarsius, Alouatta, Lagothrix, Ateles, Brachyteles,* and on the dorsal side of the fingers in *Pan* and *Gorilla*), hairs and tallow glands are totally lacking, sweat glands are hypertrophied and multiplied, and a special microscopic folding of the stratum germinativum has developed as well as a special proliferation of tactile receptors.

From this combination of characters the two main functions of the volar integument emerge. Its elastic cushions with their ridged and lubricated surfaces have adhesive properties which make the chiridia highly suitable for grasping and arboreal locomotion. Secondly, because of the proliferation and the particular topographical location of sense receptors, it is a highly qualified organ for the perception of tactile stimuli. These two functions are then the points of attack upon which natural selection has operated.

All primates possess this specialized volar integument, although there is considerable variability in its formation. The volar pads were certainly a very early acquisition of mammals. The friction skin has developed secondarily and repeatedly (i.e., convergently) in many mammalian lineages, and even in some marsupials. The following characteristics of the friction skin should be considered for taxonomic purposes: the particular ground plan of pads (primary and accessory pads of the digits and of the palm and sole respectively); the configuration of the pads (molding, limitations, fusion during embryonic and adult life); the extent of the friction skin; the course (dermatoglyphic patterns) and the quality (long or short, coarse or fine) of the cutaneous ridges etc. (Biegert, 1961).

The embryonic distribution and configuration of the pads and concomitantly the dermatoglyphic pattern can differ according to species or genus to an extraordinary degree. These embryonic pad qualities are directly responsible for the precise patterns of the cutaneous ridges because (a) the cutaneous ridges in the area of these pads form either whorls, loops or arches dependent upon the configuration of embryonic pads at the time of epidermal folding (i.e., the formation of the cutaneous ridges); and (b) the embryonically determined epidermis retains its original ridge pattern to the minutest detail for the duration of life.

From comparative studies one can construct a hypothetical, but highly probable model (subprimate stage) out of which the chiridia of modern primates have developed. This model has the following characteristics:

a) the primitive phalangeal formula $3 > 4 > 2 > 5 > 1$ with the third digit of the hand and foot as functional axis;

b) all the digits of the hand and foot are covered with arched, laterally compressed and distally pointed claws;

c) the first digits are as yet neither morphologically (no difference in size, no large nail pad, no flat nail) nor functionally (no opposability) distinct;

d) the primitive pad distribution with five primary nail pads (one on each of the terminal phalanges) and six primary pads of palm and sole (four interdigital pads, thenar and hypothenar);

e) a primitive structure of the friction skin, with short and coarse cutaneous ridges and *insulae primariae,* localized, i.e., restricted to the primary pads;

f) a primitive configuration of the pads, with flat and narrow nail pads and strongly elevated and circumscribed pads of the palm and sole;

g) in conformity with this on all the nail pads the cutaneous ridges are in the form of proximally opened loops (*sinus primarii*), while on the pads of palm and sole they are present as whorls.

Phylogenetically higher development is characterized by the following tendencies:

A quite tenacious retention of the phalangeal formula and of the third digit as the functional axis.

A very early development of a foot suited for grasping and climbing through a special strengthening and increased mobility of the big toe as a pincher with a corresponding special enlargement of its nail pad and an early transformation of its claw into a flat nail.

An increasing mobility in the remaining digits and a gradual transformation of the claw into nails.

A tenacious retention of the eleven primary pads (at least in embryonic stages) with the tendency to elevate the nail pads (i.e., a development of whorl patterns) and to flatten those of the palm and sole (i.e., a replacement of whorls by loops or arches).

An additional development of accessory pads in the region of the digits, palms and soles.

An extension of the friction skin over the entire volar side of the chiridia including the heel and a histological differentiation in the sense of a refinement and extension of individual cutaneous ridges.

Finally, a proliferation of the tactile receptors and a further differentiation of their neural relationships.

If we consider this, it can be seen that the chiridia of prosimians tend to specialize in quite different directions (see Figs. 6 and 7), while those of the simians show a trend toward higher evolutionary development. Therefore, in the chiridia of the prosimians one finds very generalized characters side by side with various specializations. The condition of the chiridia strengthens the view that the Lemuriformes, Lorisiformes and Tarsiiformes represent distinct branches on the primate pedigree which had separated quite early, and that the simians have no particularly close relationship to any of these.

Simians which have retained the generalized condition most closely, i.e., simians with chiridia of a type represented by those found in *Aotes, Callicebus* and in many respects also by the Callithricidae, cannot be derived from any of these prosimian lines, but must be traced back to an ancestral model resembling the chiridia of the Tupaiids. The Tupaiiformes can be regarded as models of an ancestral group of primates out of which the Lemuriformes, Lorisiformes, Tarsiiformes and the simians all have independently developed. The author agrees therefore completely with Remane (1961) in believing that the prosimians do not have any taxonomic unity, but that they represent a group of stages in primate evolution.

The simians, in contrast to the prosimians, show the above-mentioned trend

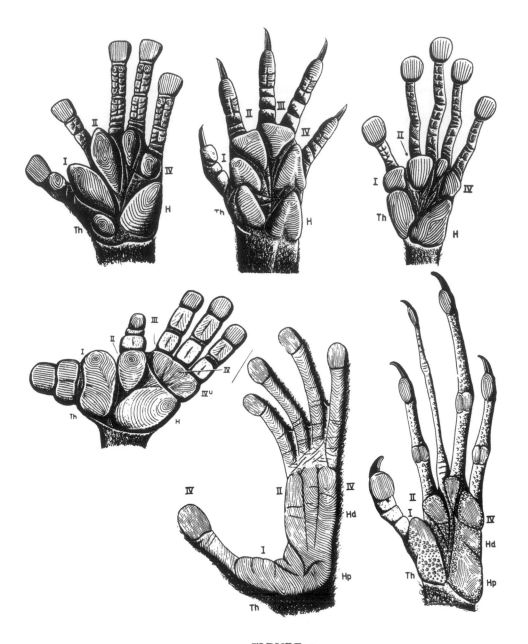

FIGURE 6

Hands of Prosimians (upper row, from left to right: *Galago, Tupaia, Tarsius*; lower row, from left to right: *Nycticebus, Indri, Daubentonia*).

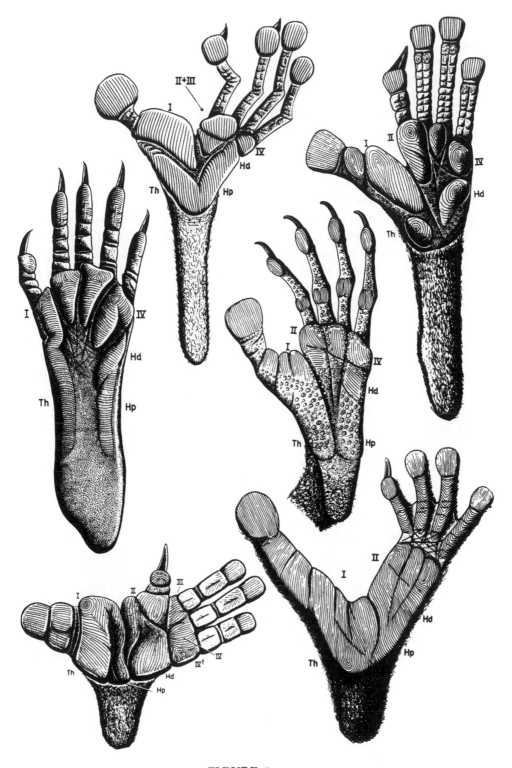

FIGURE 7

Feet of Prosimians (upper row, from left to right: *Tarsius, Galago;* middle row, from left to right: *Tupaia, Daubentonia;* lower row, from left to right: *Nycticebus, Indri*).

toward higher development of the chiridia, but there occur also, as expected, characters typical of species and groups of species as well as adaptations conditioned by locomotion. An extensive description of these specialized conditions cannot be given here (Biegert, 1961) and some few examples must suffice to demonstrate the manner in which the volar integument can assist in primate classification by either strengthening or weakening one view or another.

Two families, the Cebidae and the Callithricidae, have long been distinguished in the Platyrrhina. The number of molars and the presence of claws or nails were used as the main diagnostic criteria for their classification. Thus *Callimico*, because of its three molars (as opposed to two in the Callithricidae) was placed in the Cebidae, or in consideration of their claws in a subfamily of the Cebidae, the Callimiconinae. However, the structure of the chiridia clearly indicates that *Callimico* is a member of the Callithricidae. In contrast to the kaleidoscopic array of forms found in different cebid genera, the structure of the chiridia and the volar integument in all of the Callithricidean genera is astonishingly uniform and therefore comparatively stable.

A special and unmistakable combination of primitive characters (e.g., extent of friction skin, configuration of nail pads, claws present except on hallux) and of specialized characters (pad distribution and configuration on palm and sole, pattern of cutaneous ridges, reduction in length of opposable hallux), is typical for the Callithricidae but not for the Cebidae, even when their variability is taken into consideration. If this series of characteristics is matched against those found in *Callimico*, then there can be no doubt that it must be a member of the Callithricidae. The characters of the chiridia are therefore in this instance of considerably higher taxonomic value than the number of molars.

In the Cebidae, species differences of the chiridia are more marked, and there occur apparently four different evolutionary trends. *Aotes* and *Callicebus* remain similar to the ancestral type, which is characterized by the predominant presence of characters found in generalized primate chiridia. The other, more progressive Cebidae possess chiridia structures that can be traced back to this model. Four major evolutionary lines can be distinguished in the Cebinae (represented by *Saimiri* and *Cebus*), the Pithecinae (with *Pithecia, Cacajao, Chiropotes*), the Alouattinae (with *Alouatta*) and the Atelinae (represented by *Lagothrix, Ateles, Brachyteles*). Of these four, the Alouattinae and Atelinae—as determined through special characters such as the highly specialized prehensile tail, and a specific cutaneous ridge pattern of the sole and the palm—seems to be closer together. However, in *Alouatta* the structure of the chiridia (and particularly the dermatoglyphic pattern) is more generalized than in members of the Atelinae. *Cebus*, because of its tail and dermatoglyphic pattern, is clearly distinguished from the Alouattinae and Atelinae. It is closely related to *Saimiri*, but in many respects is more specialized.

The Catarrhina, by virtue of the development of "multi-purpose" "gripping" hands, represent a higher evolutionary stage among the simians. The development of an opposable thumb in the Catarrhina is, when compared with the

Platyrrhina, an important new attainment. Although among the Platyrrhina one finds a qualitative improvement in grasping organs (see *Alouatta, Lagothrix, Ateles, Brachyteles*), this is brought about by the development of the tail as an additional organ for touch and grasping.

Structurally the chiridia of the Cercopithecoidea are more generalized and much more uniform than those of the Hominoidea. In the Cercopithecinae in particular—because of intraspecific variability—a sharp delimitation of the species on the basis of dermatoglyphic pattern is hardly possible. In this sense the dermatoglyphic pattern is mainly group specific. In the Colobinae the species differences are more marked (e.g., *Nasalis* and *Colobus*). However, *Pygathrix* and *Presbytis* are not clearly differentiated from the Cercopithecinae. This relative uniformity of the chiridia is an outstanding trait of the Cercopithecoidea, as compared with the Ceboidea and the Hominoidea.

The structural specializations of the chiridia of the Hominoidea are by contrast well marked (Biegert, 1961) and are distinctly different from the generalized conditions in the Cercopithecoidea. Furthermore, the three families of Hominoidea are clearly differentiated in their chiridial structure. Therefore, the structural conditions found in the Hylobatidae, the Pongidae and the Hominidae are not directly related to each other, but are traceable to the generalized catarrhine model, the pattern of which is still largely retained by members of the Cercopithecinae. While the chiridia assist in constructing sharper taxonomic boundaries between the three families, there are many structural characters which are shared by the Pongidae and the Hominidae. However, *Pongo, Pan* and *Gorilla* have also developed many specialized conditions to which the structures in *Homo* cannot possibly be referred (see Figs. 8 and 9).

A classification of the Hominoidea based upon the volar integument is in agreement with their presently accepted taxonomic relationship (Fiedler, 1956). Evidence drawn from the conditions of the volar integument opposes the placement of *Pan* and *Gorilla* in the family Hominidae and supports an early, rather than a late, derivation of the Hominidae from an ancestral hominoid stock.

Since the embryonically determined dermatoglyphic pattern is not generally adaptive to functional variations in the chiridia, but reflects primarily group and species specializations, it is not a marker that can be used to determine whether the hominid chiridia have evolved directly from a generalized arboreal-catarrhine model or from a brachiation-adapted pongid model. More clearly is it possible to determine that the chiridia of the Hominidae have developed out of catarrhine-like "grasping" hands and "climbing-grasping-walking" feet with highly evolved motor and tactile facilities. A direct derivation of the Hominidae from earlier "*Tarsius*-like" forms or "protocatarrhines" cannot be supported.

For the particular problem of the taxonomic relationships and derivation of the Hominids within the Hominoidea, such specialized characters, shared with the Pongidae, as certain skeletal and soft parts, serological attributes and chromosomes are highly valuable. The fact that many structural attributes have been found to be more generalized in Hominids than in modern Pongids is not con-

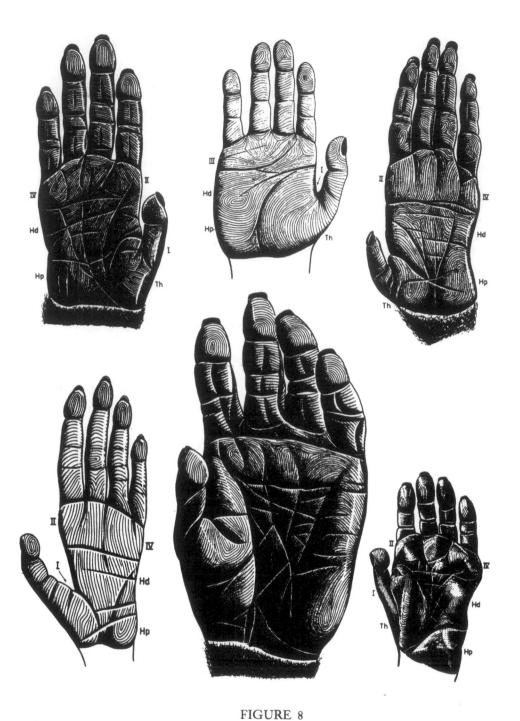

FIGURE 8

Hands of Hominoids compared with Cercopithecids. (upper row, from left to right: *Pan, Homo, Pongo;* lower row, from left to right: *Symphalangus, Gorilla, Papio*).

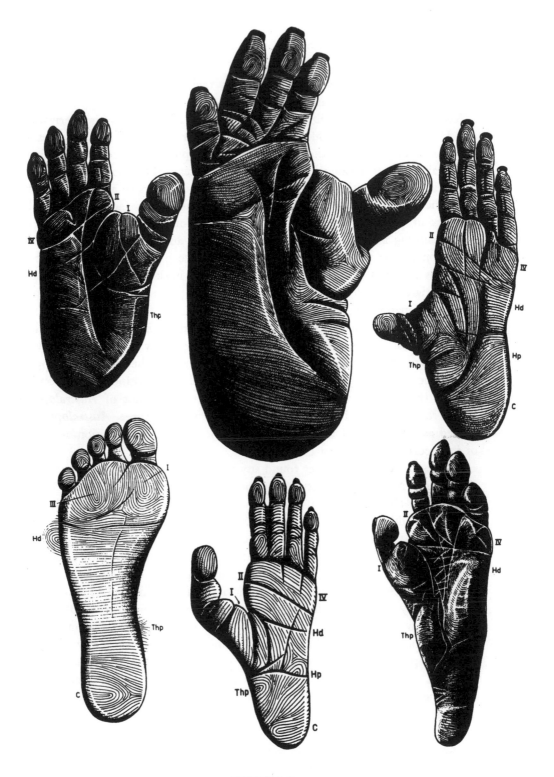

FIGURE 9
Feet of Hominoids compared with Cercopithecids (upper row, from left to right:
Pan, Gorilla, Pongo; lower row, from left to right: *Homo,
Symphalangus, Papio*).

tradictory, because the few fossil finds of *Proconsul* have already revealed that early Pongids were more generalized than their descendants in many essential points (e.g., hands, feet, and skull).

Keeping such a generalized pongid model in mind makes the derivation of the Hominids from pongid stock highly suspect. Such a view becomes more probable if one considers the fact that the development of erect posture signified the attainment of a radically new adaptive plateau. This novel manner of behavior allowed an entirely new system of selective factors to act upon the organism. The "higher development" of the motor and tactile faculties of the hand, the loss of prehensibility of the feet, a reduction in size of the canines and accompanying alterations of the lower premolars and of the jaws, the additional expansion of the neopallium, the reduction in size of the masticatory apparatus (taken as a whole) and the accompanying changes in skull structure all are adaptations initiated by exposure to this new adaptive plateau, i.e., are results of this specialized manner of hominid locomotion.

Thus viewed, it is clear that erect locomotion is the primary taxonomic criterion for Hominids. On the basis of fossil evidence we know that the specially large expansion of the neopallium, the reduction in the size of the masticatory apparatus and the accompanying changes in skull topography are distinctive taxonomic criteria of the genus *Homo* (*H. erectus, H. sapiens*) but not of *Australopithecus* which resembles Pongids in this respect still to a high degree. *Australopithecus* is classed as a member of the Hominidae on the basis of its dentition and because of its specialized pelvic-leg condition.

At present we have not yet found any undoubted Hominids earlier than lower pleistocene. Neither the posture nor the dentition of the pliocene *Oreopithecus* is hominid (Schultz, 1960). The relatively steeply inclined face, the nearly vertical mandibular symphysis and other so-called hominid characteristics of the skull are not diagnostically adequate criteria for placing *Oreopithecus* in the Hominids. To the contrary we expect that the skull topography of a really early Hominid would not differ from that of the Pongidae.

The often cited differences in the dentitions of Pongids and Hominids, which are primarily differences in the canine-premolar complex, are probably diagnostically worthless during the earlier phases of hominid evolution because it becomes more and more likely that hominid dentition was derived from the "typical" pongid dental structure. One of the leading authorities on primate dentition, A. Remane (1952), has said expressly: "Ich wüsste nicht, wie eine Ableitung der hominiden Strukturen unter Umgehung eines Pongiden-Stadiums möglich wäre. . . ."

In this context it should be noted that there is a functional interdependence in the structure of the canines and premolars. Changes in size of the canines influence the size and shape of the premolars and the size of these teeth influence the structure of the jaws (for instance the form of the dental arch, the mandibular symphysis, etc.). Any change of size of the canines results in correlative changes in the premolars and the configuration of the jaws.

In the development of erect locomotion we may find a cause for the size reduction of hominid canines. For the "armed hands" equipped with sticks and stones were far more effective for attack and defense than large canines. Thus it is believed that the attainment of the erect posture also produced a transformation of pongid dentition into hominid dentition and correlatively produced alterations in the jaws. Members of the Dryopithecinae, e.g., *Bramapithecus* and *Rahmapithecus*, because of the structure of their teeth and jaws have been diagnosed as Hominids (Gregory, Hellman and Lewis, 1938). A definitive decision, however, can be reached only if we find their post-cranial skeleton and particularly the pelvis and legs. For one would expect that the earliest Hominids would be primarily characterized by a hominid condition of the pelvis-leg complex. Therefore, it is certainly not out of the question that there are already remains of prepleistocene Hominids sitting on the shelves of museums which we do not recognize as such because we have to rely on dental structure and jaw structure alone as diagnostic taxonomic criteria.

It is an obvious hypothesis (Washburn) that the development of erect locomotion with the total emancipation of the hands was of primary significance for the evolution of a hominid form of existence. Therefore, alterations of the pelvis-leg complex into a "hominid form" are the basic distinctions of the Hominidae, whereas skull topography is only of taxonomic value for some of the Hominidae (*Homo erectus, Homo sapiens*). On the basis of our present knowledge it seems highly probable that the Hominids have evolved from a generalized pongid group which may have been similar to *Proconsul* while a derivation from "Protocatarhine" or from late Pongids of a type represented, e.g., by *Pan* is most unlikely.

BIBLIOGRAPHY

BIEGERT, J.
   1956. "Das Kiefergelenk der Primaten, seine Altersveränderungen und Spezialisationen in Gestaltung und Lage." *Morph. Jb.* 97:249–404.
   1957. "Der Formwandel des Primatenschädels und seine Beziehungen zur ontogenetischen Entwicklung und den phylogenetischen Spezialisationen der Kopforgane." *Morph. Jb.* 98:77–199.
   1959. "Die Ballen, Leisten, Furchen und Nägel von Hand und Fuss der Halbaffen." *Z. Morph. Anthrop.* 49:316–409.
   1961. "Volarhaut der Hände und Füsse." *Handbuch der Primatenkunde* II/1, Lieferung 3, pp. 1–326. Basel/New York: S. Karger.
BONIN, G. VON and P. BAILEY
   1961. "Pattern of the cerebral isocortex." *Handbuch der Primatenkunde* II/2, Lieferung 10, pp. 1–42 Basel/New York: S. Karger.

DeVore, I. and R. Lee
1963. Recent and current field studies of primates." *Folia primatologica,* 1:66–72.

DeVore, I. and S. L. Washburn
In press. "Baboon ecology and human evolution." *Viking Fund Publ. in Anthropol.*

Fiedler, W.
1956. "Uebersicht über das System der Primates." *Handbuch der Primatenkunde I,* pp. 1–266. Basel/New York: S. Karger.

Gregory, W. K., M. Hellman and G. E. Lewis
1938. *Fossil anthropoids of the Yale-Cambridge Indian expedition of 1935.* Carnegie Inst. Washington Publ. No. 495.

Hofer, H.
1952. "Der Gestaltwandel des Schädels der Säugetiere und Vögel mit besonderer Berücksichtigung der Knickungstypen der Schädelbasis." *Verh. Anat. Ges. 50. Vers.* pp. 102–113.
1953. "Ueber Gehirn und Schädel von *Megaladapis edwardsi* G. Grandidier (Lemuroidea), nebst Bemerkungen über einige airorrhynche Säugerschädel und die Stirnhöhlenfrage." *Z. wiss. Zool.,* 157:220–284.
1954. "Die craniocerebrale Topographie bei Affen und ihre Bedeutung für die menschliche Schädelform." *Homo,* 5:52–72.
1957. "Zur Kenntnis der Kyphose des Primatenschädels." Verh. Anat. Ges. 54. Vers., pp. 54–76.
1960. "Studien zum Problem des Gestaltwandels des Schädels der Säugetiere, insbesondere der Primaten." *Z. Morph. Anthrop.,* 50:299–316.

Kälin, J.
1952. "Die ältesten Menschenreste und ihre stammesgeschichtliche Deutung." *Historia mundi I,* pp. 33–98. Bern: Francke.
1956. "Zur Morphogenese des Primatenschädels." *Bericht 5. Tagung Deutsche Ges. f. Anthrop. 1956,* pp. 28–33.

Kummer, B.
1952. "Untersuchungen über die Entstehung der Schädelbasisform bei Mensch und Primaten. *Verh. Anat. Ges. 50. Vers. Marburg,* 1952, pp. 122–126.
1952. "Untersuchungen über die ontogenetische Entwicklung des menschlichen Schädelbasiswinkels." *Z. Morph. Anthrop.* 43:331–360.
1953. "Untersuchungen über die Entwicklung der Schädelform des Menschen und einiger Anthropoiden." *Abh. z. exacten Biologie,* H. 3, pp. 1–44.

Le Gros Clark, W. E.
1955. *The Fossil Evidence for Human Evolution.* Chicago: Univ. of Chicago Press.

Le Gros Clark, W. E. and L. S. B. Leakey
1951. "The miocene hominoidea of East Africa." Fossil mammals of Africa, No. 1, pp. 1–117. London: Brit. Mus. Nat. Hist.

Moss, M. L. and R. W. Young
1960. "A functional approach to craniology." *Amer. J. Phys. Anthrop.,* n.s. 18:281–292.

Penfield, W. and T. Rasmussen
1957. *Cerebral Cortex of Man.* New York: Macmillan.

Remane, A.
1952. "Der vordere Prämolar (P_3) von *Australopithecus prometheus* und die morphologische Stellung des Australopithecinengebisses." *Z. Morph. Anthrop.,* 43:288–310.

1956. "Methodische Probleme der Hominidenphylogenie III." *Z. Morph. Anthrop.*, 48:28–54.

1960. "Zähne und Gebiss." *Handbuch der Primatenkunde III/2*, pp. 637–846. Basel/New York: S. Karger.

1961. "Probleme der Systematik der Primaten." *Z. wiss. Zool.*, 165:1–34.

RENSCH, B.
1959. *Homo sapiens.* Göttingen: Vandenhoek & Ruprecht.

SCHULTZ, A. H.
1940. "The size of the orbit and the eye in primates." *Amer. J. Phys. Anthrop.*, 26:389–408.

1960. "Einige Beobachtungen und Masse am Skelett von *Oreopithecus.*" *Z. Morph. Anthrop.* 50:136–149.

STARCK, D.
1953. "Morphologische Untersuchungen am Kopf der Säugetiere, besonders der Prosimier, etc." *Z. wiss. Zool.* 157:169–219.

VOGEL, CH.
1961. "Ueber den phylogenetischen Wert von Mandibelmerkmalen bei höheren Primaten." *Z. Morph. Anthrop.* 51:275–288.

WASHBURN, S. L. and F. C. HOWELL
1960. "Human evolution and culture." In *Evolution after Darwin II*, pp. 33–56. Chicago: Univ. of Chicago Press.

WEIDENREICH, F.
1924. "Die Sonderform des Menschenschädels als Anpassung an den aufrechten Gang." *Z. Morph. Anthrop.* 24:157–190.

1941. "The brain and its role in the phylogenetic transformation of the human skull." *Trans. Amer. Phil. Soc.* n.s. 31:321–442.

WOOLSEY, C. N., P. H. SETTLAGE, D. R. MEYER, W. SPENCER, T. P. HAMUY and A. M. TRAVIS
1952a. "Patterns of localisation in the precentral and "supplementary" motor areas and their relation to the concept of a premotor area. Patterns of organisation in the CNS." *Res. Publ. Ass., Res. nerv. ment. Dis.*, 30:238–264.

1952b. "Patterns of localisation in sensory and motor areas of the cerebral cortex." *Biol. mental health and disease.* pp. 192–206.

THE CLASSIFICATION OF OREOPITHECUS

WILLIAM L. STRAUS, JR.

INTRODUCTION

OREOPITHECUS BAMBOLII is a fossil catarrhine primate from the Pontian [1] of Tuscany, Italy, dated as 10–12 million years old. Its classification has been controversial ever since Gervais described the type specimen, a mandible with teeth, in 1872. Many authors, following Schlosser (1887), have regarded it as a cercopithecoid monkey (cf. Hürzeler, 1958, 1960). Others, however, have assigned it to the anthropoid apes, some, like Schwalbe (1916), placing it in a separate anthropoid-ape family, the Oreopithecidae. Still others seem to have been inclined to view it as a sort of *forme de passage* or link between the monkeys and apes. Majority opinion, however, seems to have labeled it a cercopithecoid of some sort. All of the earlier studies dealt only with jaws and teeth—the only parts of the body then available—and most of these, unfortunately, as Hürzeler (1949; also see Patterson, 1955) has noted, were made from casts, and bad casts, at that.

The matter rested in this nebulous state until 1954, when Johannes Hürzeler rescued *Oreopithecus* from paleontological oblivion by the challenging claim, based on a detailed study of the dentition, that *Oreopithecus* is not only a hominoid (as he had stated earlier, in 1949) but, more precisely, a primitive hominid and hence a member of that evolutionary radiation which led to man.[2] Since that time, Hürzeler has recovered a remarkable group of specimens of *Oreopithecus* from lignite deposits, Pontian in age, at Baccinello (Grosseto), Italy. The remains of at least fifty individuals—including not only teeth, jaws, and skull, but also parts of the postcranial skeleton—are represented. The *pièce de résistance* is the better part of the skeleton of a young adult discovered on August 2, 1958 (Fig. 1). Roentgenograms, provided through the generosity of Dr. Hürzeler, reveal a non-united iliac crest. Otherwise, however, this skeleton appears to be fully adult in character.

All of this material has been brought together at the Natural History Museum,

---

1. The Pontian formation is regarded as Lower Pliocene by English and German geologists, as Upper Miocene by French geologists.

2. Prior to Hürzeler, only Forsyth Major (1880) seems to have recognized any resemblances between *Oreopithecus* and hominids. According to him, "*Oreopithecus* . . . zeigt in dem Verhalten seines vorderen Praemolaren Analogie mit dem Menschen."

Basel, Switzerland, where I was privileged to study it during the summers of 1957, 1958, and 1959. I am exceedingly grateful to Dr. Hürzeler for his generous permission to study the *Oreopithecus* material and to publish, without restrictions, about it. It is through his kindness, moreover, that I am able to present excellent original drawings and photographs of specimens of *Oreopithecus*. These were made for me while I was working in Basel. I wish to thank Mr. O. Garraux for the drawings shown in Figures 6, 7, 8, 9, 10, 11 and 14, and Mr. K. Rothpletz for the photographs shown in Figures 7, 8, 9, 10, 11 and 14. I also am greatly indebted to the Wenner-Gren Foundation for Anthropological Research, New York City, for generous grants which enabled me to go to Basel and which thus made my studies of *Oreopithecus* possible.

Despite this new material, which has been studied, at least in part, by a number of workers, the taxonomic position of *Oreopithecus* remains uncertain. Most recent students seem to have abandoned the notion that this animal belongs in the superfamily Cercopithecoidea and assign it, as did Hürzeler (1949), to the superfamily Hominoidea (cf., e.g., Kälin, 1955; Thenius, 1958; Butler and Mills, 1959; Schultz, 1960; Straus, 1961; and Simpson, 1962).[3] However, there is no agreement respecting its precise family assignment—whether to the Pongidae, to the Hominidae, or to a family of its own, the Oreopithecidae.

THE MAJOR MORPHOLOGICAL CHARACTERS OF OREOPITHECUS

The major structural features of *Oreopithecus* may first be considered. The following account is based not only on the August 2, 1958 skeleton, but also on other specimens discovered earlier. The ascription "WLS" refers to unpublished observations made by me. Data for other primates, except where specifically noted, are for adult animals.

BODY SIZE

I estimate the stature of the 1958 specimen of *Oreopithecus* as having been roughly about 4 feet (ca. 110–120 cm.). Schultz (1960) has reasonably estimated its trunk height (or length) to have been about 460 mm. and its body weight as approximately 40 kg.; figures which fall within the range of variation of the chimpanzee. Thus *Oreopithecus* probably was about the same size as the smaller of the two African great apes.

---

3. I am following Simpson (1945) in regarding the Hominoidea as composed of only two families, Hominidae (man and his immediate forerunners) and Pongidae (anthropoid apes). My own inclination would be recognition of two families of anthropoid apes, Pongidae (great apes) and Hylobatidae (gibbons and siamang); but, to avoid confusion, I combine them here, thus relegating the small Asiatic apes to subfamily status.

FIGURE 1

The August 2, 1958 skeleton of *Oreopithecus in situ* (ventral view). (Courtesy of Dr. Johannes Hürzeler.)

SKULL

The overall size of the 1958 skull (Fig. 2) is about that of an average chimpanzee; thus confirming the body-size estimates of Schultz (1960) noted above. The precise positioning of the foramen magnum (and occipital condyles) remains uncertain. However, from the available specimens (1958 and #40), there is good reason to believe that it was located more forward than in most pongids and many cercopithecoids; but probably not as far forward as in known hominids (WLS). Its location would thus seem to be that found in some Old World monkeys. There are relatively huge supraorbital ridges, as in the African great apes and many cercopithecoids. This, however, is not a good taxonomic character, for similarly large ridges also are found among the Hominidae—in australopithecines and some fossil men. The face, however, is strikingly short (Hürzeler, 1960; WLS). This is well exemplified by the hominid location of the anterior root of the zygomatic arch, which is above M^1 or P^4 in 1958 and #117; whereas in pongids and cercopithecoids it lies more posteriorly (Hürzeler, 1960; WLS). Indeed, the face is so shortened that the nasal bones project forward to form a partial roof for the nasal aperture; this is an essentially hominid character (Hürzeler, 1960; WLS). The zygomatic arch exhibits another hominid feature in that its anterior upper border almost attains the level of the lower border of the orbit in 1958 and #117; whereas in pongids and cercopithecoids it lies well below the orbit (WLS).

The nearly vertical mandibular symphysis regularly has neither a chin nor a simian shelf (Hürzeler, 1960; Straus, 1961). This verticality is an expression of

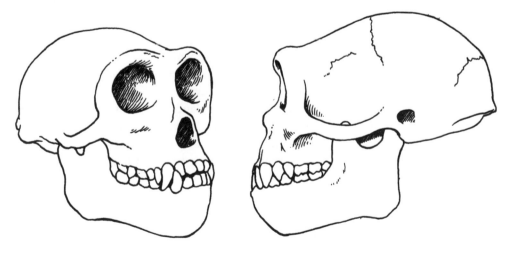

FIGURE 2

Reconstructed skull of the August 2, 1958 skeleton of *Oreopithecus*. (After Hürzeler, 1960; redrawn.)

the short face. Absence of chin and simian shelf is of no real diagnostic value; in my opinion this represents a primitive or "protocatarrhine" condition.[4] The mental foramen regularly lies at or above the middle of the corpus, as in hominids (Hürzeler, 1958, 1960). In both pongids and cercopithecoids, on the other hand, it normally is closer to the lower margin of the corpus.

The tooth rows are parallel, as in most cercopithecoids and pongids, rather than convergent anteriorly to form an arch, as in hominids (Straus, 1949).

DENTITION

The dentition of *Oreopithecus* has been described in detail by Hürzeler (1949, 1951, 1954, 1958), and has also been dealt with by Remane (1955), Butler and Mills (1959), and Simons (1960). I also have given it some attention, but cannot claim to have made anything like a thorough study. The following brief summary of its most important features largely represents my own evaluation of published data and interpretations.[5] Those who wish more detailed information are referred to the papers cited above.

The dental formula is catarrhine, $\frac{2.1.2.3}{2.1.2.3}$. The dentition is distinctly bunodont, which has led some authors to compare it with the dentition of pigs (Hürzeler, 1958).

Upper Teeth.—The upper incisors are implanted vertically and are worn off perpendicular to their long axes. This is a hominid character, probably to be correlated with the short face. The lingual surfaces of the incisors are concave; hence these teeth are somewhat shovel-shaped. At times there is a well-developed internal cusp which appears to be homologous with the so-called basal tubercle often found in hominoids. This cusp customarily is larger on I^1 than on I^2.

The upper canine is a relatively small tooth. As in hominoids, its root lacks the strong groove found in cercopithecoids. There regularly is no pre-canine diastema, although a small gap occurs in one specimen.

Both upper premolars are bicuspid. This is a regular common catarrhine feature. Other, rudimentary cusps sometimes occur, however.

The upper molar pattern finds its closest resemblance in earlier, more primitive members of the Hominoidea—namely, *Proconsul*, *Limnopithecus*, and *Pliopithecus*, usually classified as primitive pongids (Butler and Mills, 1959). Contrariwise, it bears much less resemblance to the overall pattern found in hominids, and even

4. A true chin, however, is not, as commonly believed, a feature peculiar to *Homo sapiens*; for it occurs sporadically among Old World monkeys (WLS). A true simian shelf is regularly absent in all known hominids, in hylobatines, and in early Miocene hominoids. When present, it undoubtedly represents a buttressing specialization (Straus, 1949). It is normally present in the three great apes, although variable in its degree of development; its complete absence in these animals, however, is a most exceptional occurrence.

5. I am greatly indebted to Dr. George Gaylord Simpson for calling my attention to certain features of the dentition.

less resemblance to that of cercopithecoids. In fact, it is distinctly non-cercopithe-coid. Butler and Mills (1959) regard *Oreopithecus* as probably the more primitive respecting most of the upper-molar characters in which it differs from pongids (cusps more acute; position of protocone; very small anterior fovea; groove between protocone and hypocone; lack of occlusal facet on posterior side of crista obliqua), and as more specialized in only one character (course of crista obliqua). They conclude that "In its upper molar pattern, therefore, *Oreopithecus* appears to retain primitive features that are not known in any of the Pongidae, and at the same time it resembles what are presumably the most primitive members of the family rather than the later forms. This suggests that it is a survivor, special-ized in its own way, from a prepongid stock."

Lower Teeth.—The lower incisors also are steeply implanted. They are not so markedly scooped out lingually as are the upper incisors, however.

The lower canine, like the upper, is a comparatively small tooth. A post-canine diastema is regularly lacking. As is well known, a diastema may be lacking in pongids; but this, in contrast to the situation in *Oreopithecus*, is an exceptional condition.

The lower premolars are bicuspid, two-rooted, and possess prominent talonids (that of P_4 usually being the better developed)—a non-hominid character. Their outstanding feature is their homomorphism. In this they resemble all known hominids; whereas in all known cercopithecoids and pongids these teeth are heteromorphic, P_4 being molariform and P_3 a sectorial, caniniform tooth. Indeed, this trend toward homomorphism has been regarded as one of the outstanding traits wherein the hominid line of evolution differs from the pongid and cerco-pithecoid lines.

M_1 is the shortest of the lower molars, M_3 the longest. The overall molar pattern differs in many respects from those of undoubted members of all three catarrhine families, Hominidae, Pongidae, and Cercopithecidae. Not all of these differences can be considered here. However, they include the presence of a para-conid (on M_1 and M_2 only), a well-developed mesoconid (or centroconid), a ridge connecting hypoconid and metaconid via mesoconid, and an anomalous protoconule.

The supposed dental resemblance of *Oreopithecus* to cercopithecoids will not hold water. It is true that in a few characters the lower molars of *Oreopithecus* approximate those of cercopithecoids, but their overall pattern is strikingly non-cercopithecoid. The supposed "moderately strong" (Robinson, 1956) or "partial or incipient" (Clark, 1960) bilophodontism, which has been perhaps the chief reason for the cercopithecoid assignment, has been based on the spurious trans-verse ridges produced by strong wear of the highly differentiated chief cusps. Unworn lower molars, however, in no manner suggest bilophodontism; moreover, as already noted, their detailed morphology is certainly not cercopithecoid.

Until recently, the lower molar pattern of *Oreopithecus* has generally been regarded as *sui generis*. Simons (1960), however, after a careful comparison of

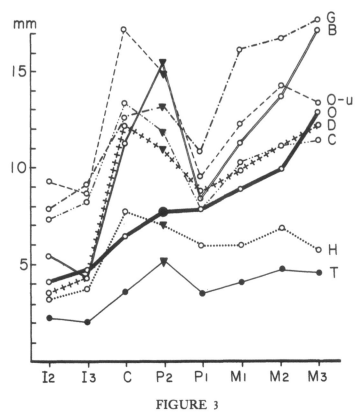

FIGURE 3

Length-proportions of the lower teeth. G, gorilla; B, Gelada
baboon (*Theropithecus*); O-u, orangutan; O, *Oreopithecus*;
D, *Dryopithecus Fontani Lartet*; C, chimpanzee; H, gibbon
(*Hylobates leuciscus*); T, talapoin (*Cercopithecus talapoin*).
(After Hürzeler, 1958; redrawn. The talapoin graph has
been added through the courtesy of Dr. Hürzeler.)

Oreopithecus with *Apidium*, has found that these two animals are practically
identical respecting the morphology of P_4, M_1, M_2, and M_3 (the only teeth
present in the single known example, a mandibular fragment, of *Apidium*). In this
he confirmed and extended an earlier observation of Gregory (1922). *Apidium*,
from early Oligocene deposits of the Fayum, Egypt, has been a controversial
animal. Originally diagnosed as a primate, it also has been variously regarded as a
primitive suid or as a condylarth. Its restudy by Simons seems to leave no doubt
about its catarrhine primate nature. According to him, "(among primates) its
greatest resemblances are to *Oreopithecus*." It scarcely need be stated that this
resemblance is of considerable significance in the taxonomic assessment of *Oreopithecus*. This will receive specific consideration at the proper place in this paper.

Length-proportions of the Teeth.—Hürzeler (1954, 1958) has shown that the
various major groups of primates (including both living and fossil forms) exhibit

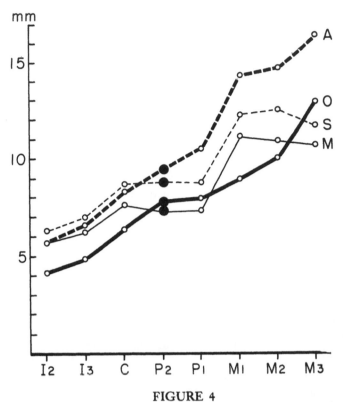

FIGURE 4

Length-proportions of the lower teeth. A, *Australopithecus*
(*Paranthropus*) *crassidens*; O, *Oreopithecus*; S, *Homo erectus*
pekinensis ("Sinanthropus"); M, *Homo sapiens sapiens*.
(After Hürzeler, 1958; redrawn.)

typical length-proportions of their teeth, which can be expressed graphically.
He showed, moreover, that the graph for *Oreopithecus* not only differs strikingly
from those for both pongids and cercopithecoids (Fig. 3), but that it shows
striking agreement with those for hominids (*Homo sapiens*, Sinanthropus, *Australopithecus* [*Paranthropus*] *crassidens*) (Fig. 4). It has been suggested to me that
these proportions merely are related to a short face and hence lack deep significance. Consequently, at my request, while I was in Basel during August, 1959,
Dr. Hürzeler kindly measured the tooth lengths of a male specimen of the short-
faced guenon, *Cercopithecus talapoin*. The resultant graph does not at all resemble those of either *Oreopithecus* or hominids; rather, it basically is similar to
those of the pongids and the cercopithecoids previously charted by Hürzeler
(1958) (Fig. 4): It can be concluded, therefore, that the length-proportions of
the teeth are not merely a reflection of facial development.

Deciduous Teeth.—Very little is known about the milk dentition. Only a few
molars, of which but two are complete, have been available for study (Hürzeler,

1951, 1958; Butler and Mills, 1959). They contribute so little to the present problem that they need not be considered here.

TRUNK

Most of the ribs of the 1958 skeleton are damaged. One of the second ribs, however, is fortunately in excellent condition (Fig. 5, II). Schultz (1960) has shown that the angle between its neck and body (54°) is relatively small, as in all of the hominoids which he studied (50°–65°); whereas in cercopithecoids this angle regularly is large (89°–100°). This definitely indicates that *Oreopithecus* had a broad and shallow trunk, as in hominoids, rather than a narrow and deep one, as in cercopithecoids. This is confirmed by the index of curvature for rib II (WLS).[6]

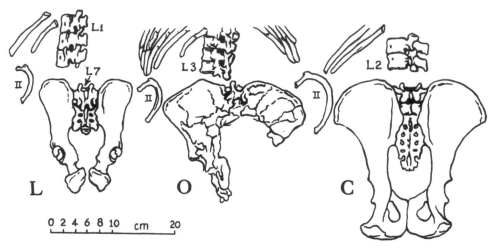

FIGURE 5

Dorsal view of pelvis, vertebrae, and ribs of the August 2, 1958 skeleton of *Oreopithecus* (center, O) *in situ*, compared with the corresponding parts of a male langur, *Presbytis entellus* (left, L), and a male chimpanzee (right, C). II, second rib. (After Schultz, 1960, redrawn.)

The lumbar vertebrae of the 1958 skeleton number 5, as normally in man and hylobatines (Schultz, 1960; Straus, 1961). Less than five lumbars, however, is the rule for the great apes. In fact, as many as five lumbars occurs in no more than 10 per cent of orangs, and never in either the gorilla or chimpanzee (Schultz and Straus, 1945; Schultz, 1961). On the other hand, cercopithecoids never possess less than six lumbars, the range of variation extending from six to nine, with six or

6. Arc length x 100/chord length for rib II = 188.7 in *Oreopithecus;* this index always is more than 150 in the hominoids and less than 150 in the cercopithecoids studied by me.

seven occurring in the vast majority of specimens (Schultz and Straus, 1945; Schultz, 1961). Thus, in the number of lumbar vertebrae, *Oreopithecus* is a hominoid.

The proportions of the lumbar vertebrae are hominoid, and not at all cercopithecoid (Schultz, 1960; WLS).[7]

On the other hand, the available vertebrae of the 1958 specimen (last T., L.1, 2, 3, and 5; L.4 is represented only by fragments of spongiosa; see Figure 5) are quite non-hominoid in having a prominent ventral mid-sagittal keel on their centra (Straus, 1961). This keel is normally well-developed in cercopithecoids, and to it is attached the ventral longitudinal ligament of the vertebral column. It also is quite evident in platyrrhine monkeys and prosimians as well as in many non-primate quadrupeds. At best, only traces of such a keel are found in pongids, and these may be entirely lacking in some specimens. It is not normally present in man.

Unfortunately, the 1958 skeleton lacks the sacrum. However, there is an earlier discovered sacrum (#50) which comprises six vertebrae (Fig. 6), as often in hominoids (WLS).[8] Cercopithecoids, however, usually have a sacrum of no more than 3 vertebrae, one with 4 being an exceptional variation (Schultz and Straus, 1945; Schultz, 1961). The *Oreopithecus* sacrum thus is hominoid in nature. It is rather narrower than is usual in hominids, however. The tapering of its lower end is perhaps suggestive of a rudimentary caudal region, as in hominoids.

EXTREMITIES

Upper Extremity.—In its overall morphology the 1958 humerus of *Oreopithecus* is that of a hominoid (WLS). The most diagnostic region is the lower end, which is distinctly hominoid in the configuration of the articular surface and in the presence of a large, projecting ulnar epicondyle (Fig. 7). It certainly is not at all cercopithecoid.[9] Two earlier specimens of lower humerus (#51 and #84) exhibit similar morphological details. The total morphological pattern found in *Oreopithecus* is a common hominoid one, for it is present not only in all living hominoid humeri but also in all known fossil ones (*Australopithecus* (*Paranthropus*), *Pliopithecus*, *Proconsul*, *Limnopithecus*). Moreover, *Oreopithecus* #51 also includes the upper part of the ulna and the head of the radius, which confirm the existence of a hominoid—rather than a cercopithecoid—type of elbow joint.

The proximal part of the ulna can be studied in detail in two specimens (WLS)

7. Schultz (1960) found the length-breadth index of L.2 to be 136.3 in the *Oreopithecus* skeleton, 111.0–161.5 in hominoids, and 70.0–100.0 in cercopithecoids. For the last lumbar vertebra I obtained a comparable index of 169.2 for the 1958 *Oreopithecus* and 147.8 for an earlier specimen (#35); 148.1 (113.0–173.1) for 21 hominoids; and 118.6 (100.0–142.3) for 17 cercopithecoids (WLS).

8. Still another, fragmentary sacrum (#35) similarly indicates hominoid affinities in number of vertebrae and general morphology (Schultz, 1960; WLS).

9. For a comparison of the lower end of the humerus in various hominoids and cercopithecoids, see Straus (1948).

(Fig. 8). It is distinctly hominoid in morphology, rather than cercopithecoid, as already noted by Hürzeler (1958). The olecranon is relatively small, as in hominoids,[10] whereas it is quite long in cercopithecoids. The joint surface for the head of the radius is in a hominoid (and not a cercopithecoid) position; and it is

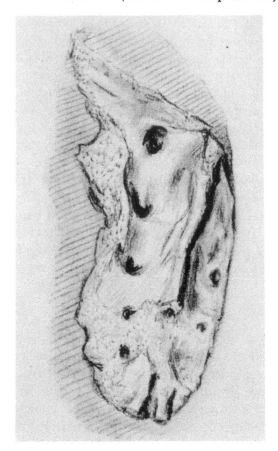

FIGURE 6
Dorsal aspect of *Oreopithecus* sacrum (#50).

not divided into two parts as customarily in cercopithecoids. Moreover, the configuration of the articular surface for the humerus resembles that of hominoids, and is quite unlike that of cercopithecoids. As in hominoids, it is divided longitudinally by a ridge; whereas, in cercopithecoids, this surface is comparatively flat. Finally, the upper ulna exhibits one distinctively hominid feature in that the areas for origin of the flexor and extensor muscles, respectively, are separated by

10. *Pliopithecus vindobonensis* is an exception in that its olecranon is relatively large (Zapfe, 1958).

only a narrow ridge, rather than by a relatively broad plateau as in pongids and cercopithecoids.

Unfortunately, the remains of the *Oreopithecus* hand skeleton available to me are too fragmentary to permit worthwhile conclusions. In the 1958 skeleton, when I saw it in 1959, metacarpale III and its basal phalanx could be identified.

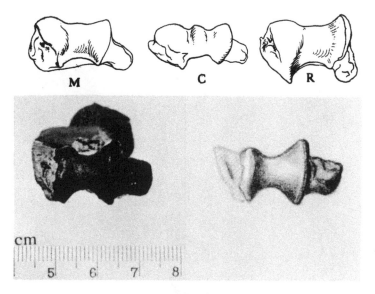

FIGURE 7

Lower end of right humerus of *Oreopithecus* (#51) (below), compared with the lower ends of the right humeri of a man (above left, M), a chimpanzee (above center, C), and a rhesus monkey, *Macaca mulatta* (above right, R). (The humeri of man, chimpanzee, and rhesus monkey are after Straus, 1948; redrawn.)

The index basal phalanx III length x 100/metacarpale III length is 72 (WLS). This ratio, however, turns out not to be diagnostic; for its range is 63–74 in hominoids (eight specimens) and 53–79 in cercopithecoids (six specimens) (WLS). The metacarpals and phalanges are slightly (but distinctly) curved, and the phalanges (especially the basals) give the appearance of being slightly scooped out ventrally. This "scooped-out" appearance is due primarily to the presence of moderately elevated ridges extending ventrally along the borders—essentially a simian condition. The "scooping out," however, is less than is usual in pongids.

Lower Extremity.—The 1958 femora have been considerably damaged, presumably by crushing, with consequent displacement of some parts, and recourse to roentgenograms has been necessary in their reconstruction. Description therefore will be limited to those features about which one can be reasonably certain.

The upper end of the femur is essentially hominoid (WLS). In two specimens (that of 1958 and #49), the neck forms a rather acute upper angle with the long axis of the shaft (being approximately 45° in the 1958 left femur) as commonly in hominoids (see Fig. 14); whereas this relationship tends to approach a right angle in cercopithecoids.

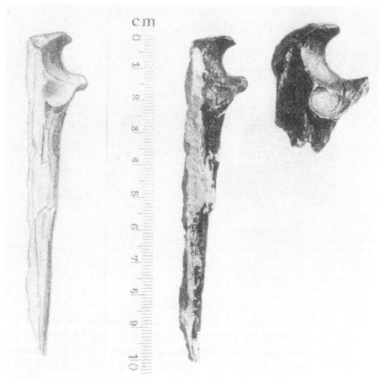

FIGURE 8
Ulnar fragments of *Oreopithecus*.

In its relative length, moreover, the neck of the *Oreopithecus* femur most closely resembles those of hominoids, especially those of the great apes and man (Schultz, 1960; WLS).[11]

11. Schultz (1960) found that the relative length of the femoral neck (measured as the greatest distance from the head of the femur to the lateral surface of the shaft, in the direction of the axis of the femoral neck), when expressed in percentage of femur length, is 25.1; in 10 Cercopithecidae, it ranges from 16.3 to 18.8; in 2 Hylobatinae, 15.0 and 19.9; in 7 great apes, 21.6–27.7; and in 3 men, 21.4–25.6. Using a somewhat different method (my "femoral neck length" was measured from the middle of the shaft to the base of the head; and my "femur length" differs slightly from that of Schultz), I found the relative length of the femoral neck in *Oreopithecus* to be 8.6; in 16 Cercopithecidae, ave. 5.9 (4.4–7.3); in 12 Hylobatinae, ave. 6.9 (5.7–8.6); in 32 great apes, ave. 10.7 (8.1–15.3); and in 12 men, ave. 10.6 (8.7–12.2).

Hürzeler (1958) has noted a resemblance of the upper end of the *Oreopithecus* femur to that of the famous Eppelsheim femur, also Pontian in age.

The femoral shaft of the 1958 skeleton is essentially uniform in width for most of its extent, broadening only in its lower part; a not uncommon condition in all three catarrhine families (WLS). The robustness index of the lower part of the shaft, 56.6, falls within the ranges of variation found in both cercopithecoids (45.7–62.9, for 37 specimens) and modern men (35.2–58.8, for 58 specimens); also, it so happens that it is identical with the index obtained from measurements of a specimen of *Australopithecus* (*Plesianthropus*) *transvaalensis* studied by Clark (1947). However, it lies outside of the pongid range of variation (32.5–51.4, for 96 specimens) (cf. Kern and Straus, 1949, for details respecting this robustness index). Transverse width of the femoral shaft midway between its two ends measures 25 mm. in *Oreopithecus*. Expressed in percentage of corrected femur length (*vide infra*), this gives an index of 9.8. This index lies within the range of variation for great apes recorded by Schultz (1960), but it is larger than any listed by him for man, hylobatines and cercopithecoids. This index of mid-shaft robustness thus is pongid; more specifically, pongine.

The angle of obliquity of the 1958 femoral shaft is about 10°, thus hominid rather than pongid or cercopithecoid (Kern and Straus, 1949). But this determination is only an approximation—and hence without deep significance—since it is based upon considerable reconstruction of the badly fractured 1958 left femur.

The relative sizes of the femoral condyles differ among living catarrhines (Kern and Straus, 1949), reflecting the weight-bearing or load line through the femur (Walmsley, 1933). In hominids and some cercopithecoids (e.g., *Macaca* and *Erythrocebus*) the area of the lateral condyle approximates or even surpasses that of the medial condyle. Among pongids, however, the lateral condyle regularly is relatively small (only in one femur of *Hylobates*, of seventeen measured, did the area of the lateral condyle equal that of the medial). Although their areas were not precisely measured, the lateral and medial condyles of two specimens of *Oreopithecus* (1958 and #66) are apparently of about equal size (Fig. 9), hence more closely resembling those of hominids and cercopithecoids than those of pongids.

Schultz (1960) has studied three patellae of *Oreopithecus*. All three of these specimens have length-width indices that fall within the hominoid range of variation, at the same time lying outside the cercopithecoid range. *Oreopithecus* thus possesses a relatively broad patella, as in all known hominoids, rather than a relatively narrow one as in Old World monkeys. It is, however, more pongid than hominid in its proportions, finding its closest resemblance in hylobatines.

Of the few foot bones available, the talus and calcaneus are of chief interest and significance here. In particular, I have made a detailed study of a talus (no #; Fig. 10) and a calcaneus (#37; Fig. 11) discovered by Hürzeler prior to

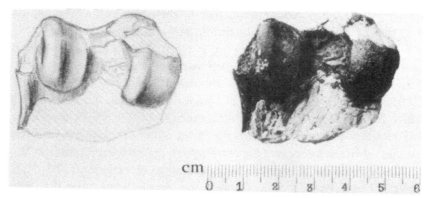

FIGURE 9

Lower end of right femur of *Oreopithecus* (#66).

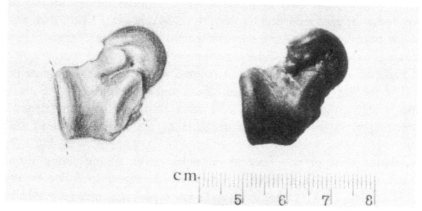

FIGURE 10

Left talus of *Oreopithecus* (no #), dorsal aspect.

1958. Both of these specimens are excellently preserved; in addition, there is an incomplete talus (#37) and an incomplete calcaneus (#92).[12]

My analysis of the *Oreopithecus* talus is based chiefly on eight characters (six indices and two angles), following the procedure of Poniatowski (1915). It resembles cercopithecoids in five characters, the Hylobatinae in two, the orangutan in four, the chimpanzee in three, the gorilla in two, and man in but one. Its overall morphology clearly is more cercopithecoid than hominoid. The slight torsion of its head and its narrowness and lowness all obviously are primitive

12. Comparisons have been made by me with series of modern catarrhines: Talus—Hominoidea, 147 specimens (102 men, 15 gorillas, 14 chimpanzees, 6 orangutans, 8 gibbons, 2 siamangs), Cercopithecoidea, 18 specimens (2 *Cercopithecus*, 1 *Cynopithecus*, 7 *Macaca*, 4 *Papio*, 1 *Colobus*, 1 *Rhinopithecus*, 2 *Presbytis*); Calcaneus—Hominoidea, 148 specimens (103 men, 15 gorillas, 15 chimpanzees, 7 orangutans, 7 gibbons, 1 siamang), Cercopithecoidea, 21 specimens (3 *Cercopithecus*, 1 *Cynopithecus*, 7 *Macaca*, 5 *Papio*, 1 *Colobus*, 1 *Rhinopithecus*, 3 *Presbytis*).

characters which suggest considerable mobility at the ankle joint. The marked medial deviation of the talar neck (Fig. 10) is indicative of a normally abducted hallux.

My analysis of the calcaneus of *Oreopithecus* rests largely on sixteen characters (including thirteen indices, using the methods of Reicher, 1913). It resembles cercopithecoids in seven characters, the Hylobatinae in six, the orangutan in six, the chimpanzee in ten, the gorilla in eleven, and man in twelve. Thus it is clearly hominoid, rather than cercopithecoid, in morphology. Its marked overall resemblance to man and the African anthropoids suggests some adaptation to progression on the ground. In four characters—relatively great width of the sustentaculum tali (Fig. 11), reflecting the marked medial deviation of the collum tali; narrowness of the corpus calcanei; absence of a lateral or external tubercle on the tuber calcanei; and strong development of the trochlear (peroneal) process—the *Oreopithecus* calcaneus is outspokenly simian and hence non-hominid. Weidenreich (1940) has shown that there is an inverse relationship between the development of the external tubercle of the tuber calcanei and the size of the trochlear process. In fact, he concluded that the external tubercle developed from a part of the trochlear process phylogenetically. Thus, all pongids and cercopithecoids regularly possess a strongly developed trochlear process; but they totally lack an external tubercle. In man, however, although the trochlear process varies in size, it normally is relatively quite small; but there regularly is a distinct external tubercle. In line with this correlation, the calcaneus of *Oreopithecus*, which possesses a prominent trochlear process, exhibits not a trace of an external tubercle.

In two other characters, however, it is peculiarly hominid and non-simian—in the exceedingly small size of the basal or plantar tubercle and in the relatively enormous plantar extension of the heel or tuber calcanei (Fig. 11). The basal tubercle usually is quite large in all cercopithecoids and pongids, and it transmits the body weight to the ground, the heel being lifted (Weidenreich, 1940). In man, however, the basal tubercle is vestigial (being represented only by the small anterior tubercle) and the body weight passes through the tuber calcanei. The basal tubercle is similarly quite small in both available calcanei of *Oreopithecus*. The extension of the heel anteriorly along the plantar surface was measured projectively from the most posterior point on the tuber calcanei parallel to the long axis of the calcaneus. This "heel length" was compared with both the greatest calcaneus length and the corpus calcanei length. For the complete *Oreopithecus* calcaneus, the calcaneus-heel index is 40.0, the corpus-heel index 53.3. Among the calcanei of living primates which I have measured, only that of one modern man has as large a calcaneus-heel index (40.3); the indices for individuals of the three great apes range from 29.4 to 37.7; and those of cercopithecoids and (especially) gibbons are considerably smaller. The same holds true for the corpus-heel index, although here the *Oreopithecus* value is also approximated in two chimpanzees (both of which have an index of 51.2).

Hürzeler (1958) has briefly described and figured an entocuneiform of

Oreopithecus discovered in 1957. According to him, it closely resembles the entocuneiform of the chimpanzee. Its articular facet for the base of the first metatarsal is similarly convex. This, together with the marked medial deviation of the collum tali, definitely indicates that the hallux of *Oreopithecus* was both abducted and opposable, as in simian primates.

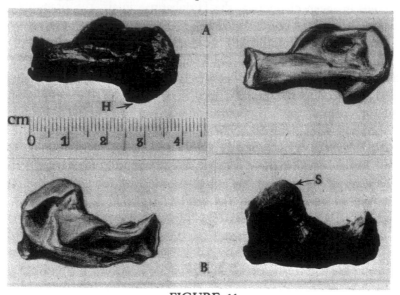

FIGURE 11

Left calcaneus of *Oreopithecus* (#37), seen from lateral (A, above) and dorsal (B, below) aspects. H, heel; S, sustentaculum tali.

Limb Proportions.—The proportions of the limb bones are of great interest and significance. Lengths of the long bones of the 1958 skeleton of *Oreopithecus* (humerus, radius, femur, tibia) were measured by me according to the technique of Mollison (1910) and Schultz (1930, 1937). These lengths, for the left side, have been estimated as follows: Humerus, 298 mm.; Radius, 283 mm.; Femur, 255 mm.; Tibia, 235 mm.[13]

It must be realized that these long-bone lengths are approximate, since all of these bones have been fractured. However, the deviations from actual lengths cannot be more than a few millimeters at most. My estimates agree closely with those of Schultz (1960), who gives length of humerus as 297 mm., of radius as 283 mm., and of femur as 243 mm. Only the femur has required any great amount

13. Neither tibia was preserved, although one almost complete tibia was present when the skeleton was recovered. Unfortunately, it fell to the floor of the mine and was shattered beyond recognition during the difficult procedure of removing the block of lignite containing the skeleton from the roof of the mine. Fortunately, however, a good photograph of the tibia and the associated femur was made before the former was lost. From measurements taken from this photograph it has been possible to make a reasonable estimate of tibia length, since the actual length of the femur on the same photograph is known.

of reconstruction before estimation of length. Roentgenograms reveal that the shaft had been fractured, followed by a twisting and overriding of the two fragments. I estimate that this overriding amounts to about 15 mm. (perhaps even a few mm. more). Hence, I have added 15 mm. to my measurement of femur length taken from the actual specimen (240 mm.), giving an approximate length of 255 mm. between tip of greater trochanter and lateral condyle, taken parallel to the longitudinal axis of the shaft. (In measuring the actual specimen, an addition of approximately 2 mm. had to be made to account for a fracture which produced an overriding of the lateral condyle on the lower part of the shaft.) This reconstruction accounts for the different femur lengths estimated by myself and by Schultz (1960). Schultz, however, recognized that a correction of about 18 mm. was in order; this would make his estimated femur length 261 mm. His measurement also differs from mine in that its upper point of reference was the femoral head, rather than the greater trochanter. In any event, as Schultz (1960) also notes, slight differences in estimation of femur length cannot significantly alter conclusions involving indices utilizing that bone.

The brachial index (radius length x 100/humerus length) is approximately 95 in *Oreopithecus*. This figure lies within the ranges of variation of chimpanzee, orangutan, and cercopithecoids, and comes close ·to the means of these two great apes and the macaque (Mollison, 1910; Schultz, 1937). The indices of man and gorilla are much lower, those of Hylobatinae much higher. As Schultz (1960) has noted, the intermediate position of *Oreopithecus* suggests a lack of specialization.

The crural index (tibia length x 100/femur length) of *Oreopithecus* is about 92 (92.1 from my estimated measurements). This falls within the ranges of variation of man, orangutan, gibbon, and cercopithecoids, and is virtually identical with the means of orangutan and macaque (Mollison, 1910; Schultz, 1937). The indices of the gorilla, chimpanzee, and siamang all are slightly lower. This index appears not to be of any great phylogenetic significance, at least not among catarrhines.

The intermembral index (humerus length + radius length x 100/femur length + tibia length), however, possesses considerable significance. It is approximately 119 (118.6 from my measurements) in *Oreopithecus*. Hence the arms of this animal were longer than its legs—a condition peculiar to habitual, outright brachiators. The *Oreopithecus* index thus is distinctly pongid—for living pongids always have an index of more than 100 and hence arms that exceed their legs in length (Mollison, 1910; Schultz, 1937). On the other hand, the legs are always longer than the arms in man and cercopithecoids—the range for man being 64.5–79.2 (Schultz, 1937), for cercopithecoids 78–97 (Mollison, 1910). However, although the intermembral index of *Oreopithecus* is pongid it actually falls within the range of variation of only the gorilla (a brachiator come to earth in adult life!). It lies slightly above the chimpanzee range, slightly below the gibbon range, and well below the orang and siamang ranges (Schultz, 1937).

That the major limb proportions of *Oreopithecus* are pongid, and those of a brachiator, is further attested by its femoro-humeral index, which approximates 117 (116.9 from my measurements).[14] This again is a peculiarly pongid character —for, among living catarrhines, only in the pongids (and, indeed, in all of them) is the humerus regularly longer than the femur (Mollison, 1910; Schultz, 1937). In fact, there are no individual exceptions to this rule in the gorilla, orangutan, siamang, and gibbon. Only the chimpanzee provides an exception; for in this animal, Schultz (1937) found that the femur was equally as long as or longer than the humerus in 31 per cent (in 30 of 96 skeletons). In both man (Schultz, 1937) and Old World monkeys (Mollison, 1910), however, the femur always surpasses the humerus in length. The value of the femoro-humeral index estimated for *Oreopithecus* falls within the range of variation obtained by Mollison (1910) for gorilla and gibbon. It lies above his chimpanzee range and below his orang and siamang ranges.

The radius of *Oreopithecus* is considerably longer than the tibia, the tibio-radial index being approximately 120 (120.4 from my measurements). In pongids, this ratio ranges from 110 to 168. The *Oreopithecus* index falls within the range of variability of the gorilla; it lies just above the range of the chimpanzee and well below those of the orangutan, siamang, and gibbon (Mollison, 1910). The radius is shorter than the tibia in the large majority of cercopithecoids and always much shorter in man; in the series of the former studied by Mollison (1910) the largest individual index was 107. The tibio-radial index of *Oreopithecus* thus is distinctly pongid.

The diameters of the caput humeri and caput femoris are 38 mm. and 29 mm., respectively, in *Oreopithecus*, giving a femoro-humeral caput index of 131 (WLS). As Schultz (1960) has pointed out, this condition is distinctly pongid. For, among catarrhines, only in the brachiators, hence in pongids, is the head of the humerus distinctly larger than the head of the femur. In the quadrupedal cercopithecoids, the two heads are approximately equal in diameter, and in bipedal man the femoral head is the larger. Schultz (1960) also expressed these diameters in percentage of trunk length; for both humerus and femur he found the index of *Oreopithecus* to be hominoid, and not at all cercopithecoid.

When humerus length is expressed in percentage of trunk length, *Oreopithecus* falls between the cercopithecoids (which have much shorter humeri) and the outright brachiators, the orangutan and hylobatines—which have longer humeri (Schultz, 1960). In this respect our fossil most closely resembles the African great apes and man. Contrariwise, the femur of *Oreopithecus*, relative to trunk length, is shorter than in all living hominoids, being as in some cercopithecoids (Schultz, 1960). It can be concluded, therefore, that the relatively high intermembral index of *Oreopithecus* results from short legs, rather than from long arms.

---

14. The femur in life obviously was considerably shorter than the humerus; and no reasonable reconstruction can make it as long as the latter bone.

Pelvis

The pelvis of *Oreopithecus*, in its total morphological pattern, is hominoid. The hip-bones are very broad—a common hominoid character (Schultz, 1960; also Figs. 5, 12). The pelvic breadth, relative to trunk length, is similarly hominoid (Schultz, 1960). Relative to hip-bone length, however, it is, to be more precise, pongid (Schultz, 1960).

But when the hip-bone of *Oreopithecus* is compared with trunk length, it is found to be relatively shorter than those of all pongids, herein resembling those of cercopithecoids and man (Schultz, 1960). The length-width index of its ilium (80), however, is distinctly pongid (being, moreover, more pongine than hylobatine); hence, although the ilium of *Oreopithecus* is relatively much broader than those of Old World monkeys it is relatively much narrower than that of

FIGURE 12

Dorsal view of caudal portion of August 2, 1958 skeleton of *Oreopithecus*, showing ribs, vertebrae, pelvis, and long bones of lower extremity.
(Courtesy of Dr. Johannes Hürzeler.)

hominids (Schultz, 1960; WLS).[15] In any event, this index, like those involving pelvic breadth, clearly attests (like the ribs), as Schultz (1960) has emphasized, that *Oreopithecus* could not have possessed a slender, cercopithecoid type of trunk. The contour of the iliac crest bears a marked resemblance to that of the orangutan (WLS).

Although it can be measured only approximately because of post mortem damage, the sacral surface of the *Oreopithecus* ilium evidently was relatively small (WLS). Using the methods of Schultz (1930), I estimate its width as 30 mm., and that of the iliac fossa as 82 mm. Dividing sacral surface width into iliac fossa width x 100, an index of 273.3 is obtained. This ratio, according to the data of Schultz (1930), distinctly resembles those of living pongids.[16] It agrees closely with his figures for gibbon, siamang and orangutan; but it is less than those of his chimpanzee and gorilla. This index is very much smaller in both cercopithecoids (in which the iliac fossa is peculiarly narrow) and man (in which both fossa and sacral surface are expanded). But the index of *Oreopithecus* cannot be considered peculiarly pongid and hence non-hominid; for similar ratios are also found in the fossil Australopithecinae, which certainly are hominids.[17] Thus, in the relative size of its sacral surface, *Oreopithecus* exhibits a general hominoid condition.

In all living pongids and most cercopithecoids there is a comparatively long space between the upper margin of the acetabulum and the lower margin of the sacral surface, so that these two structures are well separated; in other words, the lower ilium is relatively long in simian catarrhines (Schultz, 1930; Fig. 13). In modern man (Schultz, 1930) and the Australopithecinae (Straus, 1962), how-

15. From plaster casts, I estimate the length-width index of the ilium (using the method of Straus, 1929) to have been 129 for the adult *Australopithecus (Plesianthropus) transvaalensis* right hip-bone; 147.5 for the adult *Australopithecus (Paranthropus) crassidens* right hip-bone; and 122 for the juvenile *Australopithecus prometheus* left hip-bone. All of these ilia are incomplete; hence these indices represent only reasonable approximations. For modern men (244 American Whites and Negroes), this index averages about 125, with individual extremes of 107.9 and 139.5; for 13 gorillas, 91.8 (83.6–100.5); for 16 chimpanzees, 66.0 (57.6–73.7); for 23 orangutans, 73.8 (62.7–87.4); for 4 siamangs, 56.5 (52.5–61.5); for 18 gibbons, 48.7 (41.1–57.7); and for 54 Cercopithecidae (including both Cercopithecinae and Colobinae), average generic indices of less than 50, with individual extremes of 35.7 and 54.2 (Straus, 1929).

16. After Schultz (1930): 2 *Macaca*, ave. 80.9 (72.3 & 89.5); 1 *Papio*, 120.6; 23 *Hylobates*, 224.1 (166.5–293.0); 4 *Symphalangus*, 278.1 (250.0–300.0); 2 orangutans, 268.0 (254.2 & 281.8); 2 chimpanzees, 304.8 (303.5 & 306.2); 9 *Gorilla gorilla*, 393.8 (358.7–476.8); 6 *Gorilla beringei*, 352.6 (283.0–415.9); 15 men, 152.3 (125.3–190.0).

17. Using plaster casts, I have calculated the index iliac fossa width x 100/sacral surface width for three available hip-bones of Australopithecinae. Because the bones are damaged, as noted above, these indexes are only approximate estimates; but they cannot be greatly in error. For the adult *Australopithecus (Plesianthropus) transvaalensis* right hip-bone, this index is 290; for the adult *Australopithecus (Paranthropus) crassidens* right hip-bone, it is 231.5; and for the juvenile left hip-bone of *Australopithecus prometheus* it is 260. The large index of the juvenile serves to emphasize the relatively small size of the australopithecine sacral surface; for, as Schultz (1930) has shown, the iliac fossa of hominoids grows more slowly than the sacral surface, so that the index in question is largest in adult life.

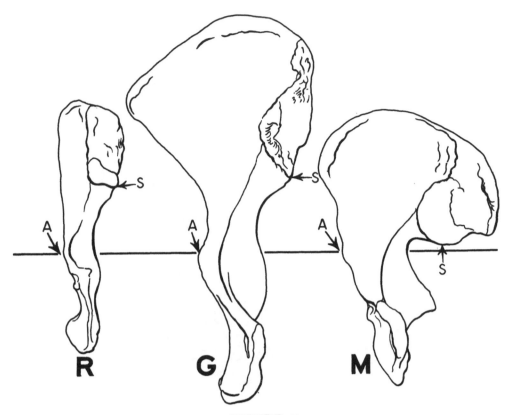

FIGURE 13

Medial views of right innominate bones of adult rhesus monkey, *Macaca mulatta* (left, R), mountain gorilla, *Gorilla beringei* (center, G), and man (right, M.) (After Schultz, 1936; redrawn.) A, highest point on acetabulum; S, lowest point on sacral surface. The bones have been reoriented so that the highest points on the acetabula lie in the same horizontal plane.

ever, acetabulum and sacral surface closely approach each other; thus the lower ilium is relatively short in hominids, this evidently being a stabilizing adaptation related to the erect bipedal posture (Straus, 1962; Fig. 13). In so far as I can determine from the two known but incomplete pelves (#49 and 1958) the distance between acetabulum and sacral surface is relatively shorter in *Oreopithecus* than in pongids—perhaps representing a more primitive or generalized hominoid condition. Indeed, if my interpretation of specimen #49 is correct, the acetabular-sacral relations closely approximate those found in hominids (Fig. 14).

The 1958 *Oreopithecus* pelvis exhibits a fairly well developed, hominid type of anterior inferior iliac spine (partly revealed on the left innominate bone in Figure 12). This structure is regarded as peculiarly related to the erect, bipedal posture in man (and australopithecines), particularly because of its association with the ilio-femoral ligament, which prevents overextension of the pelvis in the

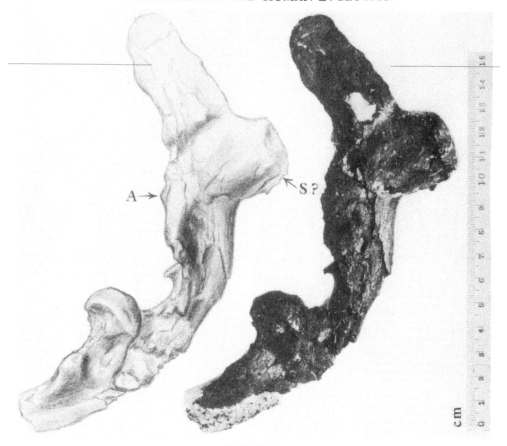

FIGURE 14

Medial view of right innominate fragment of *Oreopithecus* (#49). A, highest point
on acetabulum; S ?, lowest point on sacral surface ?. The upper end of the
dislocated femur can be seen below the acetabulum.

erect posture. This spine usually is rudimentary or relatively small in living
cercopithecoids and pongids (Straus, 1929). Even when it is moderately developed, as occasionally occurs, its conformation differs from that of hominids.

Other pelvic details of *Oreopithecus* also proclaim hominoid affinity. The
relatively short pubic symphysis (#44) is distinctly more hominoid than
cercopithecoid in structure (Schultz, 1960; WLS). The ischial tuberosity (#44)
is comparatively low and small, with rounded margins. Herein it differs markedly
from the tuberosities of Old World monkeys and Hylobatinae, finding its closest
resemblance in the tuberosities of the giant hominoids, particularly in those of
the great apes (Schultz, 1960; WLS).

In general, therefore, the pelvic morphology of *Oreopithecus* is that of a
hominoid; more specifically, perhaps, that of a generalized or primitive hominoid.
The latter is especially suggested by the relative length of the hip-bone (which

lacks the undoubtedly specialized lengthening found in all pongids) and also, possibly, by the relative length of the space separating sacral surface and acetabulum, which seems to be at least intermediate between the pongid and the hominid types, if not actually approximately hominid in character.

CRANIAL CAPACITY

Straus and Schön (1960) have estimated that the cranial capacity of the 1958 skull of *Oreopithecus* falls between 276 and 529 cc., hence within the ranges of variation of both orangutan and chimpanzee.[18] If the larger of the calculated possibilities is correct, the capacity of this particular skull of *Oreopithecus* was as much as that of many gorillas. I suspect that its actual capacity was about 400 cc., which still would leave it within the ranges of variation of all three great apes. In cranial capacity, therefore, *Oreopithecus* is wholly hominoid. Its lowest estimated capacity is considerably greater than any that has been recorded for a cercopithecoid; even large adult baboons have cranial capacities of less than 200 cc. (Schultz, 1941b, 1950b).[19] But the cranial capacity of *Oreopithecus* is not peculiarly pongine. It is true that its estimated possible maximum falls far short of the minimal capacities of both normal modern man and the most primitive known hominine, *Homo (Pithecanthropus) erectus erectus*. However, the cranial capacity of *Oreopithecus* certainly approximated, and probably was as large as or even exceeded those of some Australopithecinae, which certainly are hominids (Straus and Schön, 1960). Moreover, if we knew the cranial capacities of the Lower Pliocene ancestors of the great apes—something of which we are, unfortunately, totally ignorant—it is quite possible that the encephalic magnitude of *Oreopithecus* would be labeled hominid. For if the course of encephalic phylogeny of the Pongidae parallels that of the Hominidae—in which the brain clearly lags behind the rest of the body—one would expect the cranial capacities of Lower Pliocene pongids to be considerably smaller than those of their present-day descendants. However, this is a matter that can only be settled by future paleontological discoveries.

In relative size of brain (secured by expressing cranial capacity in cc. as a percentage of body weight in grams), *Oreopithecus* also emerges as a hominoid (Straus and Schön, 1960). Here again the Tuscan fossil may possibly be within the australopithecine range of variation.

18. The limits of variation in both orangutan and chimpanzee are almost identical with these extreme possibilities calculated for *Oreopithecus*: 276 and 523 cc. (ave. 392.2 cc.) in the combined series of 194 adult orangutan skulls measured by Gaul (1933) and Schultz (1941a); and 290 and 500 cc. (ave. 374.1 cc.) in the combined series of 94 adult chimpanzee skulls studied by Zuckerman (1928) and Schultz (1940). For 474 adult gorilla skulls, the range of variability extends from 340 to 685 cc. (ave. 504.8 cc.) (computed from the data given by Schultz, 1950a, table 7). Schultz (1962) has recently reported a cranial capacity of 752 cc for an adult male West African gorilla in the University of Zürich collection.

19. Dr. L. S. B. Leakey (personal communication), however, has a fossil baboon, *Simopithecus*, which had a brain of great-ape size.

THE TAXONOMIC ASSIGNMENT OF OREOPITHECUS

In determining the taxonomic position of *Oreopithecus* among the Catarrhina, five possibilities may be considered. These will be examined in order.

1. OREOPITHECUS IS A CERCOPITHECOID

As already noted, *Oreopithecus* long was regarded as an Old World monkey of some sort. This was largely based, as also noted, on the mistaken belief that this Tuscan fossil exhibited an "incipient" dental bilophodontism. Indeed, *Oreopithecus* exhibits but a single character which is, in our present meager knowledge of the catarrhine fossil record, peculiarly cercopithecoid:

The ventral midsagittal keel on the thoracic and lumbar vertebrae

Some other of its characters are equally shared with both cercopithecoids and pongids:

Parallel tooth rows
Talonids on the lower premolars
Structure of the talus
Brachial index

Still two other characters also occur in both cercopithecoids and hominids:

Approximately equal sizes of femoral condyles
Comparatively short hip-bone

Finally, five of its characters are also encountered in all three recognized families of catarrhines, the Cercopithecidae, Pongidae, and Hominidae:

Large supraorbital ridges
No chin
Position of occipital condyles (probably)
Bicuspid upper premolars
Crural index

Most of these characters (notably the bicuspid upper premolars, lack of a chin, vertebral keel, talar structure, brachial and crural indices, proportions of femoral condyles, and short hip-bone) probably are to be regarded as generalized or primitive catarrhine features, rather than as peculiar specializations.

It is obvious, therefore, that there is no valid reason to regard *Oreopithecus* as a cercopithecoid, not even as an aberrant one.

2. OREOPITHECUS IS A HOMINOID

The evidence that our fossil must be assigned to the superfamily Hominoidea is overwhelming. For many of its characters are common to members of both hominoid families but are not found in the Cercopithecidae:

Body size
Regular absence of simian shelf
Internal cusp on upper incisors
Form of upper canine
Shape of trunk
Number of lumbar vertebrae
Proportions of lumbar vertebrae
Form of sacrum
Morphology of lower end of humerus and elbow joint
Structure of upper end of femur
Overall pattern of calcaneus
Overall morphology of pelvis
Size of brain (cranial capacity), both absolute and relative

Many other characters, already discussed above and which will again be considered below, are found in one or the other of the two hominoid families.

What remains to be determined is the precise familial assignment of our fossil.

3. OREOPITHECUS IS A PONGID

In a few characters, *Oreopithecus* finds its parallel in the Pongidae alone. These pongid characters are:

Upper molar pattern
Proportions of patella
Entocuneiform morphology (?)
Limb proportions (intermembral, femoro-humeral, and tibio-radial indices; relative sizes of heads of humerus and femur)
A few pelvic characters (pelvic breadth; iliac length-width index; general shape of ilium; morphology of ischial tuberosity)

Some of these characters (limb proportions) seem indubitably to represent specializations connected with brachiation. In the instance of *Oreopithecus*, they probably are adaptations to an arboreal life in the Pontian swamp. A relative lengthening of the forelimb, so that the intermembral and femoro-humeral indices surpass 100, appears to be a frequent outcome of adaptation to brachiation among primates. Thus it has appeared independently in the Ponginae, the Hylobatinae, and in certain New World monkeys (*Ateles, Alouatta;* Mollison,

1910; Schultz, 1930).[20] In this instance, therefore, we are dealing with a character which appears not to possess deep taxonomic significance. The particular pelvic features and the form of the entocuneiform probably are not directly concerned with brachiation; they probably are generalized or primitive hominoid conditions. The same may possibly be true of the upper molar pattern; for it finds its greatest resemblance in early members of the Pongidae and also appears to retain primitive features not found in any known members of that family.

In view of the few peculiarly pongid characters exhibited by *Oreopithecus*, I do not believe that we are at all justified in classifying it with the Pongidae.

4. Oreopithecus is a Hominid

In a number of characters, *Oreopithecus* bears a striking resemblance to the Hominidae, and to them alone:

> Short face (including rostral location of anterior root of zygomatic arch)
> Nasal aperture has a roof
> Relations between zygomatic arch and orbit
> Vertical mandibular symphysis (associated with vertical implantation of incisors)
> Normal location of mental foramen
> Normal absence of diastemata (both jaws)
> Homomorphic lower premolars
> Length-proportions of teeth
> Overall morphology of upper ulna
> Angle of femoral obliquity (probably)
> Large heel (with concurrent reduction of basal tubercle of calcaneus)
> Form of anterior inferior iliac spine
> Relations between acetabulum and sacral surface (possibly)

Some of these features may well represent primitive hominoid characters which have been retained in the Hominidae but lost in the Pongidae—absence of diastemata; homomorphism of lower premolars; acetabular-sacral relations (?). The other characters in the above list, however, can only reasonably be regarded, in our present knowledge, as peculiarly hominid specializations.

Indeed, on the basis of these marked resemblances, I would not hesitate to classify *Oreopithecus* as a hominid if it were not for the morphology of its lower molar teeth. These were once regarded as *sui generis*. However, as I have noted above, Simons (1960) has recently confirmed the earlier observation of Gregory

---

20. The limb skeletons of both *Limnopithecus* and *Pliopithecus* definitely indicate that the Hylobatinae acquired their brachiating specializations relatively late in their evolution, after their evolutionary line had separated from the common hominoid stock. Thus these specializations represent an example of parallel development between Hylobatinae and Ponginae. In the cases of *Ateles* and *Alouatta*, however, we are dealing with an example of convergence between platyrrhines and catarrhines.

(1922) that the lower molars (and P 4) of *Oreopithecus* closely resemble those of the early Oligocene Fayum primate, *Apidium*. The pattern common to *Oreopithecus* and *Apidium* differs somewhat from that found in any known pongid or hominid, living or fossil, being on the whole more primitive. It also is more primitive than that of *Propliopithecus*, an undoubted hominoid from the same Fayum beds.[21] Yet *Apidium* and *Oreopithecus* are separated from one another in time by a period of more than 25 million years.

If the *Apidium-Oreopithecus* molar pattern is regarded as the primitive hominoid pattern, we will have to assume that the molar pattern common to all undoubted Pongidae and Hominidae evolved independently (from the *Apidium-Oreopithecus* pattern) in each family after it diverged from the ancestral hominoid stock; an instance of parallel evolution that can be matched in other characters common to pongids and hominids. *Oreopithecus*, could then be classified as an aberrant hominid which had retained the primitive hominoid pattern. This would imply that the hominid line of evolution had become independent of the pongid line at a relatively early date, probably sometime in the Oligocene. This hypothesis is not without merit. Indeed, I have previously argued for an Oligocene divergence of Pongidae and Hominidae on other grounds (Straus, 1949). On the principle of parsimony, however, this interpretation of the lower molar evidence may well be regarded by many, if not most students, as improbable. The case that can be made for classifying *Oreopithecus* as a hominid, even as an aberrant one, therefore is not clear-cut.

It also is not impossible that the *Apidium-Oreopithecus* molar pattern represents an adaptive specialization acquired independently, at different times, by these two animals. This interpretation would debar *Oreopithecus* as a lineal ancestor of man, but not necessarily from membership in the family Hominidae. I realize, however, that this interpretation is unlikely to find favor with most students of fossil teeth. For, if it be accepted, the dentition would assume a relatively insignificant role in determining the taxonomic position of *Oreopithecus*. Nevertheless, it is an interpretation which must at least be considered.

5. OREOPITHECUS BELONGS TO A FAMILY OF ITS OWN, THE OREOPITHECIDAE

Placing *Oreopithecus* in a hominoid family of its own, the Oreopithecidae— as first proposed by Schwalbe (1916); but here used with somewhat different phylogenetic implications—appears to be the only alternative to classifying this Tuscan fossil as a hominid.[22] This creation of a separate, monotypic family would rest primarily on the dentition alone. In doing so, one grants an overwhelming

21. I do not include *Parapithecus* in this comparison because the taxonomic status of this fossil, which also comes from the early Oligocene of the Fayum, is at present uncertain.

22. Kälin (1955), Thenius (1958), Butler and Mills (1959), Schultz (1960), and Simpson (1963) are among those who have assigned *Oreopithecus* to a family of its own.

priority to tooth morphology. Most paleontologists will probably agree with this procedure; for the nature of the paleontological record of necessity has granted the dentition priority in the assessment of relationships. If we had more than teeth and scraps of jaws of *Apidium* and other catarrhines for comparison, however, the molar pattern well might seem no obstacle to the classification of *Oreopithecus* as a hominid.

CONCLUSION

From the evidence at hand, it is clear that *Oreopithecus* must be included in the catarrhine superfamily Hominoidea. There exists no valid reason for regarding it as belonging to the family Pongidae. Hence it must either be classified as a member of the family Hominidae or placed in a family of its own, the Oreopithecidae. The arguments for either of these classifications appear to me as about equally sound. Whether one regards *Oreopithecus* as a hominid or as an oreopithecid is essentially a matter of taste.

If *Oreopithecus* is a hominid, its dentition in some respects has retained a much more primitive pattern than that of any other known hominid. In other respects, however, it appears to be more specialized. It must be realized that the earliest known accepted hominid is some 10 million years younger than *Oreopithecus*, a circumstance which makes the evaluation of the tooth morphology of our Tuscan fossil especially difficult.

If *Oreopithecus* belongs to its own, separate family, there must have occurred a very considerable degree of parallel evolution between it and the Hominidae— a degree of parallel evolution in some respects more striking than that which has undoubtedly occurred between the Hominidae and Pongidae.

To conclude, I am at present inclined to classify *Oreopithecus* as a primitive, aberrant member of the Hominidae. In doing this, however, I fully realize that it is quite as valid to place it in its own hominoid family, the Oreopithecidae. Any eventual firm decision between these two allocations depends upon the discovery of adequate paleontological specimens—not only cranial and dental, but, and especially, postcranial as well—not of *Oreopithecus*, but of other Tertiary hominoids.

In any event, *Oreopithecus* is an especially significant fossil in the study of human phylogeny, for it demonstrates that a number of truly hominid characters had already made their appearance within the Hominoidea by the beginning of the Pliocene, some 10 or 12 million years ago. Moreover, certain characters of its postcranial skeleton, especially certain characters of pelvis and foot, definitely suggest that the direct ancestors of *Oreopithecus* had already acquired some adaptations—prospective or otherwise—to erect bipedalism before *Oreopithecus* himself underwent adaptation to a life in the Pontian swamp.

BIBLIOGRAPHY

BUTLER, P. M., and J. R. E. MILLS
1959. "A contribution to the odontology of Oreopithecus." *Bull. Brit. Mus. (Nat. Hist.), Geol.,* vol. 4, no. 1, 26 pp.

CLARK, W. E. LE GROS
1947. "Observations on the anatomy of the fossil Australopithecinae." *J. Anat., London,* 81:300–333.
1960. *The Antecedents of Man.* Chicago: Quadrangle Books.

FORSYTH MAJOR, C. E.
1880. "Beiträge zur Geschichte der fossilen Pferde insbesondere Italiens." *Abh. Schweiz. Paläont. Ges.,* 7. (Cited by Hürzeler, 1958.)

GAUL, G.
1933. "Ueber die Wachtumsveränderungen am Gehirnschädel des Orang-Utan." *Z. Morph. Anthrop.,* 31:362–394.

GERVAIS, P.
1872. "Sur un singe fossile, d'espèce non encore décrite, qui a été découvert au Monte Bamboli." *C. R. Acad. Sci.,* 74. (Cited by Hürzeler, 1958.)

GREGORY, W. K.
1922. *The Origin and Evolution of the Human Dentition.* Baltimore: Williams & Wilkins Co.

HÜRZELER, J.
1949. "Neubeschreibung von Oreopithecus bambolii Gervais." *Schweiz. Paläont. Abh.,* 66:1–20.
1951. "Contribution à l'étude de la dentition de lait d'Oreopithecus bambolii Gervais." *Ecologae geol. Helvetiae,* 44:404–411.
1954. "Zur systematischen Stellung von Oreopithecus." *Verh. Naturf. Ges. Basel,* 65:88–95.
1958. "Oreopithecus bambolii Gervais. A preliminary report." *Verh. Naturf. Ges. Basel,* 69:1–48.
1960. "The significance of Oreopithecus in the genealogy of man." *Triangle,* 4:164–174.

KÄLIN, J.
1955. "Zur Systematik und evolutiven Deutung der höheren Primaten." *Experientia,* 11:1–17.

KERN, H. M., JR. and W. L. STRAUS, JR.
1949. "The femur of Plesianthropus transvaalensis." *Am. J. Phys. Anthrop., n.s.* 7:53–78.

MOLLISON, T.
1910. "Die Körperproportionen der Primaten." *Morph. Jahrb.,* 42:79–304.

PATTERSON, B.

1955. "The geologic history of non-hominid primates in the Old World." Pp. 13–31 in: *The Non-human Primates and Human Evolution*. Detroit: Wayne Univ. Press.

PONIATOWSKI, S.

1915. "Beitrag zur Anthropologie des Sprungbeines." *Arch. Anthrop.*, N.F. 13:1–32.

REICHER, M.

1913. "Beitrag zur Anthropologie des Calcaneus." *Arch. Anthrop.*, N.F. 12:108–133.

REMANE, A.

1955. "Ist Oreopithecus ein Hominide?" *Abh. math.-nat. Kl. Akad. Wiss. Mainz*, 12:467–497.

ROBINSON, J. T.

1956. "The Dentition of the Australopithecinae." *Transvaal Mus. Mem.* No. 9.

SCHLOSSER, M.

1887. "Die Affen, Lemuren, Chiropteren etc. des europäischen Tertiärs." *Beitr. Paläont. Öst.-Ungarns*, 6. (Cited by Hürzeler, 1958.)

SCHULTZ, A. H.

1930. "The skeleton of the trunk and limbs of higher primates." *Human Biol.*, 2:303–438.

1937. "Proportions, variability and asymmetries of the long bones of the limbs and the clavicles in man and apes." *Human Biol.*, 9:281–328.

1940. "Growth and development of the chimpanzee." *Contrib. to Embryol. no. 170*, Carnegie Inst. Wash. Publ. no. 518, pp. 1–63.

1941a. "Growth and development of the orang-utan." *Contrib. to Embryol. no. 182*, Carnegie Inst. Wash. Publ. no. 525, pp. 57–110.

1941b. "The relative size of the cranial capacity in primates." *Am. J. Phys. Anthrop.*, 28:273–287.

1950a. "Morphological observations on gorillas." *The Anatomy of the Gorilla* ('Raven Memorial Volume), Part V, pp. 227–254. New York: Columbia Univ. Press.

1950b. "The physical distinctions of man." *Proc. Am. Philos. Soc.*, 94:428–449.

1960. "Einige Beobachtungen und Masse am Skelett von Oreopithecus." *Z. Morph. Anthrop.*, 50:136–149.

1961. "Vertebral column and thorax." *Primatologia*, 4, Lief. 5:1–66.

1962. "Die Schädelkapazität männlicher Gorillas und ihr Höchstwert." *Anthrop. Anz.*, 25:197–203.

SCHULTZ, A. H. and W. L. STRAUS, JR.

1945. "The numbers of vertebrae in primates." *Proc. Am. Philos. Soc.*, 89:601–626.

SCHWALBE, G.

1916. "Über den fossilen Affen Oreopithecus Bambolii." *Z. Morph. Anthrop.*, 19:149–254, 501–504.

SIMONS, E. L.

1960. "Apidium and Oreopithecus." *Nature*, 186:824–826.

SIMPSON, G. G.

1945. "The principles of classification and a classification of mammals." *Bull. Am. Mus. Nat. Hist.*, 85:1–350.

1963. "The meaning of taxonomic statements." This volume.

STRAUS, W. L., JR.

1929. "Studies on primate ilia." *Am. J. Anat.*, 43:403–460.

1948. "The humerus of Paranthropus robustus." *Am. J. Phys. Anthrop.*, n.s. 6:285–312.

1949. "The riddle of man's ancestry." *Quart. Rev. Biol.*, 24:200–223.

1961. "The phylogenetic status of Oreopithecus bambolii." *Phila. Anthrop. Soc. Bull.*, 14:12–13.

1962. "Fossil evidence of the evolution of the erect, bipedal posture." *Clin. Orthopaedics*, 25:9–19. Philadelphia: J. B. Lippincott.

STRAUS, W. L., JR. and M. A. SCHÖN

1960. "Cranial capacity of Oreopithecus bambolii." *Science*, 132:670–672.

THENIUS, E.

1958. "Tertiärstratigraphie und tertiäre Hominoidenfunde." *Anthrop. Anz.*, 22:66–77.

WALMSLEY, T.

1933. "The vertical axes of the femur and their relations." *J. Anat., London*, 67:284–300.

WEIDENREICH, F.

1940. "The external tubercle of the human tuber calcanei." *Am. J. Phys. Anthrop.*, 26:473–487.

ZAPFE, H.

1958. "The skeleton of Pliopithecus (Epipliopithecus) vindobonensis Zapfe and Hürzeler." *Am. J. Phys. Anthrop., n.s.* 16:441–457.

ZUCKERMAN, S.

1928. "Age-changes in the chimpanzee, with special reference to growth of brain, eruption of teeth, and estimation of age; with a note on the Taungs ape." *Proc. Zool. Soc. London*, 1–42.

THE LOCOMOTOR FUNCTIONS OF HOMINIDS

JOHN NAPIER

Probably the most generally accepted definition of man at the present time is that of man-the-toolmaker, in which the concept of "toolmaking to a set and regular pattern" is implicit. Benjamin Franklin is usually credited with the idea of *Homo faber*, but it was Kenneth Oakley, of the British Museum (Natural History) who appreciated its significance as a break-through in mental development and, thus, as a basis for an objective definition of man. The definition is a good one in that it not only recognizes the importance of behavior in phylogeny but presupposes the evolution of certain functional trends such as bipedalism, manual dexterity, cerebral expansion and the possession of speech.

However while cultural toolmaking implies that such aptitudes were present, it in no way indicates the degree to which they had advanced at the time that man started to make tools to a "set and regular pattern." One can only suppose that, since this definition of man is intended to lift the progressive hominids out of the limbo of sub-humans and to place them on the side of the angels, man—at this stage—not only had hands of essentially modern form and function but in addition possessed a walking gait as purposeful and economic as that of modern man. Recent discoveries of post-cranial bones at Olduvai Gorge (Leakey 1959, 1960, 1961) suggest, particularly in the case of the hand (Napier, 1962b), that neither supposition is likely to be correct. A suspicion that has for some time been entertained in respect of the posture and gait of the australopithecines (Le Gros Clark, 1954; Mednick, 1955; Washburn, 1960; Napier and Weiner, 1962; Straus, 1962; Napier and Brook-Williams, in litt).

It is well known that animals other than Primates make use of naturally occurring objects as tools—there are numerous examples in the literature including the Galapagos Woodpecker-finches and the sea otters. Among captive Primates there are many reports such as that of Vevers and Weiner (1963) who have recorded on film the tool-using activities of a capuchin monkey in the London Zoo. Until recently however there have been no well documented reports of tool-using by free-ranging Primates. Jane Goodall who has undertaken a prolonged field study of *Pan satyrus schweinfurthi*, near Kigoma in Tanganyika, reports (1963) the prevalence of a simple form of tool-using in the group she has studied over a period of eighteen months. The activity consists of termite "fishing." The chimpanzee, after picking a number of slender stems or stalks, settles down beside a termite hill, scratches away the soil over the flight-hole with a forefinger (or thumb, as Goodall has recently observed—private communication), inserts the stick and after a few seconds withdraws it, covered with termites. The chimpanzee then passes the stick through its mouth removing the termites. Stems and stalks may be picked some distance away and brought to the termite hill—clear evidence of purposeful tool-using. If the stems are too long, they are pruned to a convenient length and if their effectiveness is marred by side branches, these are stripped off.

In these simple but nevertheless, remarkable actions in chimpanzees, one can see the beginnings of toolmaking in its simplest form—a level of activity that might be given the name of tool-modifying. There is of course, no evidence as yet that this particular aspect of chimpanzee behavior observed by Goodall is practised by any other group; it may have to be regarded therefore in schoolboy jargon as a "craze." Goodall reports (private communication) the prevalence of crazes among chimpanzees, though not necessarily concerned with the use of tools. Merfield (1956) describes the use of sticks by chimpanzees for other purposes such as collecting honey from ground nests using the "termite hill technique" described above, and Kortland (1963) discusses the evidence for the purposeful use of sticks as weapons by chimpanzees in the wild state. From what we know of the intellectual capacity of the chimpanzee and of the structural and functional anatomy of its hand, tool-modifying for an immediate purpose probably represents the zenith of its technological behavior. Furthermore, the essentially non-carnivorous diet of chimpanzees presumably does not impose a very high selection pressure on tool-making as a means of survival. These observations however, demonstrate two things: first, chimpanzees in the wild possess the rudiments of conceptual thinking, and second, the transition from tool-using to tool-making in its simplest form was probably not as complex a transition as we have supposed. Indeed if chimpanzees are intellectually capable of stripping twigs from a branch to make it more efficacious as a tool, then it seems highly probable that sub-human hominids would have been able to "modify" a naturally occurring object such as a pebble by striking a flake off it with another pebble and thus obtain a sharp cutting or chopping edge. This hypothetical procedure is in fact functionally and intellectually indistinguishable from the action of "Jacky" the London Zoo capuchin recorded on film by Vevers and Weiner (1963) which used a heavy knuckle bone to crack a nut.

The use of tool-making as a criterion of humanness carries the further implication that its evolution was a sudden event, with no prior history. Logically, normal *cultural* tool making must have been preceded by a lengthy *ad hoc* tool-making and, even earlier, of tool-modifying. Tools were undoubtedly "made" many times over using different materials ranging from wood through bone to stone, and traditions during that time must have been established, lost and found again many times over. The persistence of the tradition (or perhaps "craze" represents the situation better than "tradition") may well have come about only with the evolution of speech and language. Cultural toolmaking must be looked upon as a culminating event in a continuum of manual activity commencing perhaps as early as the Miocene and proceeding along some such lines as these:

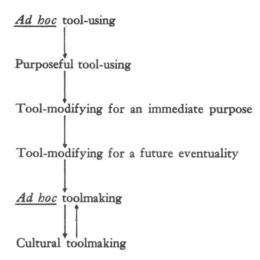

Ad hoc tool-using

Purposeful tool-using

Tool-modifying for an immediate purpose

Tool-modifying for a future eventuality

Ad hoc toolmaking

Cultural toolmaking

During the period of time covered by this continuum of manual activity, profound morphological changes must have taken place in hominid structure, some of which directly affected manual skill and some, such as improved bipedal walking, which affected it indirectly. We have perhaps given a somewhat artificial value to the concept of man-the-toolmaker by regarding tool-making as the outcome of a sudden inspired flash of conceptual thought—a bolt from the Pleistocene blue—that converted a sub-human hominid overnight into a fully fledged member of the human race.

Toolmaking skill depends on a peripheral as well as a central factor—it depends on the form and proportions of the hand as well as the size and complexity of the cerebral cortex. Washburn has emphasized that increase in brain size, characteristic of man's evolution, is more likely to have followed than preceded toolmaking. This is most likely to have been the case if toolmaking is regarded as a sort of rubicon; if, however, we look upon toolmaking, not as a sudden event, but as a continuous process it becomes apparent that both the peripheral and central factors in toolmakers and the technological level of toolmaking itself must have evolved *pari passu*. In other words that toolmaking, throughout its evolution from *ad hoc* tool-using, was equally as advanced as the hands and the brain of the toolmakers themselves.

To paleoanatomists, therefore, whose job it is to reconstruct the form of man at various stages of his evolution, the concept of man-the-toolmaker is not a very helpful one because it tells us so little about the structure of man at any particular time. What is needed is a diagnosis based on certain functional character complexes such as those that Campbell (this volume) puts forward and which others at the Burg Wartenstein conference subscribed to in large measure.

The more critical and specific these character complexes are, the more likely one is to be able to identify within the spectrum of hominid evolution the true line leading to modern man. The increasing richness of hominid discoveries, particularly of post-cranial material, makes it imperative to attempt analysis of certain aspects of function at a much more detailed level than is apparent in Campbell's diagnostic complexes. It would appear, for instance, that it is no longer enough to describe man as a toolmaking, bipedal hominid. We now have evidence from Olduvai that cultural toolmaking does not necessarily imply a hand of modern human proportions (Napier, 1962b and c) and that the bipedalism of hominids was not synonomous with the intricacies of the modern human gait. The Olduvai foot assemblage from the pre-*Zinjanthropus* level of Bed I which consists of a tarsus lacking only the posterior portion of the calcaneum, and a set of metatarsals lacking only their heads, provides evidence of a foot which on superficial inspection is extraordinarily *sapiens*-like. More detailed study reveals *inter alia* the likelihood that the axis of weight transmission in the foot differed from that of modern man (Day and Napier, in litt.).

The paleoanatomist must, within the framework of the broad diagnosis, set his sights at a high level of functional analysis. In order to do this he must select as tracer elements functional complexes that are uniquely characteristic of *H. sapiens sapiens*. Only in this way can he hope to identify the *sapiens* line within a spectrum of closely related and possibly aberrant forms. The essential requirements for such functional markers are:—

1. That they should have clearly defined structural correlates by which their presence or absence can be determined from a study of fossil remains.
2. That the functional markers so selected, should represent both the acme and the quintessence of the particular function or complex under consideration.
3. That the functional markers selected should provide clear evidence of their existence in more than one region of the body.
4. That the markers should be capable of correlation with a significant aspect of human behavior.

There are two functional markers that seem to fulfill the requirements. No doubt there are others more suitable and more valuable which will eventually suggests themselves as more fossil material comes to light. Thus: (1) Modern man strides. (2) Modern man is capable of placing his thumb and index finger together in perfect opposition. These are both familiar and simple enough concepts but each one may be regarded, in its own way, as unique and a feature of human locomotor behavior. The first is the quintessence of the modern human walking gait and the second the ultimate expression of manual dexterity. The selection of these functional markers may seem to be verging on the pedantic but as Harrison and Weiner (this volume) point out, it is important to appreciate in phylogeny of the higher Primates and hominids that the distinction between two cladistically or even patristically related forms may involve small and subtle morphological differences. It is a common experience among those concerned with hand injuries that minor *disorders* may lead to disproportionately severe *disabilities*. In a phylogenetic sense, minor inadequacies of gait which place a physiological limit on the hunting range of one group of hominids compared with another better endowed group, could well lead to extinction of the former.

To define modern man's locomotion as upright or bipedal is not enough (see also Washburn, this volume); many reptiles and mammals as well as Primates are bipedal. To define man's locomotion as a walk again is not enough, although as Washburn (1960) has rightly pointed out this facility requires a more specialized mechanism than running. Straus, in a recent paper (1962) rightly stresses the importance of prolonged bipedal standing. Walking in the sense of moving the feet alternately in such a manner that the swinging phase of the walk never exceeds the length of the stance phase, is occasionally practised by the anthropoid apes, particularly the gibbon and the gorilla. Modern man moreover can still *walk* even though his gluteus maximus is paralyzed, though to *stride* is beyond his capacity under these conditions. At the australopithecine stage of evolution it seems likely that the two known species *A. africanus* and *A. robustus* were both bipedal walkers, but it seems unlikely from available evidence that they could stride (Napier and Brook-Williams, *in litt.*). No doubt it will appear when more evidence is available that there is a spectrum of walking activity apparent even within this restricted group. The normal human gait is an extremely complex process as Saunders, Inman and Ebhart showed in their fundamental studies (1953) carried out at the University of California during World War II. They describe six major determinants of the gait all of which, interacting, direct the centre of gravity of the body along a pathway through space with the minimum expenditure of energy. The joints involved in walking are those of the vertebral column, the pelvis, the hip, the knee, the ankle and the foot; recent studies of many of these joints have demonstrated their many complex specializations (Davis, 1955, Barnett, 1954; Barnett and Napier, 1952; Hicks, 1953, 1954). The perfected gait of man which is made possible by the coordinated activity of multiple joint-muscle systems (Joseph, 1960) is clearly likely to have undergone such an extremely protracted period of evolution that, there is reason to believe, was far from complete at the australopithecine level. Striding constitutes a synthesis of the numerous specializations involved and is impossible in the absence of any one of them.

The most precise movement of which the human hand is capable, that of placing the thumb and index finger in perfect pulp-to-pulp opposition, constitutes the second functional marker. This posture, which cannot be achieved by any living primate other than man, depends on both peripheral and central factors. The main peripheral factors (Napier, 1961) are: (a) The relative proportions in length of the thumb and the digits. (b) The form of the saddle-joint at the carpo-metacarpal articulation. (c) The set of the trapezium within the carpus which is related, in turn, to the depth of the carpal tunnel. (d) The presence of broad spatulate terminal phalanges on the thumb and digits. (e) The presence of richly innervated skin on the palmar surface of the digits. (f) The presence of well differentiated thenar and other intrinsic musculature. (g) The presence of a separate long flexor of the thumb (Straus, 1939). This posture (fine precision grip) which involves only the thumb and forefinger, constitutes one extreme of the precision grip function of the hand (Napier, 1956, 1961, 1962c); at the other extreme (coarse precision grip) the thumb and the tips of *all* the digits are involved. One of the most important functional markers of the coarse precision grip is the presence of a saddle-shaped articulation at the base of the 5th metacarpal. (C. H. Barnett, private communication). The movements at this joint include an element of rotation in addition to flexion-extension, abduction-adduction: thus, it is analogous to the 1st carpo-metacarpal joint and the combined movement of flexion, abduction and rotation at the 5th carpo-metacarpal joint can legitimately be regarded as opposability of the little finger.

The Olduvai hand in functional terms was apparently capable of a power grip equal in performance but comparatively greater in strength than in modern man. There is less certainty with regard to the precision grip. This doubt relates to factors a, b and c above. While the shape of the saddle-surface of the trapezium indicates unquestionably that the thumb could be "rotated" medially about its own axis to face other digits, there is reason, in the form and set of the trapezium, for doubting that the proportionate length of the thumb to the index finger was quite as it is in modern man. Factors e and f (above) cannot be confirmed in fossil specimens but factors d and g were certainly present.

In addition to the peripheral factors already mentioned, the functional effectiveness of index-finger-thumb opposition, depends on the presence of highly evolved neuro-muscular mechanism to provide the peripheral factor for fine coordinated movements and a cutaneous sensory system sufficiently advanced to provide the afferent pathway for discriminative sensibility. These in turn depend upon central factors such as the size of the brain and intellectual capacity in general and, more precisely, on the extent of the representation of the appropriate areas of the motor and sensory cortex, and of the pre-motor and post-sensory association areas. The complex central and peripheral factors associated with manual dexterity make it clear that the evolution of manual dexterity—like that of modern walking—must have been an extremely slow process and that further discoveries are likely to reveal a spectrum of manual ability both in a horizontal and a vertical sense.

This spectrum of manual ability may possibly be apparent in the form and the quality of stone tools themselves. If this were so, the study of artifacts would provide an objective method of assessing the degree to which manual dexterity had evolved at that particular time; a method that, if properly systematized, might even provide the necessary evidence of hand evolution in the absence of appropriate bones. One can visualize considerable difficulty in determining whether the complexity and quality of a stone tool is merely a reflexion of manual dexterity or of intellectual capacity; there seem to be no a priori reasons why these two functions should have evolved at the same rate. It seems possible that one could apply the same reasoning to stone tools as one applies to fossil bones. Fossil bones possess *form* and from this form one can interpret *function;* from function one can argue certain primary aspects of *behavior.* A stone tool has form and function and from function one can legitimately deduce certain broad principles of domestic and social behavior—hunting, use of fire, fishing etc. The *form* of stone tools is largely dictated by peripheral and central factors of manual dexterity, but the *function* which is an outward expression of the purpose or idea that promoted the construction of the tool in the first place, is a reflexion of intellectual capacity. If, indeed, intellectual capacity and manual dexterity have evolved at different rates, the analysis of stone tools might reveal this fact by a discrepancy between the intended *function* of an artifact and its actual *form.* As a modern parallel one might take the case of a forearm amputee who invents a new electronic computor and sets out to build it but fails to achieve his object for physical reasons. There is some evidence of such a discrepancy in the Mousterian artifacts of the Dordogne according to Professor Francois Bordes.

There is a further consideration to be taken into account—that of workmanship. There is no reason to assume that all paleolithic toolmakers were equally good craftsmen, nor that some of the artifacts that survive are not the imitative and imperfect work of children and "trainees." This, however, is purely hypothetical. Of more immediate interest is the relation of the form and function of artifacts to the form and function of the hand. There seems good reason to believe that of the two basic patterns of the human hand—the so-called power and precision grips—the former evolved in hominids before the latter. The data on the carpals and phalanges from the pre-*Zinjanthropus* levels of Bed I, Olduvai Gorge, suggest a hand with a strong power grip but an imperfect precision grip; a similar conclusion was reached from a study of the Swartkrans metacarpal attributed to *Australopithecus robustus* (*Paranthropus*) (Napier, 1959). Stone tools on the F.L.K. N.N. site at Olduvai are of the simple Oldowan pebble-tool variety crudely flaked to form three or four faces which subtend a cutting edge. There is little doubt that such a tool could have been constructed by a hand lacking the elements of an advanced precision grip.

Since the Burg Wartenstein Conference the present author has undertaken some experiments similar to those carried out by Krantz (1960) to demonstrate that it is possible to construct "pebble-tools" and crude "hand-axes" by using a power grip only (Napier, 1962b) without the benefit of an opposable thumb. Krantz found that when using this grip he was able to produce deep flake scars, but the precisely placed light blows necessary for retouching were impossible.

There is clearly need for a critical review of the whole range of stone. tools to determine, first, how they were constructed and second, how they were employed in terms of hand function. Systematic analysis of this sort might throw some light on the evolutionary stages of the human hand, even, as already indicated, in the absence of the fossil hands themselves.

It is inevitable that *Proconsul* of the East African Miocene should be considered in the search for the Tertiary ancestors of the Olduvaian and South African Villafranchian hominids. Whatever the final status of *Proconsul* in the systematics of hominid evolution may turn out to be (see Leakey, this volume) it represents an important structural and functional stage in the phylogeny of hominid locomotion. The forelimb structure of the smallest species, *P. africanus,* is known from a single but almost complete specimen (Napier and Davis, 1959). The most striking feature of the forelimb is the combination it displays of primitive platyrrhine and generalized catarrhine features (arboreal-quadrupedal features) and specialized pongid features (brachiating features). Certain general similarities in structure and function with Colobinae were also apparent which led the authors to suggest that, in terms of locomotor habit, *Proconsul* had much in common with modern leaf-eating monkeys especially *Nasalis*. They stated that *Proconsul* represented a transitional stage of hominoid evolution in which an active quadrupedal form was developing the locomotor characteristics of a brachiator. Recently this transitional stage has been referred to as pro-brachiation (Napier, 1963). Recent work by Ashton and Oxnard (1961, 1962, 1963) and Oxnard (1963) on the functional anatomy of the shoulder region of present day monkeys has provided a certain amount of confirmation for the hypothesis proposed by Napier and Davis to account for the peculiarities of the upper limb skeleton of *P. africanus*.

The hand of *P. africanus* shows the same admixture of generalized and specialized characters as the remainder of the forelimb. The hand was undoubtedly prehensile but the thumb was not strictly opposable in the sense that the modern catarrhine thumb is opposable. It was clearly capable of independent movement but lacked the essential feature of true opposability, a saddle-shaped carpometacarpal articulation (Napier, 1961); the basal thumb joint of *P. africanus* is of the modified hinge variety, rather similar to that seen in *Cebus*. At the *Proconsul* stage of hominoid evolution there is no evidence of a truly opposable thumb, let alone the degree of finger-thumb opposability necessary to fulfill or even foreshadow the functional ideal selected as a marker for manual dexterity.

The hind-limb of *Proconsul* is incompletely known from a number of fragments (Le Gros Clark and Leakey, 1951), of the femur, the talus and the calcaneum. It is proposed to re-study these fragments in the light of recent knowledge of locomotor patterns. The fragments of the upper end of the femur (M16331 in British Museum, N.H., figured by Le Gros Clark and Leakey, 1951) shows a well rounded head, a long neck, a short greater trochanter and a backwardly directed flange-like lesser trochanter and a centrally placed fovea capitis; it corresponds, in general form, to the femur of *Semnopithecus* and is quite unlike that of *Cercopithecus*. Piganiol and Olivier who have studied the internal architecture of a series of primate femora (1958) point out the similarity of the femur of *Semnopithecus* to that of modern man. These observations are of particular interest in view of certain locomotor patterns of semnopitheques discussed below.

On morphological grounds there is little doubt that *P. africanus* was an arboreal form; on geological grounds it may well be argued that it was terrestrial or at least semi-terrestrial. The chance factors involved in the death and fossilization of any specimen make the question of the immediate environment of an individual fossil relatively unimportant. It seems unlikely that *Oreopithecus*, for instance, lived in or over a swamp. What is important is that during the lower half of the Miocene, *Proconsul* had not yet lost the clearly defined structural hallmarks that indicate an arboreal form. A lengthy discussion of the origin(s) of human bipedalism does not come strictly within the scope of this paper but a brief consideration, however, is not inappropriate. A point which appears to have puzzled some workers (Hewes, 1961) is the failure (sic) of baboons to have developed into bipeds. A number of suggestions have been put forward to account for this. Oakley (1954) suggested that the ancestors of the baboons had retained large canines when they descended from the trees whereas in the ancestors of man these structures had already been reduced during the later phases of arboreal life. Coon (1954) believed that the explanation was purely ecological. Hewes (1961) attributed hominid bipedalism to food-carrying habits. Other authors have associated it with the carriage of tools, weapons or infants. The most striking omission from this list, which thirty years ago assuredly would have been given pre-eminence, is brachiation. Hewes (loc cit.) even goes so far as to state that "the long argued issue of brachiation or non-brachiation . . . is not crucial here"! In an article in 1954 Patterson pointed out that the current attitude toward brachiation was virtually a universal rejection of a brachiating ancestry for man almost, as Patterson (1954) aptly states "as if it were a deadly sin." Thus the pendulum swings.

Gregory for many years insisted that a certain amount of ancestral brachiation was the best explanation of hominid bipedalism; that "a certain amount of brachiation" has now been given a name—semibrachiation—and the locomotor adaptations associated with it are beginning to be understood (Ashton and Oxnard, 1961, 1963; Oxnard 1963). Arm-swinging locomotion is one of the inevitable outcomes of arboreal life among the Primates and has undoubtedly evolved many times over in many different groups. Apart from the gibbons and the great apes in whom brachiation developed by parallel evolution, arm-swinging is seen in a modified form in certain of the New and Old World monkeys such as *Ateles*, *Lagothrix*, (Erikson, 1963) in *Colobus*, *Nasalis* and *Presbytis*. Not all these forms brachiate as comprehensively or as exquisitely as the gibbon, but then, neither do all cursorial forms run as fast as the cheetah. Arm-swinging locomotion constitutes a spectrum of activity ranging from the locomotion of essentially quadrupedal Primates which use the hands freely to slow up or arrest their progress when jumping gaps in the forest canopy (semibrachiators), to the fully specialized bimanes which are solely dependent on their hands for rapid and effective arboreal locomotion (brachiators).[1]

There appears to be a good reason for regarding the semi-brachiating habit of certain living Old World monkeys, e.g., certain Colobines as a "model" for the locomotor pattern that, in the ancestral hominids, was pre-adaptive for bipedalism. Unlike true brachiation, semibrachiation depends upon the specialized activity of *both* fore- and hind-limbs. A living semibrachiator such as *Nasalis larvatus* when crossing a gap in the canopy relies on the propulsive power of its hind limbs which at the moment of take-off are extended at both knee and hip joints. In flight, the forelimbs are extended at the shoulder and reach out towards the landing site to arrest the progress of the leap. At the initial stages of the horizontal leap, therefore, both sets of limbs are extended. Subsequently the hind limbs flex to assist the hands at the end of the leap. In a recent article, Harrisson (1962) describes the jumping habits of the Banded Leaf-monkeys, *Presbytis melalophus* and *Presbytis obscurus* in Malaya, thus: "The monkey . . . then literally hurls itself forward and out with an *immensely powerful thrust from the hind legs* (my italics) the arms grope wildly out ahead, the body arcs in tension, the back legs unwind and the tail flies high." In the photograph accompanying the article the leap position in mid-flight is perfectly illustrated.

During brachiation the legs are non-contributory and are more or less flexed at the hips and knees, a state of affairs which would seem to promote precisely the opposite specializations than those that are required for upright bipedal progression. It seems not unlikely, judging from the structural characteristics of the limb-bones of *Proconsul africanus*, that the pre-terrestrial arboreal phase of early forerunners of bipedal hominids was somewhat akin to that of living semibrachiators. There is no evidence that ancestral baboons went through a similar stage. No doubt ancestral baboons were equally arboreal in the sense of habitat though they must have differed profoundly from pre-hominids in the sense of habit (Napier, 1962a).

Thus, it is not to be expected that ancestral baboons and ancestral hominids having the same opportunities to adopt a terrestrial form of life but possessing different locomotor preadaptations, would pursue the same evolutionary path.

1. For a full discussion of the origins of the terms brachiator and brachiation, the significance of brachiation among living and fossil Primates, the morphological correlates in the upper limb of a semibrachiating habit, and the occurrence of semibrachiation in certain New World monkeys, the reader is referred to "The Primates" *Sym. Zool. Soc. London*, No. 10. 1963. For an experimental study of brachiation among certain apes and Old World monkeys see Virginia Avis (1962) *South Western J. Anthrop.*, 18:119–148, as well as S. L. Washburn (this volume).

BIBLIOGRAPHY

ASHTON, E. H. and C. E. OXNARD
  1961. *Proc. J. Anat. London*, 95, 4:618.
  1962. *Proc. J. Anat. London*, 96.
  1963. *Trans. Zool. Soc. London*, 29:553–650.

BARNETT, C. H.
  1954. *J. Anat. London*, 88:59.
  1962. Private communication.

BARNETT, C. H. and J. R. NAPIER
  1952. *J. Anat. London*, 86:1.

BOWDEN, R. E. M. and J. R. NAPIER
  1961. *J. Bone Jt. Surg.* 43B, 3:481.

COON, C. S.
  1954. *The story of man.* New York: A. A. Knopf.

DAVIS, P. R.
  1955. *J. Anat. London*, 89:370.

ERIKSON, G. E.
  1963. *Symp. Zool. Soc. London.* No. 10, pp. 135–63.

GOODALL, J. M.
  1963. *Symp. Zool Soc. London.* No. 10, pp. 39–47.

HARRISSON, T.
  1962. Malaya: Straits Times Press.

HEWES, G. W.
  1961. *Amer. Anthrop.*, 63.4:687–710.

HICKS, J. H.
  1953. *J. Anat. London*, 87:345.
  1954. *J. Anat. London*, 88:25.

JOSEPH, J.
  1960. *Man's posture: electromyographic studies.* Springfield, Ill.: Charles C. Thomas.

KORTLANDT, A.
  1963. *Symp Zool. Soc., London*, No. 10, pp. 61–88.

KRANTZ, G. S.
  1960. *Kroeber Anthrop. Soc.* 114–128.

LEAKEY, L. S. B.
  1959. *Nature*, Aug. 15:491.
  1960. *Nature*, Dec. 17:1051.
  1961. *Nature*, Feb. 25:649.

LE GROS CLARK, W. E.
  1954. *The fossil evidence for human evolution.* Chicago: Chicago Univ. Press.
LE GROS CLARK, W. E. and L. S. B. LEAKEY
  1951. *Fossil Mammals of Africa*, No. 1. London: British Museum (N.H.)
MEDNICK, L.
  1955. *Amer. J. Phys. Anthrop.*, 13:203.
MERFIELD, F.
  1956. *Gorillas were my neighbors.* London: Longmans Green.
NAPIER, J. R.
  1956. *J. Bone Jt. Surg.*, 38B, 4:902.
  1959. *Fossil Mammals of Africa No. 17.* London: Brit. Mus. (N.H.).
  1961. *Symp. Zool. Soc. London*, 5:115–132.
  1962a. *New Sci.*, 15:88.
  1962b. *Nature*, 196:409.
  1962c. *Sci. Am.*, 207:56.
  1963. *Symp. Zool. Soc. London*, No. 10, pp. 183–95.
NAPIER, J. R. and P. R. DAVIS
  1959. Fossil Mammals of Africa No. 16, London: Brit. Mus. (N.H.).
NAPIER, J. R. and J. S. WEINER
  1962. *Antiquity*, 36:41.
NAPIER, J. R. and P. BROOK-WILLIAMS
  In preparation.
OAKLEY, K. P.
  1949. *Man the Toolmaker*, 4th ed., London: Brit. Mus. (N.H.).
  1954. *A History of Technology*, Chap. I, vol. I. New York: Oxford University Press.
  1956. *Mem. Proc. Man. Lit. Phil. Soc.*, 98, 1.
OXNARD, C. E.
  1963. *Symp Zool. Soc. London*, No. 10, pp. 165–82.
PATTERSON, B.
  1954. *Human Biol.*, 26:191.
PIGANIOL, G. and G. OLIVIER
  1958. *Compt. rend. Ass. Anat.*, XLV.
SAUNDERS, J. B. DE C. M., T. INMAN VERNE, and HOWARD D. EBERHART
  1953. *J. Bone Jt. Surg.*, 35A, 3:543–548.
STRAUS, W. L.
  1942. *Quart. Rev. Biol.*, 17:228.
  1962. *Clin. Orthop.*, 25:9–19.
VEVERS, J. and J. S. WEINER
  1963. *Symp. Zool. Soc. London*, No. 10, pp. 115–17.
WASHBURN, S. L.
  1960. *Sci. Amer.* 203, 3:63–75.

BEHAVIOR AND HUMAN EVOLUTION

S. L. WASHBURN

THE INTEREST in human classification and evolution has been so great that it has produced a bewildering quantity and variety of terms and theories. Yet, in spite of the diversity, the major groups of primates recognized today were familiar to the scholars of the 19th century. The groups which seemed "natural" to the pre-evolutionary zoologist remain central in the thinking of the modern student of human evolution, and the purpose of this paper is to suggest reasons why this is the case. My thesis will be that the principal groups of the living primates are adaptive and that the characters by which they have been recognized are structures which are closely related to the behavior of the groups. With the addition of the evolutionary point of view, the contemporary groups are seen as the ends of adaptive radiations but the groups are the same. If this is the case, the referents of the names should be the adaptive groups. The most useful taxonomic characters will be those closely correlated with the basic adaptation. And "splitters" will be scientists who divide "natural," that is adaptive, groups.[1] Some illustrations of these points will be given before proceeding to a discussion of the Hominidae.

The primates have been found hard to define, and generalized features tend to be mentioned. Yet the order is characterized by the adaptation of climbing by grasping which is a complex specialization, separating the primates from other living mammals. It is fully developed in the prosimians and was present in the Eocene, as shown by *Notharctus*. Observation of the *quadrumana* was a factor in the recognition of the primates as a natural group, and the fundamental adaptation of elongated digits, flattened terminal phalanges with nails, and specialized palms and soles is basic to the locomotor and feeding patterns of monkeys and apes. Climbing-by-grasping is the adaptation which separates primate arboreal life from that of the tree-shrews which climb with claws. The adaptation is carried farthest in the Lorisidae and has numerous consequences. For example, in jumping primates (Tarsiidae, Galagidae) elongation is in the tarsal bones, rather than the metatarsals which is the situation in other jumping mammals. This is because the

---

1. Classifications are made by human beings with varied training and interests, and there is no reason why a particular classification should seem equally reasonable to different people. I am not suggesting that this is the only way to look at the primates, but that emphasis on behavior and adaptation is a great aid in understanding the groups which have been described and in the evaluation of the characters used in classification.

first toe must be able to oppose the others and elongation in the metatarsals would destroy the grasping adaptation of the foot.

According to this view, the basic adaptation of the primates is a way of climbing, the structure which permits that mode of locomotion, and this complex separates the order from all other groups of living mammals. Naturally, there must have been intermediate forms between tree-shrews and primates in the Paleocene, and hands and feet have evolved. But, if one compares *Notharctus* and *Homo*, it is remarkable how little the hands have changed compared to other parts of the body. In the vast majority of the living primates the original, basic loco-motor adaptation of the group is still present. If this view of the primates is correct, there is no intellectual problem in defining the order which may be done by the locomotor pattern, which is easily studied in the living forms. However, in fossils reconstruction of locomotor patterns is difficult, unless large parts of the skeleton are preserved, and emphasis will continue to be on the teeth and skull. All available evidence should be utilized, in order to get as full an understanding as possible of the earliest primates and of the trends in primate evolution. How-ever, emphasis on behavior suggests that a locomotor specialization is the basic adaptation and that the primates cannot be understood without an appreciation of this way of locomotion and its consequences.

If a special way of locomotion is the basic adaptation of the primates, then locomotor characters must be emphasized in the definition of the order. Charac-ters of the teeth and skull will be useful for identification, to determine parallelism or convergence, and to estimate the multiplicity and relations of fossil forms. But emphasis must be on locomotion.

Emphasis on behavior does give a way of evaluating structure, and this can be illustrated by comparing three pairs of taxa which have often been classed as subfamilies: Galaginae and Lorisinae, Cercopithecinae and Colobinae, Hylobatinae and Ponginae. Galagoes differ from lorises in almost every feature of their post-cranial anatomy. In locomotion the galagoes and lorises represent extremely differ-ent adaptations. The Colobinae are similar to other Old World monkeys in general structure, but differ profoundly in their viscera. The gibbons differ in no funda-mental way from other members of the Pongidae. Locomotor and visceral adaptations are the same, and there is nothing to suggest that they represent a major adaptive radiation distinct from that of the other apes. From the point of view of behavior, it is convenient to regard the galagoes and lorises as comprising two very distinct families (Galagidae and Lorisidae), each containing many genera and species. The many structural-behavioral similarities of the Old World monkeys suggest that a single family, divided into two sub-families to express the differing visceral adaptation, is appropriate. But the gibbons may be treated as a single genus composed of several species, which are small apes but not funda-mentally different from other members of the Pongidae.

Starting with three pairs of sub-families, it has been suggested that emphasis on adaptation and behavior suggests that the degree of difference is very unequal in the three cases. The adaptive situation would be more conveniently indicated by the terms: Galagidae and Lorisidae, Cercopithecine and Colobinae, and *Hylobates*. There is nothing novel about this suggestion. The names are the same as Simpson, 1945, with the exception that the contemporary gibbons are treated as a single genus which was a common opinion. All that is suggested is that the emphasis on adaptation helps in the evaluation of the classification. Further, emphasis on adaptation forces a consideration of the biology of the groups in question. It is necessary to consider many kinds of evidence and the importance of characters depends on an estimate of their importance in evolution.

These three pairs of taxa were picked to illustrate the fact that, the closer man is approached in classification, the smaller the differences which will be accepted as important. Further, these pairs were picked to suggest that emphasis on adaptation offers a way of estimating this bias, at least to some extent. The families and subfamilies of Old World primates are groups which are distinguished by major adaptations, except in the case of the Pongidae which is subdivided in a way which does not conform to any major adaptive divisions.

The Pongidae are characterized by a way of climbing which sharply distinguishes the living forms from other primates. This mode of locomotion is clearly reflected in the skeleton and many bodily proportions, and, judging from the very fragmentary fossil limb bones, this locomotor pattern probably evolved in the Miocene. If this is correct, the dentition of the apes had evolved its characteristic pattern long before the locomotor adaptation was complete, and the structure of the arm and trunk which man shares with the apes was not evolved until the Pliocene. This view is shown in the accompanying diagram, Figure 1.

The diagram is intended to suggest that the *Dryopithecus* pattern was established in the Oligocene and the separation of Pongidae and Cercopithecidae had taken place at that time. During the Miocene the apes evolved their distinctive locomotor pattern, and parallel evolution appears to have gone on in several distinct lines of brachiators. Obviously, with so few fossils such a view is only speculation, and the actual situation was much more complicated.

The locomotor pattern of the apes has been called brachiation, meaning swinging below the branches rather than walking on top as the monkeys do. This term has been widely misunderstood, partially because of the lack of behavioral data and partially because the term applies to only a small part of the locomotor pattern of the Pongidae. Apes also walk quadrupedally on the ground. None of the Pongidae, not even the gibbons, swing under branches most of the time. But in a recent study in which examples of all the living Pongidae and many kinds of the Cercopithecidae were run across the same branches, all the apes brachiated and no monkey did (Avis, "Brachiation," *Southwestern Journal of Anthropology*, 18:119–148, 1962). Old World monkeys will consistently "choose" to cross a small branch quadrupedally while the apes (Pongidae) will swing below. But, if the branch is large and stable, the apes will stay on top and move quadrupedally (*Pan, Pongo, Gorilla*) or bipedally (*Hylobates*). In a re-

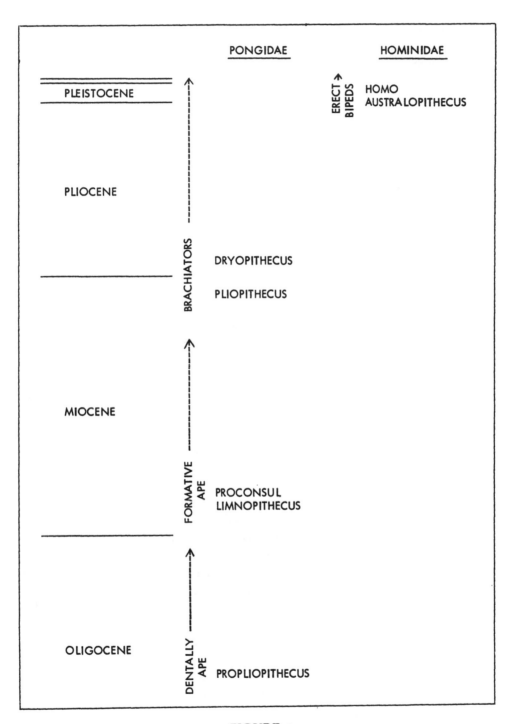

FIGURE 1

stricted sense "brachiation" applies to only a small part of ape locomotion, and, if apes are referred to as brachiators, it must be remembered that is much less descriptive of what they do than "quadrupedalism" is for monkeys.

Apes are brachiators in the sense that, in particular situations, they all swing below branches, and Old World monkeys are not brachiators in the sense that, given comparable situations, none of them swing under the branch. But for brachiation to be very useful in the description of ape behavior it must be descriptive of a much wider range of actions. These might be described as climbing and eating by reaching, a pattern of slow feeding and moving in the ends of small branches. Apes will hang by one arm and feed with the other, or hold a branch with one hand and lean out, and stretch the other hand out as far as possible to pluck food. Positively, brachiation is emphasis on long, highly mobile arms with powerful flexor muscles. Negatively, it is the reduction in importance of the lower back in locomotion. In the living Pongidae the full structural emphasis on arms may be seen and the great reduction of back motion as an essential part of locomotion. In some of the New World monkeys there appears to be a partial parallelism, both in the acquisition of brachiating structural-behavioral mechanisms and in the reduction in the back. When these are studied in more detail, they should provide a much fuller understanding of the nature of brachiation and aid in the reconstruction of the locomotor patterns of the formative apes (*Limnopithecus, Pliopithecus*) as these forms may have been in a behavioral stage very similar to that of *Ateles*.

The climbing-feeding adaptation of the Pongidae is reflected in many simple actions and postures, and these can readily be observed in zoos. For example, apes stretch by putting the arms to the side as far as possible; sit on a shelf with legs hanging down; cross arms or put one hand on the opposite upper arm; cross one leg over the other; lie flat on the back, especially with knees separated and elbows apart and hands above head. It will be noted that the positions of the chimpanzee look human and that in the list of postures above every one characteristic of the Pongidae is also characteristic of man. There is a profound similarity in the motions of the arms of man and apes, and on any playground one can see humans brachiating from bars, hanging by one hand, and exhibiting a variety of motions and postures which are similar to those of the apes. Man still is a brachiator. He is simply the one who is least frequently in the situation which calls forth this behavior. Our legs are too heavy and our arms are too weak for efficient brachiation but, when we climb, we climb like apes and not like monkeys.

The structural basis for the motions and postures which are shared by man and ape is complex, and the following listing of characters which man and ape share gives only a partial idea of the degree of similarity. Only those features are included which are importantly related to the functioning of the arms and in each case comparison is to the condition in the Cercopithecidae. The order of comparison starts with the sternum and proceeds along the arm to the fingers. In man and ape the sternum is wide, the sterno-clavicular joint stable, the clavicle long, and a large part of the pectoralis major muscle arises from the clavicle. At the shoulder, the acromion process is large and the deltoid muscle is large, permitting powerful abduction. The coracoid process of the scapula is large and the glenoid fossa nearly flat with a small supraglenoid tubercle. Trepezius is very thick in the central part, correlated with the long spines of lower cervicle and upper thoracic vertebrae. The head of the humerus is more rounded, and the bicipital groove is narrow with no extension of the articular surface into it. Flexor muscles are large relative to the extensors. Taken together these structures permit the wide range of lateral and flexor movements which are the basis of brachiation.

In the elbow and forearm are a series of adaptations which allow powerful flexor action in any part of a much greater range of pronation and supination. These are, on the humerus, a wide trochlea with a lateral ridge and a fully rounded capitulum. The proximal end of the ulna is wide, olecranon very short and the head of the radius is round. The stability of the joint depends almost entirely on the ulna-trochlea fit, and the deep olecranon fossa permits full extension so hanging is possible with a minimum of muscular effort.

In the wrist, the ulna does not articulate with the carpal bones, and this allows freer abduction and adduction necessary in swinging, where the whole force of the swing is taken in the wrist.

The hand is specialized for flexion and the sesamoid bones are lost. Compared to trunk, the hand is always long.

The structural-functional similarity of the arm of man and ape is related to less obvious similarities in the viscera and trunk. Lumbar vertebrae are few in number and the lumbar region is short. The back is not flexed in locomotion and the deep back muscles are small. In correlation with the structure of the shoulder and lumbar region, the trunk is short, wide, and shallow. This means that the viscera cannot be arranged as in the long, narrow, and deep trunks of the quadrupedal monkeys. Certainly much of the similarity in the arrangement of the thoracic and abdominal viscera is secondary to the locomotor adaptations in the form of the trunk. Even the similarity in the branching from the aortic arch and the number of rings in the trachea may be secondary.

It seems most improbable that this detailed structural-functional similarity could be due to parallelism. Certainly the fact that many other lines of evidence suggest that man and ape are related makes explanation by parallelism most unlikely. Furthermore, parallelism means that animals resemble each other because similar groups have adapted in similar ways but that the lines are genetically independent. To suggest that man evolved the structure of a brachiator by parallelism and not brachiating is to misunderstand the nature of parallel evolution.

The functioning and structure of the arm and trunk show the similarity of Pongidae and Hominidae. They are alike because of common ancestry and having lived a particular kind of arboreal life for millions of years. The divergence of the Hominidae appears to be based on a new adaptive complex in which tools and erect bipedalism are basic. Large brains and small faces followed long after the Hominidae were distinct from the apes. The earliest part of the adaptive radiation of the Hominidae now known is in the Lower Pleistocene (Upper Villa-franchian). It has become customary to indicate the human line as diverging from the Pongidae at least as far back as the Miocene. But most of the differences between man and ape seem to have evolved within the Pleistocene. Probably the only bone by which an ancestor of *Australopithecus* with a large canine tooth could be distinguished from an ape is the ilium and, possibly, the bones of the foot. The brain was not larger than that of the living apes. The hand shows many intermediate features. The ischium is ape-like, and it is tools, dentition, and ilium and foot which suggest similarity to *Homo* and the first stage in human behavior. There were at least two species of *Australopithecus* (*A. africanus* and *A. robustus*), and probably more.

The genus seems to have lasted at least 500,000 years, and possibly more than three times that. Stratigraphy and associated fossils suggest a rapid rate of evolution, but the absolute dates from the Berkeley potassium-argon laboratory suggest a much longer period. The accuracy of the method and dates is under discussion. The genus *Homo* is characterized by lower limb bones which are characteristic of modern man. It appears that the human adaptation of erect bipedalism evolved in two stages, the first of which involved primarily the ilium and the second the femur and ischium. It is possible that the first stage was primarily an adaptation to erect running, and it is only with the second stage that fully erect, long-distance walking evolved. With *Homo* appear large brains, complicated tools made by clearly defined traditions, and, probably, language. The adaptations of the genus *Homo* were so effective that these forms spread over all the Old World, except into the very cold regions and to islands where boats were needed. Conquest of the Arctic and the oceans awaited *Homo sapiens*.

It has been debated whether psychological characters should be used in classification. However, if one is interested in adaptation, there is no choice in the matter. The adaptation of the genus *Homo* is based on technical skills, language, and many attributes of mind (memory, planning, etc.), and, if complex tools and intelligence account for the radiation, they must be used in classification.

If emphasis is on behavior, it is convenient to divide the Hominidae into two genera, *Australopithecus* and *Homo*. They are distinct in locomotor adaptation, brain size, and capacity for tool making.[2] At least two species of *Australopithecus*

2. Dr. Phillip V. Tobias writes: "If generic separation between *Australopithecus* and *Paranthropus* is justified, then I feel *Zinjanthropus* is more likely to prove a third separate genus; however, I have serious doubts whether generic distinction is indeed justified between the first two forms. I am increasingly inclining to the view that the distinctions, dental and otherwise, between the two forms have been overrated and that all the South African Australopithecines should be lumped into a single genus. If this were accepted, then it seems clear that 'Zinjanthropus' belongs in the same single genus (*Australopithecus*). My study of the 'pre-*Zinjanthropus*' is not yet sufficiently advanced for me to draw any firm conclusions about its status; at the present, my *interim view* is that the scanty remains available are not incompatible with those of an australopithecine with a somewhat bigger brain than the specimens hitherto assessed; its dental features would seem to place it rather closer to *Australopithecus* (in Robinson's sense) than to *Paranthropus*, but I do not think I have yet seen any features which, individually or collectively, place it outside the probable range for *Australopithecus sensu lato*." (Personal communication)

are known, and probably there were several more, especially if the long estimates of Lower Pleistocene time are correct.

With the advent of *Homo* (possibly in the first interglacial or second glacial some 400,000 to 500,000 years ago) the human adaptation was so effective that further speciation may have stopped. The presence of the same kinds and sequence of tools in Africa, India, and Europe is evidence that there was migration over all this area during the Middle Pleistocene. The Far East appears to have been culturally distinct, and there is too little skeletal evidence to decide whether there was a distinct species of man in the Far East in the early Middle Pleistocene. At a later time, the similarity of Mapa, Solo, Rhodesian, and Neanderthal remains suggests no more than racial distinction.

DISCUSSION

The emphasis on behavior may be defended on theoretical grounds. If the direction of evolution is due primarily to natural selection, the taxa which are recognized today should be the ends of adaptive radiations. However, there are numerous practical difficulties in the application of the point of view. There are few behaviorally oriented studies. There is no agreement on the meaning of the terms, such as brachiation. The most important parts may not fossilize so that the dangers of reconstruction and speculation are *greatly* increased.

These points may be illustrated with reference to bipedal locomotion. When human bipedalism is under discussion, reference is usually made to other bipeds (birds, dinosaurs, and various jumping mammals). But human bipedalism is structurally unique, and it is not even mechanically convergent with these other forms of bipedalism. Human bipedalism is a gait which is carried on while the trunk is erect, and walking is dependent on a unique form of ilium and a pattern of muscles found in no other animal. In mammals the ischium is normally long and the main propulsion comes from pulling the leg back, primarily by the hamstring muscles. Two-joint muscles are the most important. In man the ischium is short, quadriceps extensor femoris is larger than the hamstrings, and one-joint muscles are large. A pattern of walking based on large gluteus maximus, quadriceps, and soleus is uniquely human, as is the importance of the first toe in the structure of the foot.

Considering *Australopithecus* from the point of view of behavior, it will be seen that the ilium is constantly described but the ischium is not, although both are essential to the analysis of gait. The use of the word "biped" does not distinguish the kind of bipedalism, whether like a bird, a galago, a gibbon, or recent man. And an estimate of locomotion has to contain a personal opinion on how the muscles should be reconstructed. Emphasis on behavior forces the consideration of a range of problems which are minimally considered at present. What sort of a sequence of behavioral events might lead to the ilium (and presumably the hip musculature) evolving to nearly the condition seen in *Homo* while the ischium remained almost as seen in the living Pongidae? One theory which will fit the facts is that bipedal running preceded efficient, long-distance, bipedal walking. In this stage of locomotor evolution the ischium was long (fact) and the hamstrings were relatively large (guess). From the point of view of evolution, this means that *Australopithecus* may have been able to run as fast as *Homo*. Since even some of the living apes can run faster than man, there may never have been a stage of slow, partial bipedalism. Whatever the truth proves to be, emphasis on behavior leads to making a distinction between bipedal running and walking, to postulating running as essential in the ape-to-man transition, and to see fully-erect, long-distance, bipedal walking as later. Casts of the foot from Olduvai were shown to the conference by Drs. Leakey and Napier. It appears to be even more like that of modern man than the ilium of the South African representatives of *Australopithecus*. This foot strongly supports the notion that bipedal locomotion is the adaptation which separated the Hominidae from the Pongidae.

With these behavioral distinctions in mind, it is easy to select measurements which will distinguish the Pongidae, *Australopithecus*, and *Homo*. In the Pongidae the distances from anterior superior iliac spine to the acetabulum and from the acetabulum to the origin of the hamstrings on the ischial tuberosity are both long. In *Australopithecus* the iliac measurement is short and the ischial one long. In *Homo* both are short. The dimensions may be measured in various ways and indices which identify the groups are easily contructed. But direct comparison is the most revealing. The iliac distance in a small gibbon is absolutely as long as in *Homo*, and the ischial distance in the gibbon is much longer than in man, Figure 2.

These dimensions in man are so short that they may be duplicated, or surpassed, in an ape of 1/10 the body size. Correlated with the unique locomotor pattern of *Homo*, several dimensions of the pelvis are unique and not duplicated in other living primates. In *Australopithecus* the ilium approaches the condition in *Homo* but the ischium is long, and it is suggested that this unique structure was correlated with a form of locomotion characteristic of no living primate. The interest in behavior leads to measuring the pelvis in new ways and to reaching conclusions which are different from those in the literature. The pelvis is not a structural unit, and it is very misleading to refer to the pelvis of *Australopithecus* as human. The form of the pelvis is a compromise between several sets of functional requirements, and the different sets may evolve with a considerable degree of independence.

Interest in function leads to seeing some of these complexes.

The pelvis of *Australopithecus* has been used as an example to show that interest in behavior leads to an analysis in which the emphasis is different from that in the literature. Similarity to ape or man is used to reconstruct behavior *first*, then the position of the reconstructed stage in the evolution of man is evaluated. A model of the method is: comparison of data, reconstruction of behavior, evolutionary conclusion.

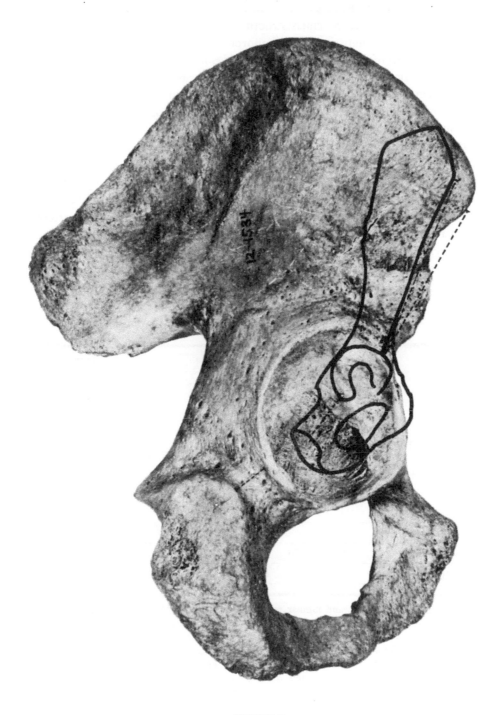

FIGURE 2

In my opinion this emphasis really makes a large difference from the traditional anthropological model in which structures are compared and conclusions drawn directly from the comparison. The importance of the difference may be illustrated by the problem of counting vertebrae. Traditionally vertebrae have been counted by the presence or absence or ribs. Using these criteria, the numbers of vertebrae in the primates have been thoroughly explored. Counted in this way, there is a gradation in the numbers of vertebrae between Cercopithecidae, Pongidae, and Hominidae.

But the vertebrae may be compared in quite a different way. The motions of the vertebrae are correlated with their joints and muscles. In rapidly-moving quadrupedal primates the back is flexed in locomotion, and sacrospinalis is large, rectus abdominis is large and long, and the lumbar region is long and composed of many vertebrae. If the vertebrae are divided by counting all ' vertebrae as lumbar which have facets allowing primarily flexion and extension, quadrupedal monkeys normally have seven cervical, ten thoracic, and nine lumbar vertebrae. Compared to the traditional method, two vertebrae have been added to the lumbar region. In Table 1, counting by ribs is compared to counting by facets.

TABLE 1

	Species	No.	Thoracic Vertebrae							Lumbar Vertebrae					
			12	13	14				4	5	6	7			
RIBS	Hylobates lar,	159	3	143	13				5	116	38				
	C. aethiops,	60	56	4						8	52				
	Papio,	11	2	9						11					
			9	10	11	12	13	14		5	6	7	8	9	10
FACETS	Hylobates lar,	159				60	97	2		59	99	1			
	C. aethiops,	60	4	56									3	54	3
	Papio,	11	1	9	1									11	

The number of thoracic and lumbar vertebrae in a sample of gibbons, vervets, and baboons, counted by the traditional method (ribs) and by the structure of the articulations (facets). This is part of a study on vertebrae started with the help of research funds provided by the University of California through its Committee on Research and continued under Grant No. G-17954 from the National Science Foundation. The writer is deeply indebted to Ralph Holloway for help in collecting the data. The gibbons were collected in one locality in Thailand on the Asiatic Primate Expedition organized by Harold Coolidge. The *Cercopithecus* monkeys were collected in Uganda under a grant from the Wenner-Gren Foundation and with the help of Dr. A. Galloway. The baboons were collected in Northern Rhodesia with the help of Dr. Desmond Clark. The gibbon collection was loaned by the Museum of Comparative Zoology of Harvard University.

When counts are made by ribs, it will be seen that many gibbons and monkeys have twelve or thirteen ribs and six lumbar vertebrae are present in many gibbons and some monkeys. There appears to be little evolution in the numbers of vertebrae. But when the counting is done by facets, a very different picture emerges. Gibbons have twelve or thirteen thoracic vertebrae and five or six lumbars, with rare exceptions. Monkeys have ten thoracics and nine lumbars, with little variation, and there is no overlapping between monkeys and gibbons in either region in even the exceptional cases (Table 1).

But even counting by facets underemphasizes the functional differences because, in the monkeys, all the lumbar characteristics are more pronounced. The neural spines are longer and the anticlinal vertebra is more pronounced. The facets fit more closely and limit motion more. The centra are longer. The transverse processes are larger. In short, correlated with large muscles, all muscular processes are larger, and the lumbar region is approximately twice the length of the thoracic. In gibbons, the opposite is the case and the thoracic region is approximately twice the length of the lumbar. The short, stocky trunk of the brachiator is not flexed and extended in locomotion as is the long, narrow trunk of the quadrupedal monkey. The number of ribs does not differ greatly between gibbons and monkeys, but the functional lumbar complex of the gibbon is approximately one-half that of a monkey of comparable weight. The condition in the monkey is primitive and part of a generalized quadrupedal complex, while the structure of the gibbon is specialized.

The specializations in the human back are basically similar to those of the gibbon, and show none of the primitive quadrupedal features seen in monkeys.

The question of the significance of the numbers of vertebrae has been elaborated because it illustrates both the importance of relating structure and behavior and the difficulty of doing so. Counting by ribs obscures the differences between Cercopithecidae and Pongidae and that hides the close similarity of Pongidae and Hominidae. If the aim is to see the relation of brachiation to the evolution of the back, the vertebrae must be counted in functional groups, but, if that is done, the counts are not comparable to those in the literature, and this is an immense loss. This then, is a general problem when interest shifts from traditional categories to behavior. I think that the following statement will be correct, but much additional material would have to be examined to prove it. Vertebral regions are most clearly defined and stable when they are important in the locomotor pattern. The Galagidae are at one extreme and the Lorisidae at the other. In the Cercopithecidae, the regions are more sharply defined and much more stable than in the Pongidae. The situation in the Hominidae is derived from that of the Pongidae, but with bipedal locomotion lumbar length has become important again. There is strong selection against a sixth lumbar vertebra, over 95 per cent of humans are at the norm, and bipedalism is so new that there has not been time for a structurally efficient lower lumbar region to evolve.

From the point of view of classification, a functional view of the vertebrae fits

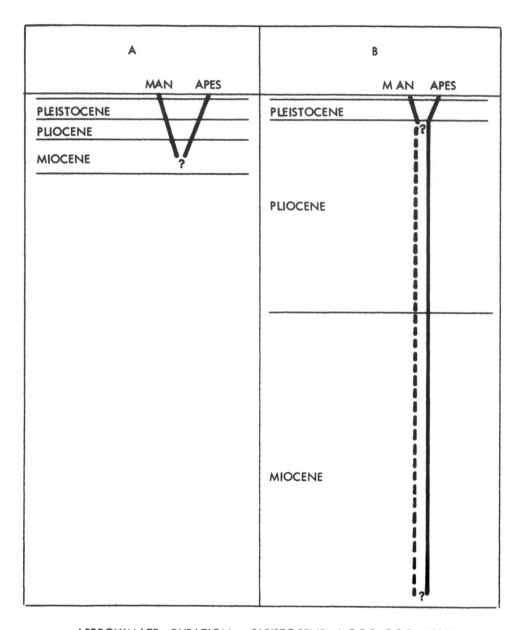

APPROXIMATE DURATION : PLEISTOCENE   1,000,000 YEARS
                         PLIOCENE   10,000,000 YEARS
                         MIOCENE   15,000,000 YEARS

FIGURE 3

the traditional categories. It suggests a much greater separation of Cercopithecidae and Pongidae than counting by ribs. It offers support for the derivation of Hominidae from Pongidae.

It has become traditional to date the separation of the Hominidae from the Pongidae to the Miocene. Earlier and later separations have been suggested, but early Miocene appears to be the commonest opinion. This view is usually diagramed approximately as shown in the accompanying Figure 3A. It will be noted that the Pleistocene appears nearly as long as the Pliocene. In Figure 3B the time scale is corrected, and it will be seen at once that all known fossils of the Hominidae are well within the Pleistocene and that a separate line has been postulated which has twenty times the duration of the known record. When Gregory and Keith pictured the separation of man and ape much as in Figure 3A, they thought that Miocene apes were essentially like living ones, that the Pliocene was little longer than the Pleistocene, and that *H. sapiens* had existed for a very long time. Figure 3A is a natural representation of this view. But now it appears that the distinctive size of the human brain evolved at the end of the Lower or beginning of the Middle Pleistocene. In the Lower Pleistocene, the ischium is long and ape-like; the ilium and hand do not yet have the full characteristics of *Homo*. Most of the characteristics of *Homo* seem to have evolved well within the Pleistocene, and there is no need to postulate an early separation of man and ape.

From the point of view of behavior, a major question is whether *Australopithecus* represents a stage of long duration, a relatively stable adaptation, or a transitional form evolving rapidly under the new selection pressures which came with the use of tools. My guess is that the second of these alternatives is correct and that tool use caused a rapid change in selection which separated the Hominidae from the Pongidae, possibly entirely within the Pleistocene.

CONCLUSIONS

It has been suggested that the "natural" groups of primates are adaptive radiations. The main adaptations were recognized by the early students of the primates, and this accounts for the agreement concerning many groupings.

Emphasis on adaptation helps in the evaluation of the structures which are used in classification, and the constant emphasis on function and behavior frequently leads to the reinterpretation of descriptive data.

MAN'S PLACE IN THE PHYLOGENY OF

THE PRIMATES AS REFLECTED

IN SERUM PROTEINS

MORRIS GOODMAN

T HE EVOLUTIONARY DEVELOPMENT of the Primates has resulted from the sum
of the processes of cladogenesis and anagenesis. Cladogenesis, or phylogenetic branching (Rensch, 1959), is the splitting of a species into two
or more species and the ensuing divergence of these species from each other.
Anagenesis, or progressive evolution (Rensch, 1959), is the process by which
the more highly organized forms of life evolved from the simpler. In the order
Primates, cladogenesis is exemplified by the prosimian phylogenies, and anagenesis
is exemplified by hominine phylogeny.

The living primate fauna can be arranged according to grades of morphological
organization into a remarkable sequence which goes from the tree shrew to
man and simulates the successive anagenetic stages of the history of the Primates.
The placement of the organisms in this *scala naturae* has guided systematists in
classifying the members of the order. For example, although man is taxonomically
grouped with the apes in the superfamily Hominoidea, he is invariably separated
from the latter by being the sole contemporary member of the family Hominidae.
In turn the apes are usually grouped together in the family Pongidae. However
it is quite possible that certain of the living apes are more closely related to man
than to the other apes. In other words certain of the apes may show a more
recent common ancestry and more genetical affinities with man than with other
apes. There are a number of such questions concerning the phylogenetic relationships of various living Primates. Indeed even the boundaries of the order are
in dispute with some systematists placing the tree shrews in the Insectivora.

To solve these problems of primate systematics we must use very dis-
criminating criteria to evaluate the genetic similarities and differences of organisms. The comparative analysis of proteins is potentially capable of providing
such criteria, for recent developments in protein chemistry and in genetics have
demonstrated that there is a close correspondence between the stuctural specificity
of proteins and the code of information in genetic material (Anfinsen, 1959;

Ingram, 1961; Crick, 1961; Chantrenne, 1961). Hence by utilizing some of the newer biochemical and serological methods for studying the structural specificity of proteins it is possible to ascertain the genetical affinities of contemporary organisms with a reasonable degree of objectivity.

Furthermore from our growing knowledge of the structure of proteins and the functional and immunological properties of proteins we can draw theoretical conclusions about the nature of the evolutionary process which led to the emergence of *Homo sapiens*. This article, following previous efforts of the author (Goodman, 1960a, 1961, will attempt to state some of the conclusions. An immunological theory of primate evolution will be presented and the results of a comparative serological study of the serum proteins of the Primates will be reviewed.[1] The data gathered in this study provide experimental evidence for the taxonomic position of man and the course of his evolution predicated by the theory.

PROTEINS AND THE EVOLUTIONARY PROCESS

A higher organism differs from a lower by the greater informational content of its genome. This is revealed by the greater biochemical complexity and more extensive molecular differentiation of the higher organism. If we compare microorganisms to lower metazoans and the latter to higher metazoans, a rough correlation is found between increases in the mass of DNA per cell and increases in the histological and morphological complexity of the organisms (Sneath, in press). However, an increase in the size or number of the DNA polymers can only provide more raw material for the coding of hereditary information. The actual development of a more highly organized living system must await the evolution of a genome whose sections or genes show an increase in the integration of their informational contents. In other words a more highly organized living system has a more highly organized code of information in its genome.

The validity of this view will become apparent after we consider some of the relationships between gene mutation, protein structure and function, and natural selection. Our inquiry will show that the process of speciation must have caused a much deeper divergence of genetic codes and a much more rapid evolution of the structural specificity of proteins among lower organisms at the primitive stages of phylogenesis than among higher organisms at advanced stages of phylogenesis. Although morphological evolution appeared to accelerate with the emergence of higher organisms, molecular evolution decelerated. The central argument we shall develop is that the possibilities for the specificity of any particular protein to vary among the members of a species decreased as the biochemical complexity and molecular differentiation of organisms increased.

1. This study is being supported by grant G14152 of the National Science Foundation. Certain phases of it have also received support from grant MY–2476 of the National Institute of Mental Health and grant 226 of the National Multiple Sclerosis Society.

A protein is composed of one or more long polypeptide chains whose complete sequence of amino acid residues (the primary structure of the protein) is under strict genetic control. The nature of this control is such that the substitution of one amino acid for another in a chain can be related to a single gene mutation (Ingram, 1961). The three dimensional architecture of a protein molecule (the secondary and tertiary structure of a protein) is determined by the coiling and specific foldings of the polypeptide chains. A typical chain may contain from 100 to 500 amino acids of up to 20 different kinds. Thus mutations can cause an exceedingly large number of permutations in amino acid sequence. A change in amino acid sequence could so alter the conformations of a polypeptide chain as to create a protein with a new tertiary structure and a new set of surface configurations. Alternatively the amino acid substitution need not alter the conformations of the chain but could nevertheless occur at a position on the chain which would alter the topography of a small portion of the surface of the protein.

It is convenient to consider the biological functioning of globular proteins in terms of three categories of surface configurations: (a) the configurations which are called the active site because they are responsible for the primary functional activity of a protein, for example an enzymatic, hormonal, or antibody activity, (b) the ancillary configurations which enable a protein to be transported preferentially across certain membrane barriers and to be taken up preferentially by certain cell types, and (c) the neutral configurations with no apparent functional activity. Obviously any mutation of a primary structure which affected the active site of a protein would also affect the survival chances of the organism. However, it is clear from the structural analysis of hormones, enzymes, and antibodies that the active sites usually occupy a small proportion of the total surfaces of the proteins (e.g., Li, 1957; Smith, 1957). Also studies of the species specificity of proteins (e.g., Tristram, 1953; Porter, 1953) establish that a protein type can vary markedly in amino acid composition and antigenic structure and still execute its characteristic physiological role. Furthermore from the x-ray analysis and other structural studies of sperm whale myoglobin (Kendrew et al., 1960) and horse hemoglobin (Perutz et al., 1960) the principle has been established that proteins and major subunits of proteins can be dissimilar in primary structure and remarkably similar in three dimensional arrangement. Thus more than one primary structure is compatible with a single tertiary structure and many of the informational bits of a genetic code can be varied without affecting a protein's active site. We may deduce that in the beginning stages of phylogenesis many permutations in the primary structure of any one protein were permissible and that a vast amount of selectively neutral genetic variability could accumulate in any stable species.

It may be argued that due to the reducing nature of the environment during the early stages of the evolution of life (Oparin, 1957) a primordial protein would have had very few if any disulfide bonds and a large proportion of its backbone polypeptide chain would have had the disordered arrangement of the random coil. Such a protein would have had a minimum of tertiary structure and for this reason alone would have lacked surface configurations which could acquire functions ancillary to the active site. Looked at in this way metazoan evolution could not begin until conditions had materialized for the synthesis of proteins with well organized tertiary structures.

A typical soluble protein in our unicellular ancestors who preceded the start of metazoan evolution probably had a well organized tertiary structure which exposed two kinds of surface configurations: a small constellation of organic groups (the active site) executing the primary functional activity of the protein, and the large proportion of remaining configurations without functional activity. The latter, in contrast to the former, must have showed extensive structural heterogeneity among the organisms of the species as the result of an accumulation of neutral genic alleles in the species. Let us refer to this protein as protein A and trace it to a descendent species in our line of ancestry much later in evolution.

The organisms in this metazoan species still produce a protein A whose tertiary structure has hardly changed since the dawn of metazoan evolution. Nor has there been any change in the active site, which may be labeled a. In addition to a the protein has other surface configurations which are uniform throughout the population. These are the configurations with ancillary effects on the functional activity of protein A. For example, the b configuration of protein A interacts with complementary configuration x of the characteristic protein of tissue X, and thus A is taken up preferentially by tissue X where it executes its primary functional activity. Any mutations which altered either the b or the x configuration would be detrimental unless a simultaneous mutation occurred which so altered the other that b' and x' were also complementary (as b and x had been). Since this would be a highly improbable event, the evolutionary stability of both b and x configurations would be relatively high. In contrast to this state of stability, configuration c of protein A interacts weakly with configuration y of plasma protein Y, which tends to divert protein A from tissue X. Consequently, mutations would be favored which altered c to either of several alternative configurations (c', c'', or c''') since none of these react with y or other accessible configurations of the plasma proteins.

Arbitrarily we shall assign to the therapsid reptilian stage of phylogenesis the species which tolerates equally well configurations c', c'', and c'''. Now we shall consider a descendent species later in anagenesis after eutherian evolution had begun. The evolution of molecular complexity in the lineage leading to this species provided the basis for such improvements as a finer regulation of physiological processes by the central nervous system, the development of a superconcentrating kidney for the conservation of water, and the establishment of a placenta which grafted fetal tissues to maternal tissues and protected the fetus from many of the vagaries of the external environment. In directing the molecular evolution underlying these improvements, natural selection had to stabilize additional protein configurations with primary and ancillary functions. Also with the increasing specificity of the stereochemical interactions of protein molecules, the thifd category of protein configurations, those without apparent function, had to be selected more carefully to insure their state of non-interference with physiological processes.

The intensity of the selection for functional and for non-interfering configurations is indicated by the large number of serologically distinct proteins found in plasma and by the high specificity of their molecular associations. For example, if red blood cells rupture, the released hemoglobin is preferentially bound to haptogobin (a distinct species of the plasma proteins). Hormones are transported in plasma by specific proteins. Thyroxine, for instance, is preferentially bound to a particular species of the alpha proteins, but also to the tryptophane rich prealbumin (Winzler, 1960; Antoniades, 1960). The selective transport of gamma globulin in preference to other maternal serum proteins across the placenta into the fetus exemplifies the high specificity of molecular associations which involve ancillary configurations of the proteins. Not only is gamma globulin transported in preference to other types of plasma proteins, but homologous gamma globulin is more readily transported into the fetus than is heterologous gamma globulin (Hemmings and Brambell, 1961). For example, when rabbit and bovine immune gamma globulins were administered to a pregnant rabbit, the rabbit protein crossed the placenta in preference to the bovine protein. Other examples including some of transport across the placenta of man, are also cited by Hemmings and Brambell (1961). Such studies show that, in addition to sites for antibody activity, immune gamma globulin has surface structures of high specificity for the selective transfer of the protein to the fetus.

In this example of the selective transport of gamma globulin across the placenta, the structural specifications of gamma globulin by the genetic code are coadapted to the structural specifications of the other plasma proteins and the placental membrane proteins. We can see from examples of this type that the process of anagenesis decreased the chances that any single gene mutation would be neutral to natural selection since any permutation in a polypeptide chain would with greater probability affect the physiological activity of a protein.

In addition to altering the structural specifications of proteins, mutations can alter the epigenetic histories of proteins. At any particular time only a portion of the DNA polymers of a cell act (through the relay system of "messenger RNA") in protein synthesis, and the portions which act depend on the environmental conditions prevailing in the cell. For instance, the presence in the cell of a substrate can activate the mechanisms for channeling specific genetic information into the synthesis of certain enzymes. Then the accumulation of the terminal product of the enzymatic reactions can repress this mechanism and activate another. Although such induction and repression can be initiated by relatively small organic molecules, the mechanisms for channeling "messages" to and from the DNA polymers must involve chains of specific reactions among macromolecules.

Ontogenetic development is in essence an epigenetic process and depends at each stage on a cascade of interactions between a changing cellular environment and genetic material.[2] At the beginning of embryogenesis only a small fraction of the total genetic information is channeled into protein synthesis. Then with each cell division a new set of environmental conditions in the descendent cells induces different portions of the genetic information to be expressed in the protein synthesis of these cells. At the completion of ontogenetic development the total amount of genetic information channeled into protein synthesis is much greater than it was in the beginning stages, but this greater sum of genetic information in the structural specificities of proteins results from addition of the activities of the various differentiated cells of the mature organism.

Natural selection has so acted on the genome that the expression of its structural information during ontogenetic development is programmed against a fixed cycle of environmental changes, one repeated from generation to generation. If the cycle is disturbed by the introduction into the fetal environment of a new factor (such as a new chemical substance) from the external environment, developmental anomalies are likely to ensue. Similarly any gene mutation which caused the early fetus to synthesize a protein with a mutated structure would introduce a new factor into the fetal environment and would tend to distabilize the subsequent course of ontogenetic development. It is not surprising that a gene mutation which acts on development from the early stages of ontogeny is more likely to be detrimental to the organism than a gene mutation which only acts at later stages of ontogeny.

A gene mutation will be selectively advantageous if it increases the capacity of the organism to withstand the disintegrating impact of the external environment and selectively disadvantageous if it decreases this capacity. In the case of selectively advantageous gene mutations the new configurations on the surfaces of the mutated proteins can be looked upon as molecular adaptations to the conditions of the environment. It will help us to pursue our inquiry if we distinguish in the higher metazoans between two kinds of molecular adaptation, the kind directed "inward" to integrate the stereochemical interactions of the proteins (and thus perfect the metabolic machinery of an organism) and the kind directed "outward" to meet the exigencies of the surroundings of an organism. With dichotomies of this type we can characterize a certain aspect of the mechanism underlying the process of anagenesis.

Let us do so by considering a paradoxical situation. It concerns the species at the forefront of the main stream of anagenesis. Compared to all other organisms the members of this species have the most elaborate endogenous makeup with the

2. See, e.g., Chantrenne (1961) and Platt (1962) who describe the gene-environmental interactions of ontogenetic development in terms of newer concepts and analogies of biochemical genetics.

greatest differentiation of parts and the largest number of "inward-directed" molecular adaptations. As a result of the superior metabolic efficiency of its members, the advanced species is supplanting competitors and colonizing the largest range of exogenous habitats. Hence the new conditions of the external environment select for new gene mutations and genetic heterozygosity since only this state can give the species a variety of new "outward-directed" adaptations. However such a state tends to disorganize the established network of highly specific protein interactions. Indeed the probability of interference with physiological processes by any mutated protein specificity is in direct proportion to the number of "inward-directed" molecular adaptations of an organism. Thus two opposing selective pressures (one favoring genetic heterozygosity, the other favoring genetic homozygosity) operate with special force on the anagenetically advancing species.

A new equilibrium is established between these two opposing selective pressures by a selection of genetic codes which increase the complexity of ontogenetic development. Such codes dissociate the rates of maturation of different somatic systems. Thus the systems which mature early in ontogeny establish the basic features of the endogenous environment, and the genes which control these early maturing systems become more uniform throughout the population. It is the later maturing systems which develop the capacity for adaptive responses to exogenous conditions and which serve as a reservoir for gene determined polymorphisms.

The ontogeny of the immunological system illustrates how the conflicting needs of an organism can to some extent be met by the late maturation of somatic systems with "outward-directed" molecular adaptations. The immunological system is one of the major achievements of vertebrate evolution. It gives the individual the capacity to respond in a highly selective manner to a fantastically large number of molecular configurations. Without the capacity to produce antibodies the vertebrate organism could not survive the invasion of micro-organisms and animal parasites. Yet with this capacity the vertebrate organism can produce autoantibodies which interfere with endogenous physiological processes and thereby lessen the chances of survival of the organism. Vertebrate evolution, however, has minimized this dilemma by selecting genetic codes which delay maturation of the immunological system and thus allow a state of immunological tolerance[3] to be acquired to endogenous proteins. As an acquired state, immunological tolerance is limited in scope. It only pertains to the proteins which contact the cellular progenitors of the immunological system during a critical stage of fetal development. Thus endogenous proteins may be autoantigenic if they appear late in ontogeny or are confined to tissues and rarely enter the

3. See Hasek et al. (1961) and Smith (1961) for a review of facts and theories on the mechanisms of immunological tolerance.

blood circulation. Furthermore, many proteins and other macromolecules which are non-antigenic within the individual synthesizing them show some degree of isoantigenicity within the species.

ROLE OF THE IMMUNOLOGICAL SYSTEM IN THE EMERGENCE OF HOMO SAPIENS

The development of placentation in the mammals initiated a new stage of progressive evolution. It provided opportunity for an intimate apposition of fetal and maternal blood vessels which could secure for the developing tissues of the embryo an abundant supply of nutrients and oxygen and an effective channel for waste removal. It also permitted fetal proteins to come in contact with maternal lymphoid cells and maternal antibodies to cross into the fetus. In this situation, a fetal protein altered by gene mutation was a potential isoantigen, injurious to the fetus by its capacity to immunize the mother. Thus in the mammalian stage of phylogenesis maternal isoimmunizations further decreased the heterozygosity of prenatally acting genes. There would also be in some lineages a selective advantage to genetic codes which delayed the appearance of proteins involved in adaptions to the external environment until after birth, since changes in the specificities of such proteins would not immunize the mother.

It was only after the transition from oviparity to viviparity in the early mammals that maternal immunizations to fetal antigens could become a factor in mammalian evolution. At first this factor was minimal. The primitive placenta of the basal stock of the Eutheria can be depicted (Hamilton, et al., 1952) as having a simple non-deciduate epithelichorial arrangement in which several avascular layers separated the fetal and maternal blood vessels. This placenta functioned with low efficiency in the transport of metabolites and would hence have blocked the entrance of fetal isoantigens into the maternal circulation. We can assume that just as small modern mammals have a relatively brief period of intra-uterine existence (Rensch, 1959) so did the small insectivore-like mammals of the Cretaceous epoch. If these animals were the size of *Sorex*, gestation may have lasted no longer than three weeks. Moreover, during this time much of embryonic development probably occurred in the lumen of the uterus before the placenta was fully elaborated. Since in mammals of any size it takes from one to two weeks for homografts to elicit immune reactions, the time during which maternal immunizations could injure the fetus was relatively short in the Cretaceous eutherian mammals. Thus isoantigens could exist in these mammals not only among the molecular structures which differentiate late in fetal development (as the gamma globulin isoantigens do in modern mammals; see Oudin, 1960; Dray and Young, 1960), but also (though to a lesser extent) among the molecular structures which differentiate early. Because there was still considerable genetic plasticity in the average populations of these basal eutherian mammals, speciation could cause extensive genetic divergencies among the radiating lineages and a rapid evolution of the structural specificities of the proteins.

The lineages which retained an epitheliochorial placenta would not be as subject to the selective action of the maternal immunological system as the lineages which evolved the hemochorial placenta with its intimate apposition of fetal and maternal blood streams. As lineages of the former type radiated into new ecological habitats they could in time acquire specializations for these habitats through the evolution of all sections of the ancestral genetic code, not just the ontogenetically late acting genes. The various molecular adaptations ("inward-directed" ones as well as "outward-directed" ones) would be selected in relation to the special features of the external environment of each lineage. A different picture would be presented by the lineages with the hemochorial placenta. In these lineages fetal proteins would have become more accessible to the maternal immunological system through the more intimate placental arrangement. If there was then a prolongation of gestation, maternal immunizations would markedly slow down the rate of divergence from the ancestral genetic code, particularly in the case of the ontogenetically early acting genes.

This immunological view of the evolution of placentation in the mammals supports the standpoint of Hill (1932) and Le Gros Clark (1959) that the hemochorial placenta is an advanced eutherian character. However, it has also been argued (Wislocki, 1929) that the hemochorial placenta is a primitive eutherian character. This argument is based on the distribution of the hemochorial placenta in contemporary eutherian orders. It is found more frequently in the orders of small placental mammals such as the insectivores (the "primitive" or lower representatives of the Eutheria) than in the orders of large placental mammals such as the ungulates. (The major exception to this pattern of distribution is the uniform presence of highly developed hemochorial arrangements in the catarrhine Primates.) There is no a priori reason to consider the presence of hemochorial placentas in many small eutherian mammals as a primitive feature of these mammals. Rather the presence of such placentas could conceivably be an example of an advanced feature which was acquired independently in many separately evolving phyletic groups at later stages of evolution. There is no rule which requires the contemporary small mammals to be primitive in all respects. The crucial question is whether the hemochorial placenta had developed in the basal stock of the Eutheria before the extensive radiation of this stock into the various orders of the placental mammals.

The immunological considerations stated in previous paragraphs provide an a priori basis for arguing that the hemochorial placenta was not present in the early stages of eutherian cladogenesis. The reduction of genetic plasticity in the basal eutherian population by maternal isoimmunizations would have been minimal in the absence of an intimate connection of fetal tissues to the maternal circulation. Conversely the genetic plasticity of this population would have been reduced by changes in the conditions of gestation. Thus the development of a placenta of the hemochorial type which increases the opportunity for immune responses to the fetus would have reduced genetic heterogeneity in the population. It is reasonable to assume that the formation of diverse orders of placental mammals resulted from the cladogenesis of a basal population with a relatively high degree of genetic plasticity rather than from one with a low degree. Therefore, a placenta of the epitheliochorial type which minimizes maternal immune responses to the fetus would have best fitted the conditions required for the first major adaptive radiations of the class Mammalia.

The immunological considerations also offer an explanation for why the hemochorial placenta is more frequently found in small rather than large placental mammals. Let us assume that in mammalian evolution there was a general trend toward an elaboration of placental mechanisms which secure optimal metabolic conditions for the developing fetus and that in the absence of maternal isoimmunizations the logical outcome of the trend was the elaboration of the hemochorial placenta. However, in the presence of maternal isoimmunizations, the trend toward the hemochorial placenta could progress more rapidly in small mammals with short gestation periods than in large mammals with long gestation periods, for the immune response of a small mammal to an antigenic challenge is no more rapid than that of a large mammal. Thus the hemochorial placenta in a small mammal with a brief gestation period would provide less opportunity for maternal immune attacks on the fetus than this same type of placenta in a large mammal with a lengthy gestation period.

These immunological considerations suggest that significant strides toward the development of a hemochorial placenta occurred in the line of human ancestry during the early stages of primate phylogeny when man's prosimian ancestors still had a relatively short period of gestation. It is not necessary to assume that these strides were equally strong in all the prosimian lineages, particularly if the adaptive radiations of the early Primates commenced shortly after the emergence of the basal members of the order. Thus the epitheliochorial placenta of the lemurs can be considered a retained primitive character rather than a regression or specialization away from a supposed ancient type of hemochorial placenta. The lemurs, however, do show advances in the mechanisms of placentation which foreshadow the advances found in higher Primates. For example, the chorion of the lemur placenta becomes vascularized relatively early in fetal development (Le Gros Clark, 1959).

Another important trend in primate evolution has been the progressive expansion of the brain. There is reason to suspect that this trend depended on the one which secured the fetus optimal conditions for its development. In contrast to the paleoencephalon, the neocortex is easily injured by oxygen deprivation (Himwich, 1952) and a rich supply of nutrients and oxygen is required during fetal life for the differentiation of this phylogenetically new part of the brain. There is perhaps a causal relationship between the rapidity by which the early embryo of man establishes an intimate connection with the maternal circulation and the remarkably expanded brain which develops in man. Although the evolution of the neocortex depended on an evolution of more efficient placental mechanisms, the acquisition of these mechanisms would not by itself guarantee that cerebral evolution would occur. Such evolution would then depend on whether the organisms in their particular environmental habitat would gain some selective advantage by acquiring bigger brains.

The arboreal environment of the primitive Primates, while favoring the preservation of such generalized anatomical features as pentadactyl limbs, selected for visual acuity and for an expansion of the higher cerebral centers controlling the motor responses to visual and tactile impressions. However, as already noted, the evolution of a more complex brain could not take place until the functional efficiency of placentation had increased. Thus we can postulate that the lineages responsible for the phylogenetic advances of the Primates were those in which an elaboration of the hemochorial placenta with its intimate apposition of fetal and maternal blood streams occurred comparatively early in the evolution of the order. The evolution of our ancestors must have then been shaped by the following immunological conditions.

Since fetal proteins could more readily contact the maternal immunological system through the more intimate placental arrangement, isoimmunizations could decrease the incidence of genetic heterozygosity and increase the homozygosity of the most commonly occurring genes. This would preserve (along with the selective action of the arboreal environment) the primitive generalized features of the Primates. However, mutant genes could alter the specificities of proteins which were adequately isolated from the maternal lymphoid cells, for such mutant genes would have escaped the selective pressure of maternal isoimmunizations. The proteins of the central nervous system depend on genes which fall in this category, as do also all proteins which do not appear before birth. Therefore, the mutant genes which act only in the postnatal organism have the selective advantage of not provoking the maternal immunological system. On the other hand a delay in the appearance of adult proteins would increase the helplessness of the offspring. Genetic codes which retarded the maturation of the new born could only be selected if there were corresponding increases in the protective care of the young by the mother or other adults. The Primates with their expanding cerebral cortex and their increasing ability for adaptive psychological responses were uniquely suited for this development.

The phylogenetic advances of the Primates which culminated in man can be considered progressive because at each stage they increased the autonomy of the individual from the external environment and gave the individual the capacity to master a larger number of external challenges. This growing capacity to master new conditions depended of course on the progressive expansion of the higher centers of the brain. The basal Primates emerging from the Cretaceous eutherian radiation appear to have had in their arboreal environment more stimulus than the basal members of the other eutherian orders for the evolutionary development of the cerebral cortex. The early members of the Hominoidea were probably still at the prosimian grade of phylogenetic development in their skeletal structure. Thus at this stage of primate evolution an arboreal environment continued to offer the hominoids the best stimulus for the elaboration of the brain. Somewhat later in the phylogenesis of the Hominoidea, after larger sized organisms had evolved, an ecological domain which was both arboreal and terrestrial offered a greater stimulus than a strictly arboreal environment for further cerebral development. Those hominoids who entered and successfully occupied this broadened ecological domain can be considered the basal members of the Hominidae. Later yet in phylogenesis, the most progressive of the Primates were the bipedal hominids who inhabited a predominantly terrestrial domain and moved over vast geographical territories. Again the broadening of the environment favored the further elaboration of the brain.

We can now recapitulate and stress the dynamics of the process depicted by our immunological theory of primate evolution. Early in the anagenesis of the Primates the hemochorial placenta developed, providing the metabolic basis for a continuing progressive elaboration of the brain. In turn this cerebral evolution, which correlated with increasing body size, allowed the Primates to invade more varied ecological domains. This led to two opposing tendencies. On the one hand the more varied conditions of a broader environmental zone selected for mutant genes and the state of genetic heterozygosity. On the other hand, as the barrier between embryonic tissues and maternal blood decreased and as the period of gestation lengthened, the maternal immunological system selected for the state of genetic homozygosity. An equilibrium was established between these contradictory tendencies by a selection of genetic codes which delayed maturation until the postnatal phase of ontogeny. The somatic systems which establish the basic features of the endogenous environment still matured before birth and were increasingly controlled by genes with a high degree of homozygosity. The somatic systems immature at birth were those with the largest concentration of "outward-directed" molecular adaptations. The expression of genetic heterozygosity then began to center in these late maturing structures. Furthermore, this process necessitated the evolution of family and social life by increasing the immaturity and helplessness of the offspring. It is not by chance that man who shows the greatest cerebral development and occupies the broadest environmental range is also the most retarded in his maturation.

Another aspect of the process of primate evolution which can be characterized by our immunological theory concerns the rate of molecular evolution in different primate lineages. The theory requires the rate of diversification of genetic codes by phylogenetic branching to decrease after each anagenetic advance. The prosimians would show the greatest rate of diversification and the most extensive divergencies in the protein specificities of the radiating lineages. There would also be marked differences among these lineages in their continuing rates of evolution. Certain groups such as the Lemuroidea who retained the epitheliochorial placenta would continue to show rapid rates of evolution and in the sanctuary of Madagascar a striking diversification of genetic codes at a fairly late stage of the evolutionary history of the Primates. In contrast, the Tupaioidea who were restricted to a narrow ecological niche, resembling the one most frequented by the Cretaceous proto-primates, must have had their genetic plasticity drastically reduced after they acquired the hemochorial placenta. This then placed a brake on the subsequent evolution of their genetic codes.

It is consistent with our theory to seek the origins of the Anthropoidea in that group of generalized tarsioids of the Paleocene in which the evolutionary advances towards the hemochorial placenta were progressing rapidly. The Platyrrhini and Catarrhini could then have independently evolved from prosimian ancestors along roughly parallel lines. They would retain despite their ancient separation a fair amount of correspondence in many sections of their genetic codes, since divergence in fetal proteins would have encountered adverse selective pressure from maternal immunizations. Due perhaps to the isolation of the South American continent during most of the Tertiary epoch, the process of anagensis was less marked in the platyrrhine Primates or ceboids than in the old world or catarrhine members of the Anthropoidea, the cercopithecoids and hominoids. According to Hill (1932) the platyrrhine placenta only attains structural efficiency at a late period of gestation. Thus, compared to the cercopithecoids and hominoids, the ceboids show the least fetal-maternal vascular intimacy during gestation. In this situation the ceboid radiations would have diversified genetic codes at a somewhat faster rate than either the cercopithecoid or hominoid radiations.

It would accord with the process of primate evolution depicted by the theory if the basal group of the Hominoidea, as distinct from that of the Cercopithecoidea, was that catarrhine lineage which, compared to the others, showed the largest degree of fetal-maternal vascular intimacy during gestation. This phyletic group would then have shown a relatively low degree of genetic plasticity. Consequently, during the radiations of the Hominoidea only small sections of the genetic codes would be diversified, and the serological specificities of the proteins of the separately evolving lineages would retain a high degree of similarity. A stage of the process would be reached with the emergence of the bipedal hominids where genetic plasticity was so reduced that the most advanced lineage (the one with the largest brain) would retain its reproductive cohesiveness and fail to speciate even though its population expanded over the surface of the earth and occupied more geographical territory than had the members of any previous primate radiation. This last and greatest expansion of the Primates can be attributed to the basal members of our own species *Homo sapiens*.

The process depicted by our immunological theory must also be affecting the current biological evolution of our species. Erythroblastosis fetalis due to the

Rh and other blood group incompatibilities is perhaps the classic example that the destructive side of the maternal immunological system is very much in evidence in the human population. It is also clear that each individual has in addition to the blood group substances a unique assortment of histocompatibility antigens in his tissues. Maternal isoimmunizations may prove to be a key factor in abortions and in the birth of premature and defective children. The process, then, of the weeding out of isoantigens is by no means over. Our thesis is also supported by the fact that some blood group substances and histocompatibility antigens do not show final maturation until the stage when the possibility of transplacental immunization has ended. We might expect even more pronounced shifts in the maturation of isoantigens to the postnatal phase of ontogeny if evolution can continue under a protective cover of an advanced technology in a society which protects its young.

However there are selective pressures apposed to further delays in maturation. These pressures emanate from the individual's own immunological capabilities. In the current stage of human evolution, the neonatal organism cannot long survive microbial invasion unless its own immunological system rapidly matures. However, if the mature immunological system is suddenly faced with "strange" endogenous molecular specificities for which it lacks tolerance, it will tend to respond to these specificities with destructive autoimmunizations.[4] Thus the threat from autoimmunizations as well as isoimmunizations has tended to further limit the evolutionary development of new molecular specificities to intracellular proteins so immobilized in stationary tissues as to normally be inaccessible to antibody producing cells. Nevertheless, the immunological threat still remains, for such immobilization cannot be absolute and tends to break down with a release of autoantigens when the tissue is under stress.

A good example is provided by the high incidence of autoimmunizations to human thyroid proteins in thyroid diseases (Roitt et al., 1958; Roitt and Doniach, 1960) and in old age (Goudie et al., 1959; Goodman et al., 1963). The auto-antibodies to human thyroid proteins cross react with thryroid extracts from catarrhine Primates (reacting more strongly with chimpanzee than with rhesus monkey) but not significantly with the extracts from non-primate mammals. In other words, the autoantigenicity of human thyroid proteins resides primarily in structural configurations which were acquired during the more recent stages of evolution. Any protein is a potential autoantigen if it does not make contact with the cellular progenitors of the immunological system in the fetus. Thus the process of anagenesis in the Primates must have caused human tissues to

4. Even if immunological unresponsiveness (or paralysis) developed the danger would exist that the paralysis would impair the general functioning of the immunological system and thus make the organism more susceptible to infections. In this connection Liacopoulos et al. (1962) found that when large doses of bovine albumin were administered to guinea pigs the immune unresponsiveness which developed to bovine albumin spread to affect unrelated protein antigens.

become rich in autoantigens,[5] for the evolution of molecular complexity created membranes such as those of the "blood-brain barrier" which effectively block the entrance of intracellular proteins into the blood stream.

SEROLOGICAL ANALYSIS OF PRIMATE PHYLOGENY

A comparative study of the serum proteins of the Primates is in progress (Goodman, 1960b, 1961, 1962a,b, in press, Goodman et al., 1960; Goodman and Poulik, 1961a,b). The principal techniques are two-dimensional starch-gel electrophoresis (Poulik and Smithies, 1958) and agar-gel precipitin testing with a variety of antisera to proteins of different Primates. These methods provide data on the phylogenetic relationships of man and other Primates. In addition, the data of the precipitin method (Goodman, 1960b, 1962a) demonstrate that certain types of proteins have evolved more slowly than other types during the radiations of the Primates. Furthermore it can be shown, though only in an oblique way, that the rate of diversification of genetic codes has been greater in certain taxa than in others. Thus it is possible to test experimentally whether or not the trends in the molecular evolution of the Primates, predicated by the theory set forth in this chapter, have occurred.

Two-dimensional starch-gel electrophoresis (filter-paper electrophoresis in one dimension followed by starch-gel electrophoresis in the other) separates the proteins of serum into 19 to 25 components. The arrangement of these components in starch-gel provides a pattern which is characteristic of the species of the organism whose serum is analyzed. Such patterns are especially useful for determining if organisms show a phylogenetic relationship at a subfamily or generic level (Goodman et al., 1960; Goodman, 1961, 1962b, in press). For example, *Perodicticus* and *Galago* (who are grouped together in the Lorisidae, but placed in separate subfamilies) have quite divergent patterns, whereas *Galago*

5. If our thesis is correct that the tissues of man are rich in autoantigens, it follows that the production of autoantibodies may play a key role in the degenerative diseases and in the process of senescence. The very extensive morphological and molecular differentiation of the mammalian organism prevents development of immunological tolerance to many intracellular molecular specificities. Thus the suggestion (Goodman et al., 1963) that autoimmune reactions cause a gradual destruction of the viable cells of the organism amounts to a version of the morphogenetic theories of mammalian senescence (Comfort, 1956). Our particular version is quite sweeping in the role that it attributes to immunological phenomena, the increase in tissue autoantigens being related to the threat to the mammalian fetus from maternal isoimmunizations.

Conceivably the science of immunobiology will be able in the not too distant future to offer a program for slowing down the rate of senescence in man. Immunobiology will first have to assess the possibilities for a molecular fetalization of our species, one which either reduces the level of autoantigens in tissues or instead decreases sensitivity to such substances by further delaying the maturation (or permanently retarding the ontogenetic development) of the immunological system. It will then be necessary to determine a means of directing such a fetalization to ensure a progressive rather than degenerate evolution of our species.

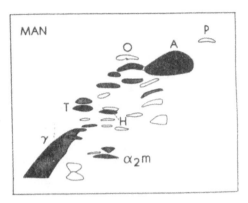
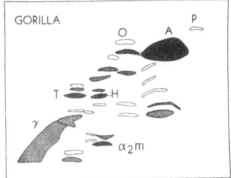
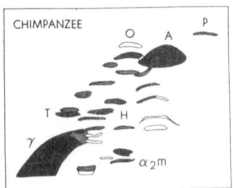
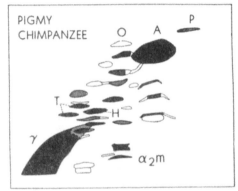
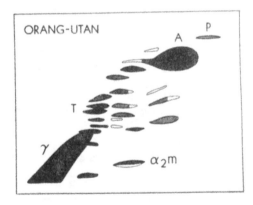
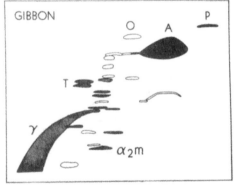

FIGURE 1

Diagrams of two-dimensional starch-gel electrophoresis patterns of hominoid sera. P = prealbumin; A = albumin; O = orosomucoid; T = transferrin; H = haptoglobin; and γ = gamma globulin. Type 1 haptoglobin is represented in the human pattern. (Inherited variations of haptoglobin occur in the human population, but appear to be minimal in non-human primate populations. Transferrin shows an extensive polymorphism among chimpanzees and perhaps as indicated by our first gorilla results, among gorillas.)

crassicaudates and *Galago senegalensis* (two species of the same genus) have almost identical patterns.

When the patterns of members of the Hominoidea (man, gorilla, chimpanzee, gibbon and orangutan) are compared with each other (Figure 1), the differences among the various hominoid types are much larger than those found between macaques and baboons or between macaques and vervets (Figure 3). Indeed the various species of the subfamily Cercopithecinae examined by two-dimensional starch-gel electrophoresis (five species of *Macaca*, one of *Papio*, and one of *Cercopithecus*) show a high degree of similarity. Although the starch-gel pattern of each hominoid type diverges sharply from the others, there is a constellation of about ten faster migrating proteins in man, chimpanzee, and gorilla (Figure 2), but not in gibbon or orangutan. The faster migrating proteins of the gibbon

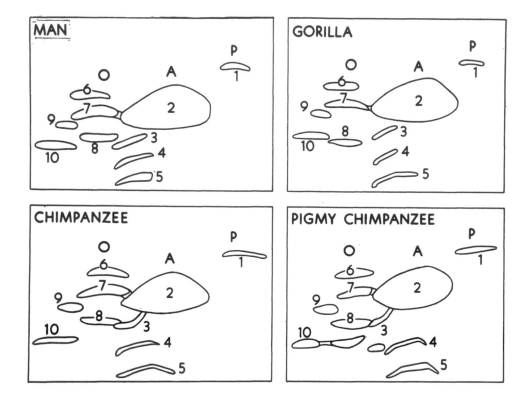

FIGURE 2

Outline diagram of ten faster migrating components in two-dimensional starch-gel electrophoresis which show similar but not identical positions in man, gorilla, chimpanzee, and pigmy chimpanzee. Note that in gorilla component 8 is to the right and beneath component 10; whereas in man component 8 is to the right and above component 10. In the chimpanzee component 8 is also to the right and above component 10, but the separation between 9 and 10 seems to be greater in chimpanzee than in man or gorilla. Note the extra unnumbered components between 10 and 4 in the pigmy chimpanzee.

and orangutan present patterns in each case which are quite dissimilar from those of the other hominoids.

Such data allow us to make educated guesses about some of the questions concerning the phylogenetic history of the Hominoidea. For example, the known hominoid fossils trace a gibbonoid line of apes back to the Oliogene epoch, but otherwise fail to establish the times of phyletic separation within the Hominoidea. However, there are fossil baboons and macaques dating from the Middle Pliocene (Le Gros Clark, 1959). Thus when the divergencies of the hominoid patterns are contrasted to the similarities of the cercopithecine patterns it can be deduced that the phyletic separations of the five major groups of living hominoids are at least as ancient as the Lower Pliocene (if not more ancient). Furthermore there is some suggestion from the divergencies of the hominoid patterns that the phyletic line which branched to give rise to gorilla, man, and chimpanzee did so after it had separated from more ancient lines leading to gibbon and orangutan.

The suggestion that *Gorilla, Homo,* and *Pan* form a phyletic unit within the Hominoidea is strongly supported by the results of the second set of experiments in which an agar-gel precipitin technique (Goodman, 1962a) was employed. This technique determines the degrees of correspondence in antigenic structure existing among the protein homologues of different species. To do so an antiserum is produced to a protein isolated from the organisms of a particular species. Then cross reactions are developed in immunodiffusion plates between the antiserum and the counterparts (or homologues) of this protein in other species. In serological terminology the immunizing protein is called the homologous antigen and the counterparts of this protein in other species are called the heterologous antigens. Since antigenic specificity corresponds closely to genetic specificity, it turns out that the cross reaction of the heterologous antigen decreases as the evolutionary separation of the heterologous species from the homologous species increases.

In the studies reviewed here a variety of antisera were used for each particular protein or antigen. This ensured that the antigenic and species specific properties of many of the protein's surface configurations would affect the measurements. For example, if a rhesus monkey is immunized with a human protein, only a small proportion of the surface configurations of the protein will be antigenic, and these will tend to be the more species specific configurations. On the other hand, if a chicken is immunized, a large proportion of the surface configurations of the protein may be antigenic, and some of these (the configurations which have been stable over long periods of evolution) will show a low degree of species specificity. Chicken, monkey, and rabbit antisera to purified human serum proteins were used to develop cross reactions with heterologous antigens from a number of different Primates (Goodman, 1962a). These cross reactions consistently demonstrated in accord with the precipitin studies of other investigators (Nuttall, 1904; Mollison, references in Kramp, 1956; Wolfe, 1933, 1939; Boyden, 1958; Paluska and Kořinek, 1960; Williams and Wemyss, 1961) the close

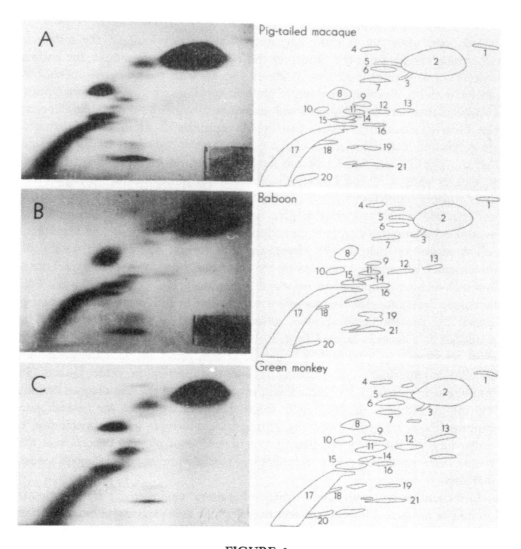

FIGURE 3

Photographs and matching outline diagrams of two-dimensional starch-gel electrophoresis patterns of Cercopithecine sera. A pig-tailed macaque (*Macaca nemestrina*), a baboon (*Papio*) and a green monkey (*Cercopithecus aethiops*) are compared. Note that a constellation of 21 components can be identified in each pattern.

phylogenetic relationship of man to anthropoid apes, his decreasing relationship first to cercopithecoids, then to ceboids, and his distant relationship to prosimians. Chimpanzee and gibbon were the first anthropoid apes to be tested (Goodman, 1962a), and chimpanzee developed the larger cross reactions.

During the past year Drs. Arthur J. Riopelle and Harold P. Klinger generously contributed for the study serum and plasma samples from several gorillas and orangutans. Many more precipitin tests have now been carried out. A large body of data (Goodman, 1962b, in press) demonstrate that man is more closely related to the African apes (chimpanzee and gorilla) than to the Asiatic apes (orangutan and gibbon). For example, hominoid albumins were compared with rhesus monkey antisera to human albumin. Gorilla albumin was quite similar to human albumin; chimpanzee albumin diverged slightly from the human protein, and gibbon and orangutan albumins clearly diverged from human albumin as well as from each other. The orangutan albumin appears to be the most divergent of the hominoid albumins. The rhesus monkey antisera also reacted with another protein (Goodman 1962a, b,) in addition to albumin. This additional reaction was afforded by the serum of man, chimpanzee, and gorilla, but not by that of gibbon and orangutan. With a capuchin monkey antiserum to human serum, orangutan and gibbon showed moderate divergencies from man, whereas gorilla and chimpanzee showed only slight divergencies from man, that of the gorilla being the smallest.

With chicken antisera to human ceruloplasmin (the copper binding protein of plasma) and rabbit and chicken antisera to human transferrin (the iron binding protein of plasma), chimpanzee and gorilla appeared identical with man, whereas orangutan and gibbon diverged from man. The gibbon transferrin appears to be the most divergent of the hominoid transferrins. With a rabbit antiserum to human alpha$_2$ macroglobulin, only the gorilla appeared identical with man; chimpanzee, gibbon, and orangutan showed small divergencies. On the other hand, with rabbit and chicken antisera to human gamma globulin, only the chimpanzee was very similar to man; gorilla, orangutan, and gibbon, respectively, were increasingly divergent. These data, therefore, not only show that the Asiatic apes are more distant from man than are the African apes, but also that all the anthropoid apes have diverged from each other.

Evidence that the chimpanzee has more recent common ancestry with man and gorilla than with orangutan or gibbon is furnished by the cross reactions of antisera to chimpanzee serum. The antisera were produced in chickens, a spider monkey, and a woolly monkey. With these antisera orangutan and gibbon showed greater divergencies from chimpanzee than did gorilla or man. Examples of the cross reactions developed by the woolly monkey antiserum are presented in Figure 4. Chicken antisera have also been produced to gorilla serum. The cross reactions of these antisera do not show any divergence of chimpanzee or man from gorilla but do show a divergence of orangutan and gibbon, the gibbon in this case showing the most divergence. It may be mentioned that Zuckerkandl et al. (1960) found that the primary structure of adult hemoglobin is very similar in man, chimpanzee, and gorilla, but somewhat divergent in orangutan. This finding provides additional evidence for the close phyletic relationship of man, chimpanzee, and gorilla within the Hominoidea.

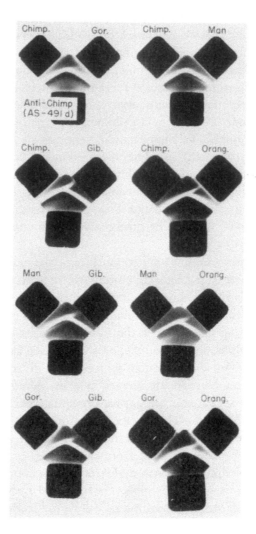

FIGURE 4

Photographs of cross reactions in type IV immunodiffusion plates developed by a
woolly monkey antiserum (AS-491d) to chimpanzee serum. The antiserum was added
in 0.1 ml amounts and the antigens (undilute sera) were added in either 0.2 ml or 0.1 ml
amounts. Chimpanzee developed a weak spur against gorilla, a very weak spur
against man, and strong spurs against gibbon and orangutan. Man and gorilla each
developed moderate spurs against gibbon and orangutan. A protein develops a spur
against its counterpart (or homologue) in another serum if it contains antigenic
configurations to which antibodies are directed that its counterpart lacks. Thus the
human and gorilla homologues of the principle protein of chimpanzee serum to
which the antibodies were directed developed larger cross reactions than the gibbon
and orangutan counterparts of this protein. (For an explanation of cross reactions in
immunodiffusion plates see Goodman, 1962a.)

All these serological findings argue for a revision of the taxonomy of the Hominoidea since in the classifications now in use *Gorilla* and *Pan* are invariably placed with *Pongo* rather than with *Homo*. A broadening of the Hominidae to include *Gorilla* and *Pan* as well as *Homo* would reflect more closely the cladistic and genetic relationships suggested by the serological data. *Hylobates* would be in the Hylobatidae and *Pongo* would remain in the Pongidae.

Placing Hylobatidae, Pongidae, and Hominidae together in the Hominoidea is supported by the cross reactions of a chicken antiserum to gibbon serum and another to orangutan serum. With these antisera, man and each of the apes yielded larger cross reactions than cercopithecoids (stump tail macaque—a member of the Cercopithecinae, and langur—a member of the Colobinae, were tested). Using chicken antisera to macaque serum, cross reactions were stronger with the langur than with the various apes and man. This supports the established primate taxonomies which separate the cercopithecines from hominoids and group them with colobines in the Cercopithecoidea. With anti-macaque sera (see also Paluska and Korinek, 1960) ceboids do not show as much relationship to cercopithecoids as do hominoids. However, cercopithecoids share some antigenic configurations with ceboids that hominoids lack. Similarly, hominoids share configurations with ceboids that cercopithecoids lack (Goodman, 1962a, in press). This type of data suggests that shortly after the dual branches of Platyrrhini and Catarrhini had formed within the Anthropoidea the catarrhine Primates branched into the Hominoidea and Cercopithecoidea. Finally, the cross reactions of various chicken anti-ceboid sera (Goodman 1962b, in press) show that the suborder Anthropoidea is a valid phylogenetic taxon. These cross reactions demonstrate that the ceboids are more closely related to the catarrhine Primates (cercopithecoids and hominoids) than to the lorisiform or lemuriform lower Primates. Thus a systematics of the higher Primates derived solely from serological data would not be much different from the established systematics which was largely derived from morphological data. A correspondence of the serological and morphological approaches is to be expected since the building blocks of morphological structures are proteins. Even the revision of hominoid taxonomy suggested by the serological data, that of grouping *Pan* and *Gorilla* with *Homo* rather than with *Pongo*, is perhaps anticipated by morphological data. Indeed, Darwin (1871) only had the evidence of morphology when he suggested that our closest relatives were the gorilla and chimpanzee and that "it is somewhat more probable that our early progenitors lived on the African continent than elsewhere."

The close phyletic relationship of gorilla and chimpanzee to man that the serological data demonstate agrees with the concept developed in a preceding section of this chapter, that the basal members of the Hominidae were the progressive hominoids who entered and successfully occupied a widening ecological domain, one which was terrestrial as well as arboreal. The gorilla and chimpanzee, who continue to occupy a terrestrial-arboreal domain, could readily have descended along with man from such a basal hominid. In turn the gibbon and orangutan, who are almost solely arboreal in their mode of life, probably trace back as separate branches (as may be deduced from the observed difference in their serum proteins) to earlier and more primitive members of the Hominoidea.

The fact that the gibbons can be traced as a separate branch of the Hominoidea back into the Oligocene, 30–40 million years ago, highlights the significance of the finding (Goodman, 1962a, b, in press) of an extensive correspondence in antigenic structure among hominoid albumins. Using chicken and rabbit anti-human albumin sera, human, gorilla, chimpanzee, and gibbon albumins appeared to be identical in the immunodiffusion plates and orangutan albumin was only slightly divergent from the others. Thus the antigenic structure of albumin has been remarkably stable during the long course of hominoid evolution. Furthermore the chicken and rabbit anti-human albumin sera developed sizeable cross reactions with cercopithecoid and ceboid albumins. (Among the Primates only lemurid, lorisid, and tupaiid albumins developed small cross reactions.) In contrast to the albumin results, the cross reactions of rabbit and chicken anti-human gamma globulin sera (Goodman 1962a, b, in press) revealed distinct antigenic differences among hominoid gamma globulins and very little correspondence between the catarrhine and platyrrhine Primates, with respect to gamma globulin. We can conclude that during the development of the suborder Anthropoidea albumin evolution has not been nearly as marked as gamma globulin evolution. The data also demonstrate that within this suborder the hominoids show the least albumin evolution and the ceboids the most (Goodman, 1962a, b, 1963). Although anti-human albumin sera detected almost no diversification of albumin within the Hominoidea, the anti-ceboid sera detected considerable diversification of albumin within the Cebidae.

These findings can be explained by the immunological theory of primate evolution developed in this chapter. The theory proposes that during the progressive evolution of the Primates, as the placental structure came to permit more interchange between fetal and maternal circulations, maternal isoimmunizations decreased the heterozygosity of prenatally acting genes and thereby increasing the evolutionary stability of proteins such as albumin which are synthesized early in fetal life. These isoimmunizations also increased the selective advantage of genetic codes which delayed maturation in ontogeny. Thus proteins which are not synthesized until after birth such as gamma globulin could evolve at a relatively rapid rate, since divergence in such proteins would not be selected against by maternal immunizations. The evolutionary stability of albumin would be greater in the Hominoidea than in the Ceboidea, since fetal-maternal vascular intimacy (and thus opportunity for selection against mutant genes by maternal immunizations) is greater in the hominoids than in the ceboids.

We can not prove that the greater diversification of albumin in the Ceboidea as compared to the Hominoidea resulted from a more rapid evolution of ceboid albumins, since the times of phyletic branching within the Ceboidea are unknown.

Possibly many of the ceboid branches are more ancient than the hominoid branches. However we can test our theory by examining other taxa. For example, the fossil record suggests that the family Bovidae first emerged during the Miocene (Simpson, 1949) after extensive branching of the Hominoidea had already occurred. Yet by the immunodiffusion plate technique chicken anti-beef albumin sera could detect divergencies even within the subfamily Bovinae, kuda albumin diverging from beef albumin (Goodman, 1962b), whereas the anti-human albumin sera failed to show any comparable divergencies among hominoid albumins. Clearly, then, albumin evolution has been more rapid in the subfamily Bovinae than in the superfamily Hominoidea. This type of finding could be predicted by our theory, for the artiodactyls have an epitheliochorial placenta which, as previously noted, minimizes the possibility of transplacental immunizations.

The serological reactions of lower Primates and insectivores also provide data upon which a theory of human evolution can be built. The data pertain to the antigenic (and thus genetic) correspondence of these organisms to man as judged by the cross reactions of antisera to purified human serum proteins. It is best to consider these data after briefly reviewing some unpublished results obtained with chicken antisera produced to the serum of lemur, potto, hedgehog, and tree shrew. It was found with anti-lemur and anti-potto sera that marked antigenic differences separate the lorisids (galago, loris, and potto) from the lemurids (lemur). However each group showed more antigenic correspondence to the other than to either the tupaiids or to members of the Anthropoidea. Thus the lorisiform prosimians and the Malagasy prosimians appear not to have separated from each other until sometime after their common ancestor had separated from the forerunners of the Anthropoidea. Nor could any special phylogenetic relationship between the Tupaioidea and Lemuroidea be discerned. With anti-hedgehog and anti-tree shrew sera, hedgehog and tree shrew appeared to show more correspondence to each other than to any of the other mammals whose sera were tested (tenrec, rat, rabbit, dog, horse, beef, elephant, man, lemur, galago, and kangaroo). But even this apparent relationship between tree shrew and hedgehog must be a very distant one, since in each case the heterologous cross reaction was many times less than the homologous. Possibly the separation of the Tupaioidea from other mammalian groups dates back to the first great radiations of the primitive eutherian mammals in the late Cretaceous epoch.

We can now consider the cross reactions of the lower Primates and non-primate mammals with the antisera to human serum proteins. When the antisera were to gamma globulin and to alpha$_2$ macroglobulin, the Strepsirhini (lemur and galago) developed larger cross reactions than the tree shrew and the non-primate mammals. However when the antisera were to albumin (Goodman, 1962a, b, in press) the tree shrew cross reactions were comparable (perhaps slightly larger) than the lorisid (galago and potto) cross reactions and clearly larger than the cross reactions of lemuroids (*Lemur* and *Propithecus*) and the other mammals tested (hedgehog, tenrec, rat, rabbit, elephant,[6] dog, horse, beef, and kangaroo). These data in confirming morphological evidence that the tree shrew has strong affinities with the Primates again suggest that maternal isoimmunizations played a role in the molecular evolution of the Primates. Such immunization could account for human albumin corresponding more to tree shrew albumin than to lemur albumin even though man appears to have a somewhat longer period of common ancestry with lemur than with tree shrew. Albumin evolution would not have been slowed in the Lemuroidea as much as in the Tupaioidea by maternal isoimmunizations, since the lemurs retained during their phylogenesis an epitheliochorial placenta while the tree shrews acquired a fairly advanced hemochorial placenta (Le Gros Clark, 1959). The serological correspondence between man and the tree shrew would seem to be due to the retention in each of very ancient or primitive genetic patterns. However the full significance of the relationship between man and tree shrew will not emerge until serological comparisons are also carried out with such groups as elephant shrews and common shrews.

SUMMARY AND CONCLUSIONS

1. Proteins play a central role in the evolutionary process. Each change in the amino acid sequence of a protein can be related to a genic mutation. In turn a change in amino acid sequence may affect the biological functioning of a protein and thereby influence the survival chances of the organism carrying the mutant gene. The biological functioning of a protein depends on the stereochemical properties of its surface configurations. Only a few surface configurations were responsible for the biological functioning of a typical soluble protein in an undifferentiated unicellular organism at primitive stages of evolution.

In contrast, many surface configurations affected the biological functioning of a typical soluble protein in a higher metazoan at advanced stages of evolution. As in the primitive organism there were the surface configurations which constituted the active site responsible for the primary functional activity of the protein. In addition there were the ancillary configurations which enabled the protein to preferentially cross particular membrane barriers, and there were also the remaining surface configurations, so contoured as to not specifically interact with any of the surface configurations of the large number of other serologically distinct proteins. As progressive evolution increased the biochemical complexity and molecular differentiation of organisms, the possibility also increased that any

6. It is also of interest that of the non-primate mammals tested with chicken anti-human albumin sera the elephant developed the largest cross reactions (Goodman, 1962a, b). This suggests that a marked deceleration of albumin evolution occurred in elephant phylogeny as in human phylogeny. In view of the elephant's exceptionally long period of gestation and the hemochorial arrangement to the zonary portion of his placenta (Mossman, 1937), maternal immune attacks on the fetus could have operated in elephant phylogeny (as in human phylogeny) to stabilize sections of the genetic code.

single mutant gene would be detrimental to the organism carrying it. This line of reasoning suggests that the process of speciation caused a much deeper divergence of genetic codes and a much more rapid evolution of the structural specificity of proteins among lower organisms at the primitive stages of phylogenesis than among higher organisms at advanced stages of phylogenesis.

2. At the phylogenetic stage of the placental mammals a new factor, maternal immunizations to fetal antigens, further restricted the divergence of genetic codes during the process of speciation. Proteins synthesized early in fetal life such as albumin showed a decreasing rate of evolution. But proteins not synthesized until after birth such as gamma globulin still evolved at a relatively rapid rate, since in this case divergence was not selected against by maternal immunizations. It is proposed that such an immunological process operated with particular force in the phyletic line leading to man. It was in this line that cerebral evolution progressed most rapidly. But the progressive expansion of the neocortex could only occur on the basis of a placental evolution which increased fetal-maternal vascular intimacy, allowing the developing fetus to obtain a rich supply of nutrients and oxygen.

As cerebral evolution progressed and body size increased, the ancestors of man invaded broader ecological domains. The new and more varied conditions of the external environment selected for a state of genetic heterozygosity. As the barrier between embryonic tissues and maternal blood decreased and as the period of gestation lengthened, maternal immunizations selected for a state of genetic homozygosity. This contradiction was resolved by a selection of genetic codes which increasingly shifted to the postnatal phase of ontogeny the maturation of somatic systems concerned with adaptive responses to the external environment, since maternal immunizations would not select against gene-determined polymorphisms in such late maturing systems. But the delay in the appearance of adult proteins increased the helplessness of the young and thereby spurred the evolution of family and social life.

3. A comparative study of the serum proteins of the Primates provides evidence for the theory of man's evolution set forth in this paper. The data show that there was a marked deceleration of albumin evolution, but not of gamma globulin evolution, during the phylogenetic advances of the Primates. This deceleration can be correlated with a placental evolution that increased the opportunity for maternal immune attacks on the fetus. Man's distant but nevertheless striking serological relationship to the tree shrew is further evidence that sections of the human genotype are extremely ancient, dating back to primitive members of the Eutheria.

The serological data provide information on the taxonomic relationships of man and other Primates. The dendrogram presented in Figure 5 summarizes the author's judgment of these relationships. A full presentation of the data concerning hominoid relationships appears in *Human Biology*, 35:377-436.

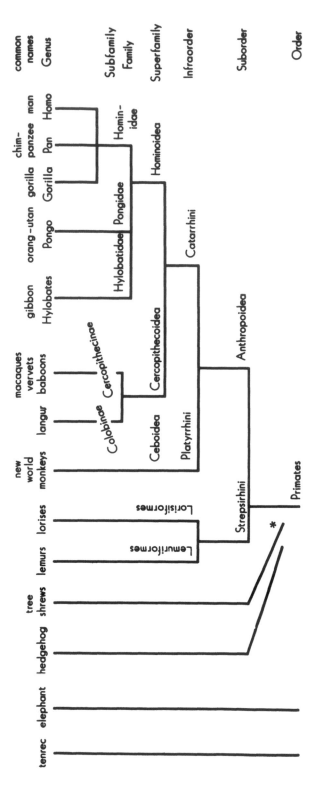

FIGURE 5

A classification from serological reactions. The line leading to elephant is interposed between hedgehog and tenrec to demonstrate that even though hedgehog and tenrec are grouped in the Insectivora by the established taxonomies of mammals no special serological relationship exists between them.

* Although serologically the tree shrews show some affinities with the Primates, they also show some with hedgehogs. Additional serological data may help identify an assemblage in which the tree shrew can be placed.

BIBLIOGRAPHY

ANFINSEN, C. B.
1959. *The molecular basis of evolution.* New York: John Wiley & Sons.

ANTONIADES, H. N.
1960. "Circulating hormones." In *The Plasma Proteins*, Vol. II. F. W. Putnam (ed.). New York: Academic Press.

BOYDEN, A. A.
1958. "Comparative serology: aims, methods, and results." In *Serological and Biochemical Comparisons to Proteins.* W. Cole (ed.). New Brunswick: Rutgers University Press.

CHANTRENNE, H.
1961. *The biosynthesis of proteins.* Oxford: Pergamon Press.

COMFORT, A.
1956. *The biology of senescence.* New York: Rinehart & Co.

CRICK, F. H. C., L. BARNETT, S. BRENNER, and R. J. WATTS-TOBIN.
1961. "General nature of the genetic code for proteins." *Nature,* 192:1227–1232.

DARWIN, C.
1871. *The descent of man and selection in relation to sex.* 2nd Ed. New York: D. Appleton & Co. (republished in 1915).

DRAY, S., and G. O. YOUNG
1960. "Genetic control of two gamma globulin isoantigenic sites in domestic rabbits." *Science,* 131:738–739.

GOODMAN, M.
1960a. "On the emergence of intraspecific differences in the protein antigens of human beings." *Amer. Nat.,* 94:153–166.
1960b. "The species specificity of proteins as observed in the Wilson comparative analyses plates." *Amer. Nat.,* 94:184–186.
1961. "The role of immunochemical differences in the phyletic development of human behavior." *Human Biol.,* 331:131–162.
1962a. "Evolution of the immunologic species specificity of human serum proteins." *Human Biol.,* 34:104–150.
1962b. "Immunochemistry of the Primates and primate evolution." *Ann. N.Y. Acad. Sci.,* 102:219–234.
In press. "Problems of primate systematics attacked by the serological study of proteins." In *Taxonomic Biochemistry, Physiology, and Serology.* C. Leone (ed.). New York: Ronald Press.

GOODMAN, M., and E. POULIK
1961. "Effects of speciation in serum proteins in the genus *Macaca* with special reference to the polymorphic state of transferrin." *Nature,* 190:171–172.
1961. "Serum transferrins in the genus *Macaca*: species distribution of nineteen phenotypes." *Nature,* 191:1407–1408.

GOODMAN, M., E. POULIK, and M. D. POULIK
1960. Variations in the serum specificities of higher Primates detected by two-dimensional starch-gel electrophoresis. *Nature*, 188:78–79.

GOODMAN, M., M. ROSENBLATT, J. S. GOTTLIEB, J. MILLER and C. H. CHEN
1963. "Effect of sex, age, and schizophrenia on production of thyroid autoantibodies." *A.M.A. Arch. Gen. Psychiat.*, 8:518–526.

GOUDIE, R. B., J. R. ANDERSON, and K. G. GRAY
1959. "Complement-fixing anti-thyroid antibodies in hospital patients with asymptomatic thyroid lesions." *J. Path. Bact.*, 77:389–400.

HAMILTON, W. J., J. D. BOYD, and H. W. MOSSMAN.
1952. "Human embryology." 2nd Ed. Baltimore: Williams and Wilkins.

HAŠEK, M., A. LENGEROVÁ and T. HRABA
1961. "Transplantation immunity and tolerance." In *Advances in Immunology I.* W. H. Taliaferro and J. H. Humphrey (eds.). New York: Academic Press.

HEMMINGS, W. A., and F. W. R. BRAMBELL
1961. "Protein transfer across foetal membranes." *Brit. Med. Bull.* 17:96–101.

HILL, J. P.
1932. "The developmental history of the Primates." *Phil. Trans. Roy. Soc. Series B.* 221:45–178.

HIMWICH, H. E.
1951. *Brain metabolism and cerebral disorders.* Baltimore: Williams and Wilkins.

INGRAM, V. M.
1961. "Gene evolution and haemoglobins." *Nature*, 189:704–708.

KENDREW, J. C., R E. DICKERSON, B. E. STRANDBERG, R. G. HART, D. R. DAVIES, D. C. PHILLIPS and V. C. SHORE
1960. "Structure of myoglobin, a three dimensional fourier synthesis at 2A° resolution obtained by X-ray analysis." *Nature*, 185:422–427.

KRAMP, V. P.
1956. "Serologische stammbaumforschung." In *Primatologia. I. Systematik Phylogenie Ontogonie.* H. Hofer, A. H. Schultz and D. Stark (eds.). Basel. S. Karger.

LE GROS CLARK, W. E.
1959. *The antecedents of man.* Edinburgh: Edinburgh University Press.

LI, C. H.
1957. "Properties of and structural investigations on growth hormones isolated from bovine, monkey and human pituitary glands." *Fed. Proc.* 16:775–783.

LIACOPOULOS, P., B. N. HALPERN, and F. PERRAMANT
1962. "Unresponsiveness to unrelated antigens induced by paralysing doses of bovine serum albumin." *Nature*, 195:1112–1113.

MOSSMAN, H. W.
1937. "Comparative morphogenesis of the fetal membranes and accessory uterine structures." *Contr. Embryol. Carnegie Inst.*, 26:129–246.

NUTTALL, G. H. F.
1904. *Blood Immunity and Blood Relationship.* Cambridge: Cambridge University Press.

OPARIN, A.
1957. *Origin of life on the earth,* 3rd Ed. New York: Academic Press.

Oudin, J.
1960. "L'allotype de certains antigéns protéid du sérum. Relations immunochimiques et génétiques entre six principaux allotypes observes dans le sérum de lapin." *Compt. rend. Acad. Sci. Paris*, 250:770–772.

Paluska, E., and J. Kořinek
1960. "Studium der antigenen eiweibverwandtschaft zwischen menschen und einigen primaten mit hilfe neurer immunobiologischer methoden." *Z. Immun. Forschung und Exper. Therapie*, 119:244–257.

Pauling, L. and R. B. Corey
1951. "Configuration of polypeptide chains." *Nature*, 168:550–551.

Perutz, M. F., M. G. Mossman, A. F. Cullis, H. Muirhead, G. Will and A. C. T. North
1960. "Structure of hemoglobin. A three dimensional fourier synthesis at 5.5. A° resolution obtained by X-ray analysis." *Nature*, 185:416–422.

Platt, J. R.
1962. "A 'book model' of genetic information-transfer in cells and tissues." In *Horizons in Biochemistry*. M. Kasha and B. Pullman (eds.). New York: Academic Press.

Porter, R. R.
1953. "The relation of chemical structure to the biological activity of the proteins." In *The Proteins* Vol. I, B. H. Neurath and K. Bailey (eds.). New York: Academic Press.

Poulik, M. D. and O. Smithies
1958. "Comparison and combination of the starch-gel and filter-paper electrophoretic methods applied to human sera: two dimensional electrophoresis." *Biochem. J.* 68:636–643.

Rensch, B.
1959. *Evolution above the species level*. London: Methuen.

Roitt, I. M., P. N. Campbell and D. Doniach
1958. "The nature of the thyroid auto-antibodies present in patients with Hashimoto's thyroiditis (lymphoadenoid goitre)." *Biochem. J.*, 69:248–256.

Roitt, I. M. and D. Doniach
1960. "Thyroid auto-immunity." *Brit. Med. Bull.* 16:152–158.

Simpson, G. G.
1949. *The meaning of evolution*. New Haven: Yale University Press.

Smith, E. L.
1957. "Active site and structure of crystalline papain." *Fed. Proc.* 16:801–804.

Smith, R. T.
1961. "Immunological tolerance of non-living antigens." In *Advances in Immunology I*. W. H. Taliaferro and J. H. Humphrey (eds.). New York: Academic Press.

Sneath, P. H. A.
In press. "Comparative biochemical genetics in bacterial taxonomy." In *Taxonomic Biochemistry, Physiology, and Serology*. C. Leone (ed.). New York: Ronald Press.

Tristram, G. R.
1953. "The amino acid composition of proteins." In *The Proteins*, Vol. I, A. H. Neurath and K. Bailey (eds.). New York: Academic Press.

Williams, C. H. and C. T. Wemyss, Jr.
1961. "Experimental and evolutionary significance of similarities among serum protein antigens of man and the lower Primates." *Ann. N.Y. Acad. Sci.*, 94:77–92.

WINZLER, R. H.

1960. "Glycoproteins." In *The Plasma Proteins*, Vol. I F. W. Putnam (ed.). New York: Academic Press.

WISLOCKI, G. B.

1929. "On the placentation of the Primates, with a consideration of the phylogeny of the placenta." *Contr. Embryol. Carnegie Inst.*, 20:51–80.

WOLFE, H. R.

1933. "Factors which may modify precipitin tests in their applications to serology and medicine." *Physiol. Zoo.*, 6:55–90.

1939. "Standardization of the precipitin technique and its application to studies of relationships in mammals, birds, and reptiles." *Biol. Bull.*, 76:108–120.

ZUCKERKANDL, E., R. T. JONES and L. PAULING

1960. "A comparison of animal hemoglobins by tryptic peptide pattern analysis." *Proc. Nat. Acad. Sci.*, 46:1349–1360.

THE CHROMOSOMES OF THE HOMINOIDEA

HAROLD P. KLINGER, JOHN L. HAMERTON, DAVID MUTTON
AND ERNST M. LANG

THE ROLE of chromosome cytology in primate evolution is reviewed and discussed thoroughly by Chu and Bender (1961 and in press).

In our study the data on primates of the above mentioned authors and others (Young et al., 1960; Chiarelli, 1961; Hamerton et al., 1961) have been supplemented with chromosome studies of some species of the Hominoidea which have never been studied or for which the previous data are incomplete. We repeated some species because it was thought important that the material all be processed in the same laboratory under similar technical conditions. This should increase the reliability of comparative studies.

We have started at the top of the evolutionary scale and plan to work backwards. This somewhat illogical order was dictated by the fact that it was at the top of the scale, primarily among the living Hominoidea, where the available data are most scanty.[1]

The animals studied are listed in Table 1:

TABLE 1

Species	Sex	Number of Animals
Pan troglodytes troglodytes	male	1
	female	2
Pan troglodytes paniscus	male	1
	female	2
Gorilla gorilla gorilla	male	1
	female	1
Pongo pygmaeus	male	1
	female	1
Hylobates lar	male	1
	female	1

1. Since this paper was presented Chiarelli (1962) has published data on some of the great apes which are to be discussed.

235

Chromosome studies in all cases were made from short term cultures derived from blood and skin samples. Technical details will be published elsewhere. (Also see appendix to this article, page 239).

The results of karyotype analyses are summarized in Table 2. The chromosomes have been divided according to their morphological characteristics, i.e., size and centromere position, into four groups. These groups are somewhat arbitrary but an attempt was made to be consistent within groups. In the case of some chromosomes difficulty is encountered in placing them in a particular group. However, the error which may have been incurred would not alter the findings and in particular the conclusions significantly.

It is important to point out that the arrangement of the chromosomes in groups as shown in Figure 1 and Table 2 implies a great deal more than we really know. Particular chromosomes from two different animals may look exactly alike but we do not know if they are homologous, i.e., carry the same genetic information. Until such homology can be determined our studies will remain partly on a conjectural basis.

OBSERVATIONS

Since most is known about human chromosomes we will compare those of the other Hominoidea with them.

From Table 2 and Figure 1 we see that all the apes examined except for *Hylobates lar*, have 48 somatic chromosomes as compared to 46 for man.

The chromosome set of *Pan troglodytes troglodytes* most closely resembles that of man. Its prime difference rests in one extra pair of acrocentric chromosomes. Several of the other chromosome pairs differ somewhat from those of man in the same size sequence. This also applies to the other apes but these differences will be neglected in this report and, for the sake of simplicity, only major divergences will be pointed out.

All the chromosomes of *Pan troglodytes paniscus* seem to be identical with those of his larger relative except for the 22nd pair which is acrocentric in the latter but metacentric in *paniscus*.

Next in order we would place the gorilla who differs from man primarily with four fewer small metacentries and six more large satellited acrocentrics. The Y is metacentric in the gorilla as compared to the acrocentric Y of man and the chimpanzees.

The orangutan has two large and four small metacentrics less, ten more large and two less small acrocentrics than man. The centromere location of the Y is probably subterminal in the orangutan (indicated as [M] in Table 2 since in this table, for the sake of simplicity, we have called all chromosomes metacentric which are not clearly acrocentric).

The gibbon has 44 somatic chromosomes of which in the female all are metacentric. The Y is minute in the male gibbon and probably acrocentric. It is inter-

FIGURE 1

Representative karyotypes of *Homo* (Ho), *Pan troglodytes troglodytes* (Pa.t.), *Pan troglodytes paniscus* (Pan.p.), *Gorilla gorilla gorilla* (Go), *Pongo pygmaeus* (Po) and *Hylobates lar* (Hy.1.). The Y and X sex chromosomes are in the second line of each karyotype at the extreme right.

esting that in all the species studied the long arms of the X are about twice as long as the short arms and in all the X is about the 5th to 7th chromosome in size (Fig. 1). The nuclei in the cells of oral mucosal smears and fibroblast-like cells grown in culture were sex chromatin positive in all the females of the species studied and negative in all the males. The ape nuclei more often had other chromocenters than human nuclei do. However, these could not be confused with sex chromatin because they are generally smaller, irregular in shape and rarely situated against the nuclear membrane as is more common for the sex chromatin of man.

TABLE 2

Species	2n †	Autosomes				Sex chromosomes	
		Metacentrics		Acrocentrics			
		large	small	large	small	X	Y
Homo sapiens	46	24	10	6	4	M	A
Pan troglodytes troglodytes	48	24	10	8	4	M	A
Pan troglodytes paniscus	48	24	12	8	2	M	A
Gorilla gorilla gorilla	48	24	6	12	4	M	M
Pongo pygmaeus	48	22	6	16	2	M	[M]
Hylobates lar	44	[24] *	[18] *			M	A

† 2n = Diploid number of chromosomes.
M = metacentric   A = acrocentric   [] = tentative.
* The separation into small and large metacentrics in this species is most difficult (Fig. 1) nd therefore arbitrary until further data become available.

DISCUSSION

As pointed out previously we do not know if chromosomes with morphological identity are really homologous. However, the striking resemblance of the chromosomes of man, chimpanzee and gorilla strongly suggests a close evolutionary relationship. The orangutan seems to be less closely related to man. The gibbon is quite different. Although a valid basis of comparison is not yet available one gains the impression that the gibbon differs as much from the orangutan as he does from man and the other great apes.

It is evident that more detailed studies will be required to obtain data of full comparative value. Measurements of relative chromosome arm lengths may provide some further information on the number of morphologically similar chromosomes of these species. Determinations of the deoxyribonucleic acid content of interphase nuclei will indicate if all these species have the same amount of this nucleic acid; the carrier of most of the genetic information. Such studies are now in·progress and should help decide if the observed differences in the chromosome morphology of the great apes is due to a loss or gain of chromosome material or simply to rearrangements of chromosome segments.

CONCLUSIONS

Chromosome morphology indicates that the African great apes are more closely related to man than the Asiatic apes. The chimpanzee seems to be closest and then the gorilla. The karyotype of the orangutang suggests a more distant relationship to man and perhaps to the other apes. The karyotype of the gibbon is completely different and resembles more that of the specialized Cercopithecoid monkeys of the genera *Macaca, Papio* and *Cercocebus*. Perhaps in view of this data, a taxonomic revision of the group, particularly in relation to the position of the gibbons, is called for.

These interpretations are in good agreement with those of Goodman (1962 and this volume) which are based on data of a completely different method.

ACKNOWLEDGEMENT

We thank Professor B. Grzimek, Director of the Zoological Gardens in Frankfurt and his coworkers for permitting and helping us to obtain material from their pygmy chimpanzees.

APPENDIX

It has been suggested that we add in some detail the method of collecting skin and blood from animals for chromosome studies. We are glad of this opportunity and hope this appendix will serve a double purpose: first, to illustrate to directors of zoological gardens and others the harmless nature of the procedure (perhaps they will then more readily allow us to take material from their charges); second, we hope that workers in far removed areas will be willing to collect material for us. The following instructions should then prove of help. We provide a kit containing everything needed. The instruction designations therefore refer to articles in this kit.

PROCEDURE

For animals which cannot be held easily and safely during the following procedures we use a tranquilizing drug called Sernyl (Parke & Davis Co., Detroit). Extensive experience by others and ourselves on many different primates has convinced us of the complete safety of the drug. An accidental several-fold overdose has no ill effects. The animals just sleep a bit longer. (The drug should be used with caution in animals other than primates). It can be given orally, subcutaneously, intramuscularly or intravenously. When given orally it can be mixed with any liquid the animal will drink such as milk, fruit-juices, Coca-Cola, etc.

Sernyl Dosages:

Oral: 4–5mg. per kilogram of body weight.

Injected: 1 mg. per kilogram. If sedation is inadequate 0.5 mg./kg. may be supplemented.

When given orally the first effects are noticed after 20–45 minutes. When injected after 5–15 minutes. Small animals weighing less than 1000 grams require about 50 per cent or more of the drug on the weight basis. Sernyl and sterile disposable 2 ml. plastic syringes are enclosed in the kit for removing the drug from the ampule and for transfer to a liquid or for direct injection.

When possible we prefer to have a skin and a blood sample. The former is more important and is usually easier to obtain.

Skin Biopsy

Since the skin fragment will be grown in a culture medium capable of supporting the growth of bacteria and molds it is important that it be removed under aseptic conditions. For this reason please use only the sterile equipment provided in our kit and follow these directions as closely as possible.

1. On an extremity of the animal cut off the hair with scissors from an area about 8 cm. long and 5 cm. wide.
2. Soak with "Soap Solution" using sterile gauze swab from container marked accordingly.
3. Shave the cleared area with the razor provided.

FIGURE 2
Method for taking skin biopsy.

4. Wipe again with another sterile swab and soap solution wipe center of shaven area first then periphery. Use this procedure in following steps as well.
5. Wipe with sterile swab soaked with "ether-alcohol."
6. Repeat step 5 twice more, each time with new sterile swab. Then wipe only center of area with dry sterile swab.
7. With sterile forceps from glass tube with red cap (remove from tube without

touching front of forceps) grasp the skin so as to produce a skin fold about 1 cm. long and protruding about 2 mm. above the forcep blades as shown in Figure 2.

⁻8. With freshly unpacked sterile razor blade, grasped only at the side as shown in Figure 2, cut off protruding skin quickly using a slight sawing motion and forceps blade margins as guiding rails for razor.

9. Grasp cut off piece of skin with surgical forceps from tube with blue cap and take care not to touch it with any non-sterile objects. Place in screwcap bottle marked "For skin biopsy." Do not use forceps used for holding skin to grasp cut-off piece. When opening and closing the bottle do not touch bottle lip or inside of cap or allow these to come in contact with non-sterile objects. Do not leave bottle open longer than is absolutely necessary. If skin biopsy does not float off forceps into solution use another sterile forceps to detach.

10. Mark bottle at once with species of animal, sex, age and locality captured, using glass writer included with kit.

11. If skin has been cut through and edges separate (occurs only in small animals) then place two or three stitches with instruments and suture material provided. Dust with aluminium disinfecting powder provided and leave uncovered. Animals will generally remove sutures themselves at the right time.

12. Place bottle with biopsy in insulated container provided. If possible add 3 to 5 cubes of ice (do not use dry ice, solution should not freeze) and expedite as quickly as feasible. The skin dies off if too long in transport so please use fastest transportation means available. All shipping expenses will be reimbursed.

Venous Blood Sample

1. Prepare blood set by cutting open plastic bag containing plastic tube with needles at both ends and air vent. Pull air vent out of plastic protector and push through rubber top of bottle marked "For Blood." Remove plastic protector from short needle on tube and insert likewise in bottle. Leave plastic protector on long needle till just before insertion in vein. The caliber of the needle may seem large but experience has shown that this size works best. Most primates have sufficiently large veins to allow its introduction. The vein may be even slightly thinner than the needle. For very small primates use the 10 ml. disposable plastic syringe with finer needle. In this case insert only air-vent into bottle and inject blood from syringe into bottle by pushing needle though rubber top.

2. Select a vein on animal. In our experience one of the veins on the inside of the elbow is most suitable. Other good veins course along the lateral and dorsal surfaces of the lower leg. To find vein apply rubber hose provided in tourniquet-like fashion on proximal portion of extremity and secure hose with clamp which is likewise included. Veins should stand out, if not, massage extremity some or readjust tension of hose. Do not apply too much pressure or arterial flow will also stop.

3. Cut away hair, shave and disinfect a small area above vein (if vein is selected first and skin prepared accordingly the biopsy may be taken at the same site but preferably not directly over vein).

4. Insert needle in vein. When blood appears in tube, suck on air vent. Allow bottle to become about ⅔ full (about 8 ml. of blood). Shake bottle continuously while blood is flowing into it to mix thoroughly with anticoagulant. Blood should not clot. If flow is good, transfer air vent and needle of tube to a second bottle and fill likewise.

Remove tourniquet and then withdraw needle from vein. If during procedure blood ceases to flow massage extremity. Releasing rubber tube and reapplying it also often starts flow again. If blood clots in tube take new set, but this is very rare. A minimum of about 4 ml. of blood should be obtained (about ⅓ of bottle) but we prefer much more.[2]

5. Label bottle with species name, age, sex and locality captured and ship together with skin biopsy as fast as possible.

BIBLIOGRAPHY

CHIARELLI, B.
   1961. "Chromosomes of the orang-utan (*Pongo pygmaeus*)." *Nature*, 192:285.
CHIARELLI, B.
   1962. "Comparative morphometric analysis of primate chromosomes. I. The chromosomes of anthropoid apes and of man." *Caryologia*, 15:99–121.
CHU, E. H. Y. and M. A. BENDER
   1961. "Chromosome cytology and evolution in primates." *Science*, 133:1399–1405.
   (In Press). "Primate Chromosomes." In *Evolutionary and Genetic Biology*. J. Buettner-Janusch (ed.). New York: Academic Press.
GOODMAN, M.
   1962. "Evolution of the immunologic species specificity of human serum proteins." *Human Biology*, 34:104–150.
HAMERTON, J. L., M. FRACCARO, L. DE CARLI, F. NUZZO, H. P. KLINGER, L. HULLIGER, A. TAYLOR and E. M. LANG
   1961. "Somatic chromosomes of the gorilla." *Nature*, 192:225–228.
YOUNG, W. J., T. MERZ, M. A. FERGUSON-SMITH and A. W. JOHNSTON
   1960. "Chromosome number of the chimpanzee *Pan troglodytes*." *Science*, 131:1672.

2. After we remove the white cells for culture purposes the red cells and plasma are sent on to other workers for haemoglobin, serum protein and blood group studies. The sample is thus utilized completely and helps many complementary investigative projects.

PERSPECTIVES IN MOLECULAR ANTHROPOLOGY

EMILE ZUCKERKANDL [1]

INTRODUCTION

THE IDEA OF STUDYING EVOLUTION at the chemical level is by no means a new one. The first important synthesis in this field was published nearly twenty years ago by Marcel Florkin. In his book *Biochemical Evolution* (1944) Florkin lists many suggestive chemical differences between organisms, including quantitative changes in inorganic and qualitative changes in various organic molecules. Such changes certainly illustrate evolution. But seriations of morphological characters seem to be more informative markers of evolution and to do more for the establishment of evolutionary affinities between organisms than seriations of chemical characters involving inorganic or most types of organic molecules. Although morphological characters of living matter undoubtedly are among the most complex effects in existence in the universe and as such should be among the poorest analytical tools, they can be revealing of fundamental relationships in that they express the interplay of a great number of the most specific organic constituents.

There are, however, in living matter two types of compounds that are far more informative than any other characters, namely certain kinds of nucleic acids and their products, the proteins. These are the specific constituents just mentioned of which morphological characteristics are an overall expression. Nucleic acid molecules contain the coded information that is used, in collaboration with the internal and external environment, for making an organism what it is. Proteins are transcripts of this information, and they are produced and carry out their functions so as to confer on an organism all of its properties in response to the environment.

The polypeptide chains in proteins contain about twenty different amino-acids in specific sequences, whereas the corresponding specificity in the deoxyribonucleic acid molecules that constitute the genes on the chromosomes is obtained by a linear succession of only four different kinds of purine and pyrimidine bases. The structural gene is here defined as that segment of deoxyribonucleic acid that contains the information for one polypeptide chain. The distinction between

1. Contribution No. 2995 from the Division of Chemistry and Chemical Engineering, California Institute of Technology.

polypeptides and proteins is nowadays based on the fact that a protein is often made up of more than one polypeptide chain, and that different polypeptide chains are controlled by different genes, so that the production of one protein can be controlled by one to several structural genes.

Thanks to the knowledge that has been gained recently about the relationship between proteins and genes, the study of amino-acid sequences in proteins is now able to achieve the most precise and the least ambiguous insight into evolutionary relationships and into some of the fundamental mechanisms of evolution. Forces of evolution operate of course at many levels other than the molecular level, but nucleic acids and the corresponding proteins furnish in the last analysis the specific raw-materials for all these forces.

MAIN METHODS

The analysis of proteins is an indirect but accurate way (although the accuracy is not necessarily 100 per cent) of analyzing that part of a given genome that is composed of structural genes in a state of synthetic activity. Inasmuch as other types of genes exist, or the same type in other states, the structural analysis of proteins rather than of the genes themselves bears out an important distinction between genes and is, therefore, of theoretical interest. However the choice of protein rather than of nucleic acid molecules as the objects of chemical analysis in evolutionary studies is not based today on theoretical considerations, but on the availability of the required biochemical techniques.

In order to establish evolutionary relationships one may initially rely on assaying degrees of differences between homologous proteins whose precise structure remains unknown. The pioneer work in this field has been done by Reichert and Brown (1909). As long as this approach is satisfactory, immunological techniques, as applied most recently by Goodman to Primates (cf. Goodman's contribution to this volume), are the most powerful ones since they allow one to gather a relatively large amount of information in a short time. Concomitantly it is necessary to pursue on a much smaller number of proteins the more cumbersome elucidation of their amino-acid sequence. This is, indeed, at present the only way by which the information carried by active structural genes can be defined completely and in absolute terms.

Goodman has presented in this volume the methods pertaining to the immunological approach. I may, therefore, limit myself to mentioning briefly the techniques of "absolute" structural studies. Amino-acid sequences can be established either by the physical method of X-ray diffraction (Kendrew et al., 1961) or by the stepwise chemical degradation of a chain of amino-acids, starting at one end of the chain, and making the successive amino-acid residues available for separate analysis (Fraenkel-Conrat et al., 1955). An alternative chemical method is to break down the protein by a wide array of proteolytic enzymes that split the chains of amino-acids at different places, each enzyme being used in a separate experiment. After analyzing the amino-acid composition of the different pieces obtained with each enzyme, the original sequence can in theory be deduced from a comparison of the pieces. Each of these chemical methods has its practical limitations and complete sequence analyses have so far only been achieved by various combinations of the two. As to the analysis by X-ray diffraction, it is at present of a much more restricted applicability than the chemical methods.

A technique that yields results intermediary between the relative and the absolute type is that of peptide pattern analysis, which has been called "fingerprinting" (Ingram, 1958). It consists of digesting the protein with a highly specific proteolytic enzyme, i.e., one that splits the chain of amino-acids only next to a few strictly defined types of amino-acid residues. Trypsin is most commonly used. The mixture of the products of such a partial breakdown of the protein is spread two-dimensionally over a piece of filter-paper by a combination of electrophoresis and chromatography. Thereby a characteristic pattern is obtained for each protein and each specific enzyme. The structural information given by such a pattern remains qualitative and relative as long as the various spots that compose it are not analyzed further. It is easy, however, to push tryptic peptide analysis one step further and to gather a limited amount of absolute structural information—an amount that may in some instances be critical and illuminating (Ingram, 1959).

SOME CONSIDERATIONS ABOUT THE EVOLUTION OF PROTEINS

A. Amino-acid substitutions

How do proteins and structural genes evolve? A priori, there are several ways in which either may change. They may grow terminally; they may split up; they may lose or gain whole sequences anywhere along the chain; they may duplicate. In fact, the most common change, or at least the event most commonly retained by evolution, consists in the substitution of one single amino-acid for another in the protein, of one single base-pair for another in the structural gene. This is amply illustrated by what has been learned, since the original work of Ingram (1959), of different types of hemoglobin polypeptide chains, particularly in relation to normal and abnormal human hemoglobin chains. Hemoglobin molecules of higher Vertebrates are made up of two types of polypeptide chains, controlled (at least in man) by two non-linked structural genes, and each molecule contains two chains of each type. Thus, there is a total of four chains per molecule, of nearly equal length (slightly less than 150 linearly arranged amino-acids), nearly identical conformation in space, and a striking sequential homology in spite of numerous differences in sequence between the two types of chains (Braunitzer, 1961).

Homologous proteins in closely related species usually differ at most by a few single amino-acid substitutions, which most likely have occurred independently, whereas many such substitutions may differentiate species far removed from each other on the evolutionary scale. This was illustrated two years ago by a survey of different animal hemoglobins by tryptic peptide analysis (Zuckerkandl et al., 1960). By this method gorilla, chimpanzee, and human hemoglobins were indistinguishable (Fig. 1). Identical looking peptide patterns are compatible with the existence of a moderate number of differences in actual amino-acid sequence. But the similarity was meaningful in view of the differences that were on the other hand found between human hemoglobin and hemoglobins from other species. Whereas the pattern given by orangutan hemoglobin differed slightly from the patterns characterizing man, chimpanzee and gorilla, that of the Rhesus monkey *Macaca mulatta* differed more significantly, although the similarities predominated largely.

In Artiodactyles, one group of peptide spots was found in the same places as in humans, and another important group in new positions. Much further down the line, in different groups of fish the differences between the hemoglobins seemed to extend to every tryptic peptide, although this does not exclude that

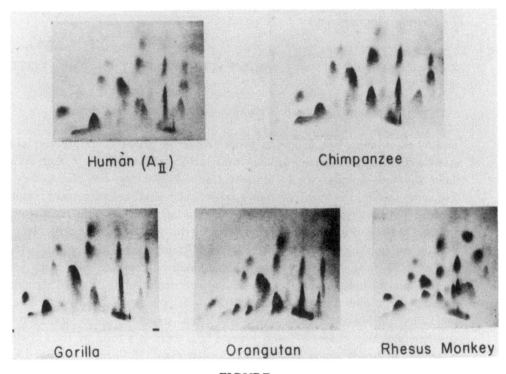

FIGURE 1
Tryptic peptide patterns obtained with adult hemoglobins from different Primates (from Zuckerkandl et al., 1960).

important sequences of amino-acids may still be identical or similar. It does indicate however that mutational changes have spread over the whole molecule. Hemoglobin studies of this type have also been carried out by other authors, notably C. Muller (1961). A survey of Primate hemoglobins has recently been made by R. Hill and J. Buettner-Janusch (personal communication).

A study of the amino-acid sequences in the hemoglobin chains from an adult gorilla is under way, in collaboration with W. A. Schroeder (Zuckerkandl and Schroeder, 1961). It has shown in one of the polypeptide chains, the so-called α-chain, one well established difference in amino-acid sequence by comparison with the human α-chain, namely the replacement of a glutamyl residue, found in man, by an aspartyl residue, probably in the twenty-third position from the amino-end of the chain. There is a second, more doubtful, difference, one seryl residue being perhaps missing in the gorilla chain. In the β-chains, the second type of chains that make up human and gorilla adult hemoglobin, there is only one apparent difference between the two species: an arginyl residue is missing in the gorilla chain and is probably replaced by a lysyl residue.

Assuming that there is only one real difference in sequence between human and gorilla α-chains, the differences found in the case of both polypeptide chains are the smallest possible. They are of the same order—namely one single amino-acid substitution—as the differences that have so far been found between humans with normal and abnormal hemoglobin chains. Therefore, from the point of view of hemoglobin structure, it appears that gorilla is just an abnormal human, or man an abnormal gorilla, and the two species form actually one continuous population. It is possible that a gorilla hemoglobin chain exists as a mutation in some humans. It would not be easily detected because its electric charge, decisive in electrophoretic analysis, would be identical with and its functional properties probably similar to that of the corresponding normal human hemoglobin chain. The reverse—the existence of a human hemoglobin chain in some gorillas—is equally possible in principle, but less probable because of the comparatively small number of gorillas in existence.

B. THE CONDITIONS UNDER WHICH A MUTATION MAY BE EVOLUTIONARILY EFFECTIVE

The fixation of mutational changes in a protein may be restricted both by physico-chemical properties of the corresponding structural gene and by natural selection. At the level of the mutational event itself there may be regions of the deoxyribonucleic acid molecule that mutate more readily than others. The existence of mutational "hot-spots" has been demonstrated in the phage T-4 by Benzer (1961). It is possible, as pointed out by Pauling (Zuckerkandl et al., 1960) that the thermodynamic stability of mutational substitutions varies and that mutant alleles often display a greater mutation rate than the normal alleles at the site affected by the mutation. There are indications that amino-acid substitutions in abnormal human hemoglobins may occur at preferential sites, although these sites would be spread over large regions of the molecule. If so, it would not be established that the sites in question are mutational hot-spots. We might alternately be dealing with effects of natural selection.

From the functional point of view mutations undoubtedly are tolerated at certain sites of the molecule better than at others. Comparing the tryptic peptide patterns of pig, cow and human hemoglobins (Zuckerkandl et al., 1960), it was found that all three had a certain set of apparently identical tryptic peptides (peptides obtained by digestion with trypsin), whereas those that differed in any of the two, often differed in the third again as compared to the first two. The impression was gained that there were regions more stable, perhaps not to mutation, but to evolutionarily effective mutation, i.e., to mutations whose results have been retained by natural selection. This is confirmed by the work of Braunitzer and Matsuda (Cullis et al., 1962) on horse hemoglobin, which shows that of the seventeen differences in sequence between the horse and the human α-chains, nine differences, i.e., 50 per cent, are crowded in two stretches totaling twenty-two amino-acid residues out of 141, i.e., 16 per cent.

For any polypeptide chain a distinction may be made between amino-acid residues that change relatively frequently during evolution, residues that change rarely and others that never change as long as the polypeptide in question is recognizable in terms of its main characteristics. The residues that may not change are those whose physico-chemical and steric properties are essential for the function of the whole. In general they represent probably a relatively small percentage of the total (see below). In hemoglobin they are those directly implicated in the binding of the heme group, in the maintenance of the overall conformation in space which is again essential in relation to the specific binding and position of the heme, and perhaps some of those that are critical for the relation between one polypeptide chain and its partner chains. Possibly it is the restrictions imposed in the case of hemoglobin by this last category of residues that account for the apparently greater variability of myoglobins in closely related species (Stockell, 1961). Myoglobins, which are the muscle hemoglobins, are indeed similar to hemoglobins in basic function and in spatial conformation (Kendrew et al., 1960), except that they exist as single polypeptide chains instead of forming molecules composed of four such chains. It will consequently be of great interest to compare the myoglobins from humans and from anthropoid apes.

A second category of amino-acid residues will be implicated more indirectly in the function of the protein. Their substitution will also have an effect on function, but it will not be an all–or–nothing effect. A change in specific functional properties without a loss of the general functional characteristics may or may not be selected for, depending on the kind of change the exterior or interior environment of the organism is undergoing at the time of a given mutation. Amino-acid residues of this type will be subject to evolutionary change, but not very frequently. Examples would be found for instance among amino-acid residues decisive for interchain relations and, furthermore, for intermolecular relations, whereof solubility is one manifestation.

Finally, there are the nearly indifferent substitutions. It may well be that no substitution exists that does not have some effect either on the functional properties of the molecule, or on properties that define its relation with the intracellular milieu and are thus indirectly also functional. But a number of possible substitutions will have only slight effects in these respects. They relate to amino-acid residues in positions of particular "tolerance," or to others with little tolerance, but where some particular substitution will not create much disturbance. In the case of many or most residues the permissible maximum number of substitutions will, however, not be fixed. In proteins composed of several polypeptide chains such as hemoglobin this number will depend on changes taking place in partner chains and on questions of mutual fit of identical chains; in all proteins, including those composed of a single polypeptide chain, this maximum number of different substitutions will depend on the possibility of finding an evolutionarily effective way to get over intermediary unfavorable substitutions (unfavorable from the point of view of a given function) which represent bridges between favorable ones. This possibility in turn depends often on environmental conditions, although cases of absolute impossibility no doubt do exist.

The situation is, indeed, complicated considerably by the fact that the consequences of one mutation often depend on the occurrence of others. A given amino-acid substitution may be incompatible with function, but this incompatibility may cease if a second change occurs at another site of the same or of a different polypeptide chain. Conversely, a favorable substitution may cease to be favorable when a second substitution occurs somewhere else in the molecule. The greater the number of changes required to render an initial change compatible with function, the less likelihood there is of course that these later changes occur. In energetics we learn that certain molecules may be in several different states of low energy, i.e., of stability and high probability, but that the spontaneous passage from one of these states to the others may be unlikely because it requires a high energy of activation: a hump must be passed between two valleys. In evolution, similarly, getting from one favorable amino-acid sequence to another favorable one may require the passage over a hump of functional incompatibility.

Functional incompatibility of an amino-acid substitution is often not absolute, and may sometimes even be selected for in the heterozygous state when the disadvantage of a change is compensated by an advantage of another kind. A case in point is that of the sickle cell condition, first recognized by Pauling et al. (1949) to be a molecular disease attributable to a mutational change in the hemoglobin molecule. This change, a single amino-acid substitution, has later been pinpointed by Ingram (1959). In spite of the fact that the sickle cell gene causes a deadly disease in the homozygous state, sickle cell hemoglobin has been spread in certain regions of the world by natural selection. A correlation has been established between a relative resistance to the agent of the most severe form of malaria, *Plasmodium falciparum*, and the presence of the sickle cell gene in the heterozygous state (Allison, 1957). Thus a mutation that is unfavorable from the point of view of the specific function of a cell can nevertheless be selected for in the presence of certain factors in the environment. Thereby the possibility is left open that a second mutation may occur that would fully restore functional compatibility and might even open up a new level of functional possibilities that could not have been reached otherwise. The probability of such a second favorable mutation is low, because it must no doubt be of a very specific kind. The likelihood of its occurring would be negligible if the first "unfavorable" mutation did not spread considerably in the population. But this may well happen a number of times during the evolution of a given lineage of organisms. In this way molecular disease may actually be a powerful agent of evolution (Zuckerkandl and Pauling, 1962).

Since the genetic code has been tentatively deciphered and is established in some of its principles (Nirenberg and Matthaei, 1962), we know that a single mutational event cannot change a given amino-acid into any other amino-acid. Only certain substitutions are theoretically possible. Therefore, at a given site of a protein molecule and in a given environment, amino-acid C may be functionally better suited than amino-acid A, but in order to go from A to C it may be necessary to pass through the unfavorable substitution B. Whether the passage from A to C actually occurs will depend on how serious a molecular disease amino-acid B causes and on whether or not it may be selected for in the heterozygous state because of certain factors present in the environment. Evidently, if state B represents a selective advantage, it is possible that a passage to state C will result in the loss of this advantage. Although an improvement with respect to a specific function, state C may therefore not be selected for, unless the environmental factor that turned state B into an advantage disappears. Whether this situation obtains will depend on the balance between the types of advantages that are involved. But we must recognize that the passage from an initial state of amino-acid sequence to a final state via an intermediary state will in a number of cases require a simultaneous succession of changes in the environment.

We conclude accordingly that there are many restrictions to evolutionarily effective amino-acid substitutions but that, on the other hand, through the channel of molecular diseases endowed with a selective advantage, evolutionarily effective transformations of proteins may occur that would be otherwise impossible.

C. The coexistence of multiple homologous genes: isogenes

Beside amino-acid substitutions, another feature of fundamental interest to evolution is the existence within single organisms of a multiplicity of proteins that are closely similar in structure but differ by more than one character of amino-acid sequence. Such homologous proteins are normally traceable to non-allelic genes. Hemoglobin will again serve here as an example that one may believe to be representative.

When a total hemoglobin preparation from any animal is subjected to refined chromatographic procedures under optimal conditions, heterogeneity is invariably found. One to several components are always present that deserve to be called major, plus a number of minor components. It has been shown for human hemoglobin that most minor components are probably secondary modifications of major components and as such do not depend on distinct structural hemoglobin genes (Jones, 1961). Minor components under a separate genic control do however exist. Some are minor during certain phases of development of the organism and major during other phases. This is exemplified by the human β-chain, which is a minor component during early intrauterine life and a major component during adult life, and by the γ-chain for which the reverse holds. Two β-chains together with two α-chains form adult hemoglobin (Hb A), two γ-chains together with two α-chains form fetal hemoglobin (Hb F). On the other hand Hb A₂, composed of two δ-chains and two α-chains, is never better than a minor component. It is probably undetectable during the greater part of fetal life and forms in the normal adult only about 2 per cent of the total hemoglobin present.

Minor components that are temporarily major will be subjected to natural selection along with components that are permanently major, as the human α-chain. But components that remain permanently minor cannot play a significant role in relation to the function carried out by the major component. If their function is of a different kind and specific, they may be treated by natural selection as major components in their own right. But such another specific function may not exist, and in that case all intrinsically stable mutations that occur may be retained, as long as the gene is not bodily eliminated, even if at some point the mutations cease to be compatible with gene expression. One may, thus, expect minor components to be often more variable than the corresponding major components and thereby to be of special interest to evolutionary studies in closely related forms, such as man and the anthropoïd apes. However, the remarkable apparent structural stability of Hb A₂ in humans, as judged from electrophoretic data only, has been emphasized by Prof. Neel (personal communication) and must be mentioned here.

Prof. Dobzhansky (personal communication) states that a functionless gene will be bodily eliminated in a short evolutionary time. It does not follow that a gene that has ceased to express itself to a significant extent in polypeptide language will necessarily be eliminated. Genes may lose their specific function because their protein product becomes quantitatively negligible (in minor components) or null (in the case of the hypothetical dormant genes [Zuckerkandl and Pauling, 1962]), and yet they may continue to carry on another type of function that confers upon them a positive selective value, for instance, the maintenance of the topological status quo in certain regions of the chromosome. The distance of a gene from a controlling element may not be unimportant. And inasmuch as the elimination of a gene A, assumed to be functionless from the point of view of polypeptide synthesis, would change this distance for the adjacent genes B, C, D, natural selection might favor the retention of the "functionless" gene A. The rate of activity in protein synthesis of a gene may be adversely affected if its neighbor is bodily eliminated.

Minor component genes may constitute a gene reserve permitting the rapid adaptation of populations to significantly modified conditions, since an increase in the rate of synthesis of a protein that is preadapted to a novel situation may take a smaller number of mutational steps than would the required structural modification of the major component.

Three cases should be distinguished in relation to minor component-genes:

(a) A structural gene for a minor component may have no function related to its base sequence. As a result, this gene may be rapidly eliminated, but perhaps not necessarily so, as has been pointed out. All mutational events that affect such a gene, except perhaps its elimination and certain deletions, will be expected to be retained by evolution to the extent the inherent stability of the mutations permits.

(b) A structural gene for a minor component may possess a specific function linked to its base sequence but different from the function of the homologous major component-gene (e.g., in the case of hemoglobin, from the function of oxygen transport). Natural selection will in this case limit the number of evolutionarily effective mutations. This number will probably be greater than in the corresponding major component during the period in which the new specific function of the minor component becomes established.

(c) If it is found that during a period of evolution the structure of a minor component-gene is about as stable as that of the major component-gene, and if no specific function of the minor component protein product can be detected, the hypothesis might be considered that the structural gene has a regulatory function with respect to other structural genes—that the structural gene is at the same time a regulator gene in the sense of Jacob and Monod (1961). It is in keeping with experimental results to assume that the structural specificity of a regulator gene is decisive for its action. In such a case the minor protein component might only be a byproduct of the gene's specific function.

The following interpretation of genic multiplicity in relation to a given type of polypeptide chains has been proposed by Ingram (1961) and, independently, by Prof. Pauling and myself (1962), on the basis of chemical evidence on the structure of human hemoglobins. The picture is in effect an application of concepts due to the work of the geneticists Bridges (1935), Metz (1947) and Lewis (1951).

It is assumed that genes occasionally duplicate, either by unequal crossing over, or by chromosome duplication, or by some other mechanism. When gene duplication does not occur through chromosome duplication, the duplicate genes may be at first not only identical, but adjacent. They will mutate independently. Their identity and vicinity may be selected for under certain circumstances, namely when the increase in quantity of a certain gene product is of advantage. When this is not the case, divergent mutations in the duplicate genes may be retained. One of the genes may eventually be transported by translocation to another part of the genome and thus come under the influence of another controlling element. Alternately, in the case of chromosome duplication, a mutation in a controlling element on one of the duplicate chromosomes may be selected for, whereas the homologous controlling element on the other chromosome derived from the duplication may be maintained invariant by natural selection. Different controlling elements will often repress gene activity at different times of development, hence, one may presume, the succession in time of the activity of similar, but not identical genes, which is so striking in hemoglobin synthesis.

The α-, β-, γ- and δ-chain genes are believed to have a common gene ancestor. We shall not go into details here, except to point out, as already emphasized by Ingram (1961), that the relatively small number of differences between the hemoglobin β- and δ-chains (less than ten) indicate that they represent the most recent duplicates. This seems to be in keeping with observation. Kunkel et al. (1957) found no minor hemoglobin component with the electrophoretic mobility of Hb A_2 in cattle, horse, pig, dog or rabbit. On the other hand, they found such a component in the chimpanzee and in the New World monkey, *Cebus*. It was however absent from the Old World monkeys *Macacus*, *Cynomolgus*, and *Cerocpithecus*. As long as structural studies are not performed on hemoglobin components from representatives of these latter genera we cannot be sure that components more closely related to human Hb A_2 than to human Hb A, but of different electrophoretic mobility, are not found in Old World monkeys also. The same applies to mammals in general. It would therefore be premature to draw conclusions from the distribution of Hb A_2-like components about the evolutionary relationship between Old World monkeys, New World monkeys, anthropoid apes, and man. But taking present evidence at face value, it would seem that the δ-chain locus had not been in existence for a long time before the beginning of the evolution of the Primates.

It appears that minor components must have arisen from major components. A minor component such as Hb A_2 could not have originated with the specific properties it possesses as an oxygen carrier except if it, or its ancestor, had been selected for this function. We assume, however, that this selection could take place only in a major component, since the minor components must be considered ineffective from the point of view of this function. Thus the minor component gene must have had an ancestor that differed from it at least in the magnitude of its synthetic activity. Whether this ancestor existed before the last gene duplication or whether even after this event one of the duplicates remained a major component gene before turning into a minor one cannot be said with certainty. But the latter possibility may obtain when the functional properties of the minor component appear to be "good," as in Hb A_2 (Meyering et al., 1960). Such properties could, indeed, not be expected to be maintained over long evolutionary times if certain mutations were not eliminated by natural selection, although they might be eliminated, as mentioned above, if the minor component gene has a regulatory function. Accordingly it is permissible to hypothesize that a Hb A_2-like hemoglobin was once a major component, instead of or along with the equivalent of Hb A, which is the major component in adult man today. Perhaps a hemoglobin homologous with Hb A_2 is a major component in some of the monkeys that seem to lack Hb A_2.

The picture of gene duplication is almost inescapable if on the one hand one accepts evolution as a fact and on the other takes note of the extraordinary homologies between the amino-acid sequences of polypeptides of a certain type controlled by distinct structural genes. These extraordinary homologies become ordinary through the hypothesis of the existence of a common gene ancestor.

The existence of several distinct structural genes of a given type within one individual—let us call them isogenes—may well be a widespread phenomenon. This is suggested by the occurrence of so-called isozymes, structurally different enzymes of a given type that are produced in different tissues of one organism (Markert and Møller, 1959; Kaplan et al., 1960) and perhaps sometimes in different cells of the same tissue (Bakemeier, 1961). Isozymes may be controlled by different duplicates of a given gene which are placed in different regions of the genome (Zuckerkandl and Pauling, 1962). Since in different cellular environments different parts of the genome are active in protein synthesis, one will expect different gene duplicates to be active in different tissues.

After discussing some of the features of protein structure and multiplicity that relate to evolution, it is appropriate to examine the foundations on which to base the claim that the structural analysis of proteins may greatly contribute to anthropology in particular and to the study of evolution in general.

THE POTENTIALITIES OF THE MOLECULAR APPROACH TO ANTHROPOLOGY AND TO THE STUDY OF EVOLUTION

The study of anthropology at the molecular level is as yet little more than a project. What are the advantages and the disadvantages of this approach? [2] Let us first examine the difficulties and disadvantages.

2. This topic was discussed during the meeting at Burg Wartenstein by a restricted committee composed of Drs. B. Campbell, Th. Dobzhansky, M. Goodman, G. A. Harrison, H. P. Klinger, E. Mayr, G. G. Simpson and myself, and, after extensive agreement was reached, was rediscussed thereafter in a plenary session. Observations of the various participants have been taken into account in the following presentation. The abbreviation "c.c." following the reference to a discussant stands for "communicated at conference."

A. Difficulties

(1) The amount of information on molecular structure relating to human evolution is at present very limited. This is obviously only a temporary condition.

(2) In one important respect, however, the lack of information may well constitute a major obstacle; for the molecular approach is at present only possible in relation to living organisms, and of value for historical anthropology only inasmuch as inferences about fossil forms can be drawn from data about living forms. Perhaps it will prove possible to extract proteins or significant protein fragments from skeletal remains. Professor Hodge, California Institute of Technology, with whom I have discussed this possibility tells me that he has had a similar idea relating to collagen in bones. Projects of this kind are threatened by two main potential difficulties. One is that so many amino-acid residues may be transformed, and so many peptide bonds broken, that the organic products of extraction would not be recognizable in terms of the original polypeptide chain. The extent of breakdown may vary a great deal depending on the age and the condition of preservation of the remains. Second, it may happen that those few types of proteins one would hope to find in human or prehuman remains will prove to be of the structurally more stable kind, like insulin. In this case, they would be of little or no value as biochemical markers during the relatively short span of evolutionary history that is under scrutiny in anthropology. R. C. Leif, of California Institute of Technology, points out (personal communication) that hair would be a favorable object of studies in molecular anthropology, since the obvious differences among the human races in relation to hair strongly suggest a high degree of variability in the structure of hair proteins.

(3) This leads us to a third point, namely that, even with respect to living forms, studies of amino-acid sequence in homologous proteins meet with specific difficulties of interpretation at both extremes of the scale of comparisons, namely in comparisons bearing on very closely or on very distantly related species.

In closely related species, and still more in the case of subspecies and races, even in the most variable types of proteins only few differences will probably be found. Therefore many different proteins will probably have to be investigated to make possible precise quantitative statements about the degrees of difference between the organisms concerned.

Between distantly related species a significant number of differences in amino-acid sequence will easily be found, but a difficulty will arise in the interpretation of the results in terms of phyletic distance. This will be best understood by examining a concrete case.

As already mentioned, Braunitzer and Matsuda (Cullis et al., 1962) showed that there are seventeen differences in amino-acid sequence between the human α-chain of hemoglobin and the horse α-chain. With respect to the hemoglobin α-chain horse may therefore differ from man seventeen times more than the gorilla (see above). This figure may be expected to vary considerably according to the polypeptide chains under consideration.

Among the factors that intervene in this variation will be (a) the length of the polypeptide chain, (b) the tolerance to amino-acid substitutions with respect to a given function, (c) the selection pressure in favor of the maintenance of that function, (d) the difference in functional requirements between the species to be compared, differences that may be small for some functions and large for others, (e) the question as to whether the polypeptide chain belongs to a major or to a minor protein component, and (f) intrinsic differences in mutability, in the stability of certain mutations in different regions of the structural gene that controls the polypeptide.

If horse differs from man seventeen times more than gorilla with respect to the hemoglobin α-chain, how different is seventeen times? The degree of difference may be expressed as a percentage figure relative to the maximum theoretically possible difference in sequence that does not lead to a total loss of the specific functional properties of the protein. If it is assumed that the comparison of the amino-acid sequences of the human hemoglobin α-chain, the horse hemoglobin α-chain, and the sperm whale myoglobin chain gives us a nearly correct idea of the number of amino-acid residues that may not be changed without this type of chromoprotein losing completely its properties of function and of conformation in space, we find that of the 141 amino-acids of the α-chain, 20 cannot be changed. Of these, one lysyl residue is changed to an arginyl residue in the human γ-chain, characteristic of fetal hemoglobin, according to the structure of the γ-chain as worked out by W. A. Schroeder, R. Shelton et al. (personal communication). This leaves a maximum of 19 amino-acid residues, i.e., about 13 per cent of the total number of amino-acid residues, that probably may never be changed, and these residues are distributed over nearly the total length of the chains while clustering in certain regions. Actually the percentage of amino-acid residues that may not be changed in hemoglobin proper (as distinct from myoglobin) is presumably larger. If we accept the above figures then, comparing horse and man, 17 out of 122, i.e., 14 per cent of the number of amino-acids are substituted that may be changed without the general type of chromoprotein under consideration losing completely its functional characters. In other words, the difference between human and horse hemoglobin is about 1/7th of the maximum *measurable* number of differences between vertebrate heme-containing respiratory pigments. This is a sizable fraction if we consider that two mammals are being compared.

It indicates that, as the comparison will bear on much more distantly related forms, the observed number of differences will be significantly smaller than the actual number of evolutionarily effective amino-acid substitutions that have occured, i.e., of substitutions that, during a time and in a given line of descent, have been retained by natural selection. Indeed, the greater the proportion of observed substitutions in a polypeptide, the greater the chances that a number of amino-acid residues have been changed more than once since the time of the common ancestor of the two chains that are being compared. The probability of two or more successive and successful mutations relating to the same amino-acid site can, of course, be calculated, and a correction be introduced for the probable actual, as opposed to the observed, number of evolutionarily effective amino-acid substitutions, neglecting possible mutational events of other types. (The latter appear to be evolutionarily effective in only a small proportion of cases).

To give two arbitrary examples, if 29 differences in amino-acid sequence were found between two homologous polypeptide chains containing 147 amino-acids each, one might expect that a total of 31 evolutionarily effective substitutions have actually taken place at the 29 sites; if 131 differences in sequence were observed, then 173 actual evolutionarily effective substitutions may be expected to have taken place at the 131 sites.[3] Thus, while the number of evolutionarily effective substitutions that have occurred is likely to be accurately represented by the number of observed substitutions as long as this number represents a small fraction of the number of permissible changes, the discrepancy between the observed and the actual number of presumed mutational events of evolutionary significance will increase rapidly as this fraction becomes large.

Instead of referring the number of observed amino-acid substitutions to the number of residues that may be exchanged for at least one other specific amino-acid without a radical loss of functional properties, one might have chosen to refer it to the total number of amino-acids present in the chain, arguing that the loss of functional properties does not change the relations of descent between polypeptides and therefore should not be taken into account in phyletic studies. However, if such a functional loss occurs in a major component of, say, hemoglobin, i.e., in a component that is undoubtedly subjected to natural selection by virtue of its known function, then this mutation will be lethal and therefore evolutionarily ineffective. In a major component (and probably, as we saw, at least in some minor components) the total number of amino-acid residues is therefore *not* "on the market" for evolutionarily effective substitutions.

The introduction of a statistical correction as proposed above will make the comparison between very different homologous polypeptide chains of use for the evaluation of phyletic distance, as long as the hypothesis of their homology remains reasonable. The results, however, might be erroneous if a sizable fraction of the observed changes were due to events other than single amino-acid substitutions. The same comparison cannot fully serve in the establishment of phyletic lineages, since the nature of substitutions that have been effaced by subsequent ones occurring at the same sites remains of necessity unknown. To establish kinships beyond distance evaluations, it would be necessary to analyze homologous polypeptides in forms intermediary between the two distantly related species that are being compared and their common ancestor. These

3. I am indebted to Professor Linus Pauling for this calculation.

intermediary forms, however, most likely do not exist any more. At best, inferences as to these intermediary forms may be drawn from the structure of the homologous polypeptides of those of their living descendants that are presumed or known to have changed the least. Phyletic relationships will become better established as more data become available; i.e., the number of different polypeptide chains subjected to analysis increases. Different polypeptides of an organism may preserve evidence of a relationship with different ancestral forms, some closer and some further removed in time.

(4) A fundamental restriction with respect to the use of protein analysis in the study of evolution resides in the fact that such studies do not lead to the elucidation of the causality of evolutionary trends. They are not fit for determining, without the use of other information, what causes evolution to proceed in certain directions rather than in others. The first step in evolution is always made by a gene and therefore by a change in the structure or in the quantity of a polypeptide chain. Yet which of the great number of such steps is going to be evolutionarily effective, molecules alone do not decide. Molecules make proposals, and these proposals are taken up or dropped at other levels. To be sure, this restriction applies not only to the molecular level, but to any other level considered individually. It is likely that the analysis must be carried out simultaneously at a number of levels, and the results coordinated, in order fully to define an evolutionary event. Yet, when the molecular, supra-molecular, cellular, tissular, organic, systemic, individual, and further the ecological, sociological and psychological levels are considered, it appears that the key to the determinism of evolutionary trends is found at the higher levels of integration more than at the lower ones. This appears to be so because natural selection acts on functional characters and functions are carried out by coordinated wholes.

(5) Lastly, it is evident that protein analysis tells us nothing about the openess or closedness of a given gene pool. This notion, as Professor Mayr has been emphasizing for many years, is all important to the notion and to the reality of biological species. Inasmuch as the species concept is basic in taxonomy and speciation a basic evolutionary process, the analysis of amino-acid sequences is therefore irrelevant to taxonomy. On the other hand, immunological incompatibilities will be understood in terms of amino-acid sequences and the analysis of these sequences will thus contribute to the elucidation of the process of speciation.

B. Advantages

Turning to the positive qualities of the molecular approach to anthropology and evolution, it may be useful to begin by pointing out that the molecular level, in the case of proteins, cannot be treated as a single one. Three main molecular sub-levels must be distinguished. They are closely interrelated, but must be kept apart because the events at one of them often are not exclusively determined by the events at the next lower sublevel. These sublevels are, in their hierarchical order: (a) the amino-acid sequence of the polypeptide chain, defining its so-called primary structure; (b) the way the string of amino-acids is wound around in space, defining the so-called secondary and tertiary structures; and (c) the direct interaction of polypeptide chains with others or with different types of molecules to form stable or unstable higher molecular units, and their indirect interaction in reaction chains. These distinctions are of importance in relation to several of the topics to be discussed below.

Let us now briefly consider the characteristics of the molecular approach to the assessment of phyletic relationships with respect to the following eight topics: (1) the discontinuity of characters; (2) the arbitrariness of character definition; (3) polygenic effects; (4) pleiotropy; (5) necessary correlates; (6) convergence; (7) the distinction between types of genes or of gene action; and (8) environmental effects.

(1) Dr. Campbell (c.c.) pointed out that in the recently developed techniques of "numerical taxonomy" (Sneath and Sokal, 1962) the treatment of continuous characters as discontinuous represented a serious hazard to the result. Characters of amino-acid sequence are discontinuous, and may prove well suited to this kind of taxonomic investigation. Professor Simpson (c.c.) does not however consider that they have any important advantage over many morphological characters. Let us only remark that to the extent to which proteins are uniquely defined chemical species, the discontinuity of characters of amino-acid sequence renders finite the total number of characters of a given organism, instead of infinite, or rather indefinite, in the case of morphological characters. The fact that the building stones of proteins do not vary fractionally but only by whole units would be compatible with a "quantized" but otherwise continuous variation of protein structure around a mean. To establish that such a variation does not in fact obtain in the organism has been one of the achievements of recent biochemistry.

(2) At the molecular level phyletic studies may also suffer somewhat less from the bias introduced by the arbitrary choice of taxonomic characters. Granted, since we cannot hope to analyze the sequence of all proteins to be found in a given organism, not even of all their major components, the proteins to be analyzed must be chosen, and this choice, whether subjective or imposed by circumstances, may introduce a bias in the establishment of phyletic relationships. However, once the proteins are chosen, the delimitation and definition of their characters is not left to any arbitrary decision, and this would appear to be of considerable advantage over characters of bones.

(3) Professor Dobzhansky (c.c.) and Professor Mayr (c.c.) have particularly emphasized the importance of polygenic effects. "Every character is affected by numerous genes and is therefore polygenic; virtually every gene a poly-

gene. . . . In view of the similar action of the factors in a polygenic set, the same phenotype can be produced by many different gene combinations" (Mayr, in preparation [4]).

This phenomenon will tend to lead to underestimates of phyletic distance when the measurement of this distance is based on morphological characters. On the other hand different gene combinations may not in general be thought, in the light of present evidence and theory, to contribute to the primary structure of any given protein. This theory does not go unchallenged (Commoner, 1962), but remains the most probable. Secondary modifications of amino-acid sequence, which possibly do occur to a minor extent (H. Borsook, personal communication), inasmuch as they are under genetic control, may be a mild manifestation of polygeny at the level of primary protein structure. With respect to the tertiary structure and to the function of proteins the situation is different, since these factors are acted upon by the intracellular environment, which in turn results from the collaborative action of many genes. This remark however does not primarily concern us here, and one may state that, from the point of view under consideration, proteins are a "cleaner" material for phyletic investigations than morphological characters.

(4) So they are with respect to the question of pleiotropy. There is in principle no pleiotropy at the level of amino-acid sequence in proteins, since one structural gene is supposed to determine the primary structure of one kind of polypeptide chains only. In practice one can only say that pleiotropic effects are here very much reduced: the occurrence of a low percentage of "errors" in the transcription of the genic nucleotide code into polypeptide language (Pauling, 1957), although it has not been demonstrated, remains probable. All available results on primary structure of proteins concur in making it likely that such pleiotropic effects at best affect only a small proportion of the population of molecules that is synthesized under the control of a given structural gene. Again the situation must be more involved in the case of the tertiary structure of proteins (their conformation in space), and hence their functional characteristics. Some polypeptides probably will not change significantly their conformation in space within the limits of physiological variations of the intracellular milieu. Their tertiary structure is in this case unequivocally determined by their primary structure. Other types of polypeptides and proteins, however, may possess within the limits of physiological conditions more than one stable conformation; the one that is realized initially, at the moment of synthesis, may be relatively protected from changing over to other conformations by the high energy of activation necessary for such a switch.

(5) While macroscopic morphological characters can be ascribed to an almost hopelessly complex interweaving of causes and effects, different characters being partly affected by the same causes and one given character by different causes,

4. I am indebted to Prof. Ernst Mayr for offering me the opportunity to read a chapter of his book "Animal Species and Evolution" before publication.

this network is relatively disentangled at the molecular level. It is probably very much disentangled at the level of primary protein structure; and probably less so at the level of tertiary protein structure. The latter structure, as the former, is, of decisive importance for the interrelations between molecules. These interrelations cannot be neglected in the assessment of the characteristics of an organism and in inquiries about phylogeny. Therefore the reduction of ambiguity in the relation between a character and its ultimate determinant, the gene, is only a partial one at the molecular level. Even such a partial reduction of ambiguity represents of course a substantial gain.

If one desires to establish the amount of difference between organisms, aminoacid substitutions in functionally correlated protein molecules immediately pose a problem of judgment. Let us take the case of two polypeptide chains that must interact in order for a certain reaction to proceed. It is evident that if chain A changes, it may be necessary for chain B to change also, and in a specific way, if A and B are to continue to interact approximately as before. Thus in a sense it would be erroneous to add the difference in amino-acid sequence found in chain B to the difference to chain A for the purpose of assessing the overall difference between two organisms. Indeed, once an amino-acid is substituted in chain A, the organism that produced the two chains may change less, phenotypically, if an adequate substitution occurs in chain B also than if chain B remains invariant. In the latter case an important biochemical pathway may be lost, and if it is, multiple genotypic consequences will soon result. Thus genotypically as well a greater stability of the organism is insured if the mutation affecting chain B follows the mutation affecting chain A.

Necessary correlates, i.e., distinct characters varying together because they are necessarily correlated, must be recognized and taken into account when inquiries into phylogenetic relationships are made (Cain and Harrison, 1958, 1960; Campbell, 1962). Three types of such correlates may be distinguished:

(1) A type not involving gene action: this applies to the mathematically or logically necessary correlates (Cain and Harrison, 1958).

(2) A type involving correlated structural genes. Here we deal with functionally or ecologically necessary correlates that have arisen by the successive mutation of independent genes, or by drawing into the correlation, after the occurrence of a given mutation, other pre-existing genes. Natural selection, as long as it maintains a certain function, imposes the correlation. An example quoted by Cain and Harrison (1960) may be pertinent here, namely the correlation between the shape of the posterior face of the lower canines and the anterior face of the upper canines in mammals in which these teeth act together as a shearing device.

(3) A type involving the action of structural genes, but no necessary correlation between them, some characters being automatically dependent on others without any ad hoc structural genic specification. A probable example is one mentioned by Professor Washburn (c.c.), namely characters of the hominid

palate that change as an apparently automatic consequence of a change in size of the canine teeth. Here the presence or absence of a certain morphogenetic process occurring in one region and attributable to gene mutation may determine the presence or absence of morphogenetic processes in adjacent regions, without involving in the tissues occuping the latter genic modifications other than regulatory.

This type of correlation may be indistinguishable from pleiotropic gene effects, not only in practice but also theoretically, if one understands by pleiotropy the effect on more than one character of a single gene mutation. But it then becomes apparent that the notion of pleiotropy covers two different types of processes: (a) the mutation of a structural gene leads to the intercellular transmission of a stimulus that may produce secondary reversible genic modifications of a regulatory type in adjacent tissues; (b) the mutation of a structural gene, through a change in primary structure of an enzyme, the gene product, has a basically intracellular effect on more than one totally or partly distinct biochemical pathway, the effect being irreversible (except through the occurrence of a second reverse mutation).

Type (1) of necessary correlates is easily taken care of at the level of the primary structure of proteins, and the third type does not occur at that level. However type (2), very definitely, reappears here. In favor of the molecular approach as compared to other approaches we again register only a partial improvement of the situation and certainly not the elimination of all difficulties.

(6) As is well known, structural resemblance does not necessarily express evolutionary relationships, since it can also express functional convergence. This source of ambiguity is likely to exist also at the level of protein structure. But it appears probable, from the limited available evidence as well as a priori considerations, that evolutionary relationships, at this level, will not be seriously blurred by such effects. The sequences of amino-acids most directly implicated in specialized functions of a protein appear to comprise only a relatively small fraction of the total number of amino-acids, as Dr. Goodman also mentions in his contribution to this volume.

It is not easy to discover which sequences reflect more faithfully true evolutionary relationships, the more critical ones with respect to function or the less critical ones. In fact the more critical ones, although a minority, may be the most unlikely to arise by convergence. First, it will be realized that the appearance of, say, only fifteen specific amino-acids in specific positions, out of a total of 150, is already an improbable event on a pure chance basis, and before selection can operate, chance must furnish it with the raw materials. Second, these fifteen specific amino-acids in specific positions make sense from the functional point of view only if a number of other conditions of sequence are fulfilled. These other conditions are of the "coadaptive" type in that there is no absolute requirement of specificity for any amino-acid residue, but the nature of any one determines a requirement with respect to the nature of others. Several "solutions" to a "problem" of sequence are possible, but the character of each amino-acid residue must comply with the particular solution that is adopted. An example is afforded by the problem of how to form a sharp angle in a given region of a polypeptide chain. Proline residues are used in angle formation. The comparison of the structures of hemoglobin and myoglobin chains by Perutz, Kendrew and their collaborators has shown that proline can in this respect be replaced by another amino-acid. Surely, however, the amino-acid sequence in such a region cannot be just of any kind.

Therefore, in order to obtain a molecule with the general characters of conformation in space and of function that pertain to hemoglobin and myoglobin, a much greater number of conditions of sequence must be fulfilled than is apparent from the consideration of the most "critical" amino-acid residues, recognized as such by the fact that they are the most invariable. Considering from evidence on hemoglobin the rate at which evolutionarily effective amino-acid substitutions appear—the order of magnitude is one in ten million years (Zuckerkandl and Pauling, 1962), it seems very doubtful that enough time was available since the beginning of vertebrate evolution to form by convergence two nearly identical polypeptide chains of 150 units. All the intermediary phases during such a convergent transformation would indeed have to be functional (unless the gene under consideration were temporarily dormant), although not functional in terms of the final type of molecule. If we add to this, as mentioned above, that at a given site of the molecule a given amino-acid is not free to change directly into any other because of restrictions imposed by the genetic coding mechanism, the number of obstacles in the way of convergence of whole polypeptide chains may well be insuperable. A marked divergence in amino-acid sequence, while general function, critical amino-acid residues and tertiary structure are preserved, appears more likely than a convergence of the same magnitude. Once the function and the tertiary structure are acquired and the residues are in place, all changes in sequence that are compatible with them can be tried out and the primary structure of the molecule be transformed progressively. This is why mammalian hemoglobins and myoglobins probably have a common molecular ancestor, as Ingram (1961) also believes, in spite of the fact that between the two types of molecules the differences in sequence are more numerous than the similarities. The chances that the similarities have arisen by chance are the smaller as the non-similarities are also restricted in kind. A polypeptide chain of both the spatial conformation and function of myoglobin and of a hemoglobin chain may not have arisen twice during the history of evolution.

Within the limits of a general type of spatial arrangement and of function of a polypeptide chain, convergence may play a role in evolving particular modalities of the function as a response to the challenge of a common environment. It will thus affect the less "critical" amino-acid residues and true phyletic relationships will remain recognizable at least on the basis of the more critical residues, even though they may be a minority: it would indeed not be a minority without widespread "support."

If the same amino-acid residue establishes itself secondarily at a given molecular site' in two different but homologous polypeptide chains, we may be dealing either with a case of convergence or with what should be more properly termed a coincidence. The term convergence applies when the amino-acid residue has a similar function in the two molecules that are being compared. This need not be the case. The functional effect of a given amino-acid substitution depends on the remainder of the protein molecule. If the differences in the remainder of the molecule are such that identical amino-acid residues in corresponding positions actually have different functional effects, the identity is better described as a coincidence than as convergence.

General functions may well be compatible with a number of different spatial arrangements of polypeptide chains and general molecular solutions, and in this sense functional convergence can undoubtedly arise at the molecular level; but the similar function and/or convergent morphology does then not imply a convergence in amino-acid sequence, except perhaps for a restricted number of residues localized at the active center. A similar comment has also been made by Dr. Goodman.

In conclusion, convergence is likely to intervene to a lesser extent at the level of molecular than at that of macroscopic anatomy. Therefore a purely phenetic classification, i.e., one based on differences and similarities without judgment of evolutionary relationships (Cain and Harrison, 1960), will here be closer to a phyletic classification; i.e., one that is based on such relationships.

(7) Most morphological characters appear each to be determined by numerous distinct genes, the so-called polygenes. Professor Dobzhansky (c.c.) insisted on the important function of polygenes which, he pointed out, is not sufficiently recognized by biochemists. Quantitative variations of characters are indeed mostly explained on the basis of the hypothesis of polygeny (Sinnott, Dunn and Dobzhansky, 1952). Professor Mayr (in preparation; see footnote to p. 000) shows the role this concept is to play in the understanding of the internal cohesion of the gene pool and its important consequences for evolution.

On the other hand Professors Dobzhansky and Mayr do not believe that a special category of controlling genes exists: a structural gene may control the rate of activity of another gene, and every rate controlling gene may at the same time be structural in its own right.

It follows that polygenic effects are considered to be attributable to structural genes only. "There is no evidence for a separate class of polygenes" (Mayr, loc. cit.).

Comparing certain anatomical structures that differ in apes and man, Professor Schultz (c.c.) states: "The essential differences seem to be ontogenetic ones which also produce the marked variability in these structures." In the cases discussed by Professor Schultz, whatever mutations of structural genes have occurred may indeed be rendered responsible for rate controlling effects on the remaining genes. It is nevertheless suggestive that, as Professor Schultz reveals, many anatomical differences between man and the anthropoid apes can be reduced to a matter of differential growth rates. A special class of rate controlling genes may well exist. "Not in a single species" says the same author "occur only retardations . . . nor exclusively accelerations, since every single feature can independently shift its place in the sequence of ontogenetic processes in either direction."

When the product of a structural gene is present, it is potentially able to carry out its function at once. Why then does the development of certain characters start earlier or later in some species than in other related ones? The proponents of the view that there exists only one type of genes would have to answer, it seems: because in some species certain products of structural genes inhibit certain sets of other genes for a longer time during development than in other species. And why this difference in timing of the action of these controlling gene products? Because, again, of the intervention of still further structural genes. Then, why the change in timing in the activity of the latter? Pursuing this reasoning ad infinitum one would, of course, rapidly drive it ad absurdum. Something beside structural genes must act in the regulation of time of activity, and this something, whether "plasmagene" or nuclear gene (in a broad sense of the word) is heritable.

It is true that structural genes may be regulatory at the same time, as Lee and Englesberg (1962) have indeed shown. But it remains to be seen whether all regulatory genes also act as structural genes. Since mutations of regulatory genes are known (Fincham, 1960), one would expect, in case they also function as structural genes, to note a double effect of such mutations, one with respect to the rate of synthesis of one enzyme, and another with respect to the structure of another enzyme. Negative evidence is insufficient, but as the known cases of rate controlling genes, without any known simultaneous structure controlling activity, multiply, it may appear increasingly likely that special rate controlling genic elements exist that are not also controlling the structure of a polypeptide. If so, do they nevertheless control the structure of a ribonucleic acid?

Be this as it may, local properties of chromosomes that are independent of synthesizing activities seem to participate in determining the time sequence in which linearly arranged genes become activated, as illustrated by the findings of Gall and Callan (1962). These authors show that manifestations of what appears to be gene activation extend linearly along the loop of a lampbrush chromosome at certain stages of development. It is likely that if a segment of this loop were removed by translocation to another part of the genome, the structural genes on the segment would not be activated at the same developmental time as within the loop.

Granted, it may be that effects produced by structural characters of the chromosome are ultimately also the expression of "structural genes" whose products control the physico-chemical and morphological properties of the chromosome itself. If so, two types of genes may again be distinguished, those presiding over the formation of cellular organels such as the chromosomes and the others —unless polygenic and pleiotropic effects intervene at this level also, whereby such a distinction would be blurred and may lose its foundation. Due respect must therefore be payed to our present ignorance.

It remains that a purely quantitative change in gene action, which apparently does not involve any mutation in a structural gene, may greatly alter a number of morphological and other characters of an organism. A trisomy in the human chromosome #21, one of the smallest human chromosomes, results in Down's disease, a congenital disease more commonly known as mongolism (Lejeune, Gautier and Turpin, 1959). Cases of this kind tend to show that both the proportions of substances produced and the organization of living matter may be changed by a simple variation in the dosage of gene products without the introduction of any new types of genes.

Under these circumstances, what distinction, if any, does the analysis of amino-acid sequence in proteins, i.e., the indirect analysis of structural genes, allow one to carry out? A distinction between quantitative and qualitative characters? The answer seems to be no, since quantitative characters may be polygenic and their variation involve mutations of structural genes, as do variations in qualitative characters.—A distinction between structural and rate controlling genes as separate categories, although both acting through molecular products? This remains possible but is not at present establised.—A distinction between structural genes and genes involved in establishing chromosomal characteristics that preside over the temporal and perhaps quantitative regulation of the activity of structural genes without the intervention of special molecular products? This also represents a possibility.—Finally, a distinction between pure dosage effects of genes (as in Down's disease) and structural effects (as occurring in mutations of structural genes)? Here a positive answer is in order, although the analysis of amino-acid sequences will, of course, detect structural effects only to the extent to which genes express themselves structurally in a polypeptide product.

Much clarification is necessary before these questions can be settled, or even properly put. But it appears that in one way or another the analysis of the primary structure of numerous polypeptides from two organisms that are to be compared is likely to take into account automatically important differences in gene action and perhaps in the nature of genes. Especially, extensive investigations of protein structure in relation to one type of organisms would allow one in principle to establish the degree to which quantitative changes in characters are associated with mutations in structural genes.

This leads us to the last point to be discussed: that of environmental effects, which express themselves in quantitative variations of gene action, barring mutagenic effects.

(8) As emphasized by Dr. Harrison (c.c.), the elimination of environmental

effects, inasmuch as they alter the characters used in taxonomy, is a prerequisite for establishing a phyletic classification. Professor Schultz (c.c.) has shown the confusing extent of intraspecific variability in certain hominoids, and this must, of course, be due in part to environmental effects. While some minor protein components appear to result from secondary intracellular modifications which may be influenced by the environment, the amino-acid sequence in major protein components may be said to be immune to such influences and to constitute, from this point of view, the best type of characters for phyletic investigations. This does not apply to the same extent to other properties of the protein molecules. An increasing number of results tend to show that the tertiary structure, the specific winding of the strings of amino-acids in space, arises spontaneously as the result of a specific amino-acid sequence (Mizushima and Shimanouchi, 1961; Haber and Anfinsen, 1962). As already mentioned, different types of polypeptide chains must, however, be expected to differ in the degree of strictness of this determination. A given amino-acid sequence will lead to n probable conformations of the polypeptide chain within the limits of physiological conditions and n will vary between one and an unknown number which, in extreme cases, may be quite large. The greater n, the greater the potential effect of the environment on the functional activity of a given polypeptide chain. The necessary activation energy for the passage from one conformation to another might be furnished, for instance, by the successive adsorption and desorption of a globular protein to a cellular interface.

Dr. Harrison (c.c.) has summed up the situation by pointing out that the further away we get in the series of integrated biological levels from the gene level, the more disturbance is caused by environmental effects with respect to the unambiguous expression of the structure of a given gene.

So far we have dealt with the analysis of protein fine structure as a means of establishing phyletic relationships, without distinguishing the two types of information required for a completely defined phyletic classification, namely, (a) the degree of similarity (the inverse of the degree of difference) between two organisms, convergence effects being excluded (the "patristic" affinity of Cain and Harrison, 1960), and (b) the distance in time (number of generations) between the forms to be compared and their common ancestor—the "cladistic" affinity of Cain and Harrison, (1960).

It results from the foregoing discussion that the amino-acid sequence of proteins constitutes the most satisfying character for the establishment of patristic affinities. To what extent can the same type of information be used for determining the time at which two evolutionary lines became divergent?

Here data on molecular structure are insufficient per se. An example of how such data can be tentatively used if independent paleontological dating is available for establishing an average rate of appearance of evolutionarily effective amino-acid substitutions is given by Zuckerkandl and Pauling (1962). As these authors emphasized, calculations of such a type are beset with pitfalls. They furnish interesting indications, but cannot be relied upon in the absence of confirmatory evidence from a different source. Beside other limitations of the accuracy of this approach, the time at which two homologous protein molecules have derived from their common ancestor does, of course, not have to coincide with the time at which the lineages of the organisms as a whole have become distinct, i.e., at which speciation occurred. More importantly, the variation of the rates of evolution along different lines of descent and within any given line at different periods may cause serious errors. Only average values for rates of evolution can be used and for this reason, as Professor Simpson (c.c.) has pointed out, the proposed method of evaluation of the time of common ancestry of homologous proteins could not be applied to short lapses of evolutionary time.

When technical progress will have made it possible to perform sequence analyses on a significant number of proteins from many organisms, the following procedure may be of value to the study of the living forms and their evolution:

(1) Measure the phenetic affinity between two organisms by using exclusively data on protein fine structure;

(2) Make similar measurements separately at various other levels of biological integration: the subcellular, cellular, tissular, systemic, individual levels, the levels of physiology, ecology, ethology, psychology etc. The results of any number of these levels may be provisionally combined as long as the amount of available information is scarce.

(3) Compare the results of operation (1) with the average of the results of operation (2).

If the difference between corresponding proteins are on the average greater than the differences found on the average at other levels of biological integration, convergence has taken place, and the difference between the two differences may be taken as a measure of the amount of convergence. The organisms are genetically more different than is apparent phenotypically.

If on the contrary the differences between homologous proteins are on the average smaller than differences found at other levels, several interpretations may be possible. One would be that the changes are due to mutations in rate controlling genes more than in structural genes.

At any rate, a comparative study in quantitative taxonomy (Campbell, 1962) as applied to distinct hierarchical levels of organisms will allow one to characterize them in a much more refined way than heretofore.

SUMMARY

(1) After briefly surveying the main methods of protein fine structure analysis, some aspects of the evolution of proteins are considered, related to the questions of single amino-acid substitutions and of the multiplicity of homologous polypeptides within a single organism.

(2) The difficulties in using sequence analyses of amino-acid in proteins for anthropological studies are discussed: (a) the provisional difficulty in securing the required large number of results; (b) the fact that protein studies on fossil remains can at best be made to a very limited extent, if at all; (c) difficulties in reaching correct conclusions when comparing very closely or very distantly related forms; (d) the incompetence of molecular studies with respect to the elucidation of the causality of evolutionary trends, and (e) the incompetence of molecular studies in relation to the application of the species concept.

(3) The advantages of sequence analyses of amino-acids in proteins for the study of evolution are examined in relation to the following questions: (a) the discontinuity of characters, (b) the arbitrariness of character definition, (c) polygenic effects, (d) pleiotropy; (e) necessary correlates, (f) convergence, (g) the distinction between types of genes or of gene action, and (h) environmental effects.

(4) The general outlines of a plan of study in quantitative taxonomy are indicated.

ACKNOWLEDGMENT

I want to express my great appreciation to Dr. Bernard Campbell for his many helpful criticisms and suggestions in relation to the present article.

BIBLIOGRAPHY

ALLISON, A. C.
1957. "Malaria in carriers of the sickle cell trait and in newborn children." *Exptl. Parasitol.*, 6:418–447.

BAKEMEIER, R. F.
1961. "A possible cellular explanation of the multiplicity of steroid reductases." *Cold Spring Harbor Symp. quant. Biol.*, 26:379–386.

BENZER, S.
1961. "On the topography of the genetic fine structure." *Proc. Natl. Acad. Sci.*, 47:403–415.

BRAUNITZER, G., R. GEHRING-MÜLLER, N. HILSCHMANN, K. HILSE, G. HOBOM, V. RUDLOFF and B. WITTMAN-LIEBOLD
1961. "Die Konstitution des normalen adulten Humanhämoglobins." *Hoppe Seyler's Z. physiol. Chem.*, 325:283–286.

BRIDGES, C. B.
1935. "Salivary chromosome maps." *J. Heredity*, 26:60–64.

CAIN, A. J. and G. A. HARRISON
1958. "An analysis of the taxonomist's judgment of affinity." *Proc. Zool. Soc. Lond.*, 131:85–89.
1960. "Phyletic weighting." *Proc. Zool. Soc. Lond.*, 135:1–31.

CAMPBELL, B.
1962. "The systematics of man." *Nature*, 194:225–232.

COMMONER, B.
1962. "Is DNA a self-duplicating molecule?" In, Kasha and Pullman, eds. *Horizons in Biochemistry*, pp. 319–334. New York: Academic Press.

CULLIS, A. F., H. MUIRHEAD, M. F. PERUTZ and M. G. ROSSMANN
1962. "The structure of haemoglobin. IX. A three-dimensional Fourier synthesis at 5.5 A resolution: description of the structure." *Proc. Roy. Soc.*, A, 265:161–187.

FINCHAM, J. R. S.
1960. "Genetically controlled differences in enzyme activity. *Adv. Enzymol.* 22:1–43.

FLORKIN, M.
1944. *L'évolution biochimique.* 1st ed. Paris: Masson & Cie.

FRAENKEL-CONRAT, H., J. I. HARRIS and A. L. LEVY
1955. In: David Glick, *Methods of Biochemical Analysis*, vol. II, p. 359. New York: Interscience.

GALL, J. G. and H. G. CALLAN
1962. "H³ uridine incorporation in lamp-brush chromosomes." *Proc. Natl. Acad. Sci.*, 48:562–570.

HABER, E. and ANFINSEN, C. B.
1962. "Side-chain interactions governing the pairing of half-cystine residues in ribonuclease." *J. Biol. Chem.* 237:1389–1844.

INGRAM, V. M.
1958. "Abnormal human haemoglobins. I. The comparison of normal human and sickle-cell haemoglobins by 'fingerprinting.'" *Biochim. Biophys. Acta*, 28:539–545.
1959. Abnormal human haemoglobins. III. The chemical difference between normal and sickle cell haemoglobins." *Biochim. Biophys. Acta*, 36:402–411.
1961. "Gene evolution and the hemoglobins." *Nature*, 189:704–708.

JACOB, F. and J. MONOD
1961. "On the regulation of gene activity." *Cold Spring Harbor Symp. Quant. Biol.* 26:193–211.

JONES, R. T.
1961. "Chromatographic and chemical studies of some abnormal human hemoglobins and some minor hemoglobin components." Ph.D. thesis, California Institute of Technology.

KAPLAN, N. O., M. M. CIOTTI, M. HAMOLSKY and R. E. BIEBER
1960. "Molecular heterogeneity and evolution of enzymes." *Science*, 131:392–397.

KENDREW, J. C., H. C. WATSON, B. E. STRANDBERG, R. E. DICKERSON, D. C. PHILLIPS and V. C. SHORE
1961. "The amino-acid sequence of sperm whale myoglobin. A partial determination by X-ray methods and its correlation with chemical data." *Nature*, 190:666–670.

KUNKEL, H. G., R. CEPPELLINI, U. MÜLLER-EBERHARD and J. WOLF
1957. "Observations on the minor basic hemoglobin component in the blood of normal individuals and patients with thalassemia." *J. clin. Invest.*, 36:1615–1625.

LEE, N. and E. ENGLESBERG
1962. "Dual effects of structural genes in Escherichia coli." *Proc. Natl. Acad. Sci.*, 48:335–348.

LEJEUNE, J., M. GAUTIER, and R. TURPIN
1959. "Les chromosomes humains en culture de tissus." *Compt. Rend. Acad. Sci.*, 248:602–603.

LEWIS, E. B.
1951. "Pseudoallelism and gene evolution." *Cold Spring Harbor Symposia Quant. Biol.*, 16:159–172.

MARKERT, C. L. and F. MØLLER
1951. "Multiple forms of enzymes: tissue, ontogenetic and species specific patterns." *Proc. Natl. Acad. Sci.*, 45:753–763.

METZ, C. W.
1947. "Duplication of chromosome parts as a factor in evolution." *Am. Naturalist*, 81:81–103.

MEYERING, C. A., A L. M. ISRAELS, T. SABENS and T. H. J. HUISMAN
1960. "Study on the heterogeneity of hemoglobin, II." *Clin. Chim. Acta*, 5:208–222.

MIZUSHIMA, S. I. and T. SHIMANOUCHI
1961. "Possible polypeptide configurations of proteins from the viewpoint of internal rotation potential." *Adv. Enzymol.*, 22:1–27.

MULLER, C. J.
1961. "A comparative study on the structure of mammalian and avian haemoglobins." Ph.D. Thesis, University of Groningen.

NIRENBERG, M. W. and J. H. MATTHAEI
1962. "The dependence of cell-free protein synthesis in Escherichia coli upon naturally occurring or synthetic polyribonucleotides." *Proc. Natl. Acad. Sci.*, 47:1588–1602.

PAULING, L.
1957. "The probability of errors in the process of synthesis of protein molecules." In: *Arbeiten aus dem Gebiet der Naturstoffchemie, Festschrift Arthur Stroll*. Basel: Birkhauser, pp. 597–602.

PAULING, L., H. A. ITANO, S. J. SINGER and I. C. WELLS
1949. "Sickle cell anemia, a molecular disease." *Science*, 110:543–548.

REICHERT, E. T. and A. P. BROWN
1909. "The differentiation and specificity of corresponding proteins and other vital substances in relation to biological classification and organic evolution." *Carnegie Inst. Washington*, No. 116, 338 pp.

SINNOTT, E. W., L. C. DUNN and TH. DOBZHANSKY
1952. *Principles of genetics*. 4th edition. London: McGraw-Hill.

SNEATH, P. H. A. and R. R. SOKAL
1962. "Numerical taxonomy." *Nature*, 193:855–860.

ZUCKERKANDL, E., R. T. JONES and L. PAULING
1960. "A comparison of animal hemoglobins by tryptic peptide pattern analysis." *Proc. Natl. Acad. Sci.*, 46:1349–1360.

ZUCKERKANDL, E. and L. PAULING
1962. "Molecular disease, evolution and genic heterogeneity." In *Horizons in Biochemistry*. Kasha and Pullman, eds. New York: Academic Press. pp. 189–225.

ZUCKERKANDL, E. and W. A. SCHROEDER
    1961. "Amino-acid composition of the polypeptide chains of gorilla haemoglobin."
    Nature, 192:984–985.

Comment by Dr. Mayr: "The majority of the workers probably would agree with the following statement: The evidence of chromosome configuration, hemoglobin and serum proteins agrees essentially with the results of the best morphological studies. African apes are closer in all these characters to *Homo* than they are to orang or the gibbons. The gibbons, the most different group of anthropoids on the basis of the morphological evidence, seen to have branched off earliest according to the limited available paleontological evidence, are also the most distinct in their chromosomes and serum proteins."

SOME PROBLEMS IN THE ANALYSIS

AND COMPARISON OF MONKEY

AND APE BEHAVIOR

K. R. L. HALL

INTRODUCTION

IN THE EARLY BEHAVIOR STUDIES of captive monkeys and apes, for example in the accounts of Romanes (1889), Hobhouse (1901), Thorndike (1901) and others, the object of comparison was the "mentality" of these animals and that of "Man." Hence, observation, experiment, and interpretation all were oriented for many years toward discovering the intellectual limits of the few representatives of the very few species that were studied, while, at the same time, some authors, notably Romanes (1889), tended to over-emphasize the almost-human mental attributes of individual animals.

We may describe this as the "intellectualist" era of comparative psychology. Darwin's (1872) study was the only one, until very recently, in which some attempt at compiling the primate emotional repertoires was made, while no systematic data at all became available on the ecology and social organization of these animals in the wild until the early 1930's. Although this is a matter of historical reference, it is cited because the validity of experimental laboratory findings obtained from captive individual monkeys and apes must, for comparative purposes, be much in question where they cannot be related to the natural behavior repertoire and natural social setting. Hence the comparative significance, either amongst the monkeys and apes themselves or phylogenetically, of much of the problem-solving data, on tool-using, detours, delayed responses, and discrimination learning has not been at all clear. As an example, Harlow's (1951) careful evaluations of relative achievements in terms of learning sets by chimpanzees, cebus and rhesus monkeys may need reviewing in the broader context of naturalistic study, while Klüver's (1933) outstanding laboratory work on, amongst others, instrumentation problems given to monkeys of the *Cebus* genus calls in question the whole basis of intellectual gradings amongst the monkeys and apes by demonstrating that, in *Cebus* at least, their tool-using performances may differ qualitatively from, but not thereby be inferior to, those achieved by Köhler's chimpanzees.

A precise evaluation of the way in which one or more kind of laboratory "test" attainment by such animals relates to the total picture of the animal's natural adaptability becomes necessary, for there are, as is well known, many examples of genuine tool-using in birds (Thorpe, 1956) where learning is involved over and above a simple facilitation of some species-specific manipulatory tendency. A similar complex of factors is involved in the comparisons of delayed response or detour performance which, if taken out of a setting of interest to the animal or totally unrelated to any natural experience, can provide measures of little or no validity for comparative purposes.

In spite of inadequate behavioral knowledge, in laboratory or in natural environment, of even a single species of Old or New World monkey, so strong was the pull of the current evolutionary model that Yerkes (1943, p. 195) was impelled to say:

Looking from the chimpanzee monkeyward, there is gulf which has tended to widen as knowledge became more inclusive and reliable. In the behavioral categories symbolized by the terms sense, perception, memory, imagination, the superiority of the chimpanzee over all observed Old and New World monkeys is definitely indicated by naturalistic and experimental data. Furthermore, creative imagination, learning processes, linguistic and emotional expressions are far more varied and highly developed in the chimpanzee. Especially conspicuous are the contrasts between ape and monkey in readiness and degree of response to training, general adaptation to experimental and other requirements, ability to learn to cooperate with each other or with man. In each of these the ape has a long lead.

Yerkes here goes beyond the intelligent to the emotional and social characters of the chimpanzee in what is, in effect, a superordinate classificatory summary of behavioral attributes that distinguish this animal from all monkeys. While he may turn out to be right, at the time the statement was made, and still today nineteen years later, there are no clear-out observational or experimental grounds for it. Yerkes knew a very great deal about chimpanzees, and about representatives of the other anthropoid apes, but one suspects that a sufficient familiarity with some at least of the monkey species might have led him to a more cautious approach.

A similar kind of men-oriented criterion for comparison is that of Warden et al. (1936, p. 420): "The capacity to adjust to the complex human mode of life is, perhaps, the best indication we possess of the superior intelligence of the higher primates." This, however, is a possibly confusing standard requiring considerable elaboration, for the rhesus monkeys of the cities and temples of India might seem to have gone much farther towards it than any of the living anthropoid apes, and the baboons of Africa survive remarkably well the changes in their habitat brought about by cultivation, and by visitors to nature reserves and parks.

This brief review of some of the difficulties involved in trying to use for behavior ratings amongst the contemporary monkeys and apes a variety of concepts unsupported by systematic experimental or field observation may serve, so it seems, to point the need for a caustic and critical approach to the basic problems of method in drawing out comparative statements. Even at this early stage, it is probably worth guessing that the greatest achievements in comparative primate behavior study will come as research workers learn to synthesize new concepts from field and laboratory data after eliminating or radically altering many of those in current use.

Because the general problems of evolution of contemporary primates and of the possible taxonomic use of behavioral data as an aid to morphological classification involve attention to particular problems of observation and method, some of these will be reviewed prior to their illustration mainly from field study of baboons. The likelihood that data on behavioral similarity and differences, combined with structural data, might eventually lead to modification of the phylogenetic picture derived from morphological data was envisaged by Carpenter (1958), and the possibility that behavior data may be of classificatory value at even the inter-specific level cannot be excluded if a level of analysis such as that outlined by Mayr (1958) and Hinde and Tinbergen (1958) for lower vertebrates and invertebrates turns out to be feasible.

PARTICULAR PROBLEMS

The problems of behavior comparisons of the monkeys and apes are not, it seems, distinguishable from those which the ethologists have had to surmount in their studies of sub-primate animals, except perhaps in two major respects. The first is the relative difficulty that observers sometimes have found, in spite of the derision of anthropomorphism over many years, of adopting standards of objectivity when dealing with these animals which they would find easy when dealing with lower vertebrates. The second is that wild, free-ranging groups of monkeys and apes offer an observational problem of practical difficulty and complexity such as is not usually afforded by, say, gulls or geese.

1. CHOICE OF CHARACTERS

The principle of taxonomic relevance, emphasized by Le Gros Clark (1959) with reference to structure, can be assumed to be crucial also for behavioral comparison. As Nissen (1958, p. 183) pointed out: "For a phylogenetic comparison of behavior, the problem of classification, of identifying the most meaningful and useful dimensions and descriptive units of behavior, is of basic importance. Indeed, one may say it is *the* problem of psychology generally." To the fallacy of making comparisons on the basis of experimental performance of doubtful validity is added that of making a comparative judgment in terms of an insignificant variable, such as conditionability in the classical Pavlovian situation. Hence, Nissen suggests that ". . . some orderliness in behavioral evolution will appear if, instead of considering each particular character (or trait, or mechanism) separately, we group together those characters which are behavioral equivalents or alternatives" (p. 201).

The problem of choosing the kind of behavior most relevant for a classificatory or evolutionary scheme is basically one of deciding empirically on traits of greater or lessor comprehensiveness, some or all of which may serve with varying degrees of usefulness. For example, Le Gros Clark (1959) speaks of "adaptability" as a general trait characterizing primates, but it is obvious that wild, and captive, monkeys and apes differ enormously in the extent to which they can modify their behavior to meet varying circumstances, whether experimentally arranged or naturally arising through alterations of habitat. Here again baboons and macaques may show, in one sense, a higher degree of natural adaptability than any of the anthropoid apes in coping with human encroachment, but chimpanzees seem capable of learning, or being taught, more humanoid tricks such as give them a high entertainment-rating for intelligence. But it must be remembered that seals, dolphins, or elephants are also excellent humanoid tricksters, and the comparative problem is not greatly advanced by this kind of bias in choice.

At the other extreme, the task would be to try to discover small differences in the behavior patterns of threat, mating, or social gestures which could reliably distinguish, say, *Macaca mulatta* from *M. fuscata* or *M. radiata*, or *Papio ursinus* from *P. anubis* or *P. hamadryas*. Whereas the detailed behavioral analysis necessary for such comparison has been proved feasible in studying various gull or duck species, it is possible that intraspecific variability in the monkeys and apes will prove too great for interspecific comparisons within a genus to be reliably determined even at this basic behavior-repertoire level. Nevertheless, such studies, on wild and confined animals, should be thoroughly carried forward, for *negative* results in the sense of failure to find distinctions in, for example, the repertoires of communicative expressions or gestures, except possibly of a quantitative rather than a qualitative kind, might indicate where revisions in the morphological taxonomy could be necessary.

The search for behavioral data relevant for comparison and for criteria with which to judge the levels of complexity or integration giving the best distinguishing features is likely to resolve itself as the amount of systematic observations of Old and New World species, of terrestrial and arboreal and intermediate types, and of the anthropoid apes, increases. The problem of choice is emphasized, however, because it tends to be associated with the problem of validity of general comparative statements. Thus, when Yerkes and Yerkes (1935, p. 994) said that baboons were "far below the howler in communal tendencies, group cohesion, and extent and variety of important social behavior," with the chimpanzee best of all the non-human primates in respect of these behavioral characters, the wording is such as to convey, not a speculation or even a hypothesis for testing, but a definite comparative judgment. It so happens that such judgments or generalizations can only now begin to be derived from a sufficiency of observation, and it is obvious that no behavior characters at any level can be chosen for their phylogenetic and taxonomic relevance unless they are so derived.

While this kind of error goes with unsupported generalization at the sociological level, another kind of error may come into the search for significant character at the behavior-component level of analysis. For example, it may become clear, after prolonged observation, that component X varies in its species-meaning with the context of other expressive behavior accompanying it, and that the repertoire of a species must receive its detailed examination within the whole framework of age, sex, and social differences within the natural groups before it can provide a meaningful mosaic for comparison. Nevertheless, there is reason to hope, from the accumulation of data on a few species in the wild and in captivity, notably in the Macaque and Baboon genus, that analysis of behavior patterns into component sequences, at the communicative or purely locomotor and maintenance level, may aid in classification provided that the methodological safeguards are sustained.

There are two further points which need careful consideration here. The first concerns not so much the "level" of behavior character that may be relevant and useful, but the "normality" or "usualness" of it. Thus, situations sometimes occur, man-contrived or natural, which drastically alter the ecological set-up of the wild monkey group. These then produce behavior which is almost certainly characteristic of the species, and perhaps of the genus, only it is not commonly seen. Two examples will illustrate the point. Rescue-operations from the flood-formed islands of Kariba have shown that baboons and vervet monkeys not only swim readily to avoid capture, but that vervets in particular dive and swim underwater. Baboons, perhaps because of their heavier coats, dive much less readily, although one young female dived and swam so deeply under water that she was not recovered. Reactions to the circumstances of captivity may likewise be of some comparative interest, although the range of individual difference within a species is likely to be very great.

The second point concerns the question whether only the "innate" behavior character should be used for comparison and classification or whether the modifications due to ecological and social influences should be included. Perhaps the question is now dead, but it should be mentioned because of the favored isolation method as a way of extracting uncontaminated repertoire data. There is, of course, a continuity of grades of environmental variation, ranging from maximal social and physical isolation through the contingencies of the natural group or captivity mother-infant relationship, to the kind of super-normally attentive setting provided for Viki by the Hayes. One would nowadays hesitate to exclude any kind of developmental data from possibility of comparative use. If it can be shown that the "fear-grimace" or the posture of "presenting" occur on first occasion on which an experimental isolate is associated with another animal of the same species, we can conclude that neither expression owes anything to learning.

If, on the other hand, a home-reared or laboratory-trained animal achieves performances of a kind not observed in its wild species, it is reasonable to assume that these are learned from the special environment provided.

Both the extreme forms of data are of very doubtful actual or potential comparative value for a variety of practical and methodological reasons, and one may suppose that the data obtained in the natural group, the zoo or compound group, and the minimal subgroupings achieved in most laboratories, are likely to have significance, in that order, in any classificatory scheme using behavioral information. One may well propose, as a substantial hypothesis, that the extent and significance of early learning in the primates is proportional to the time-scale in the achieving of physical maturity, as one evident criterion, but no way of "weighting" the influences of learning as against those of other factors in determining the complex behavior patterns of these animals has yet been worked out, so that we should take the data as it comes.

There seems thus no particular justification, in the light of present inadequate knowledge, for especially selecting from the most complex (social) level, from intermediate levels (mating pattern), or from basic movement levels (facial gestures, intention movements, and so on). Reliable evidence from any or all levels would seem to hold equal promise as aids in filling out and validating the taxonomic distinctions made on morphological grounds.

2. Definition and Terminology

Adequate descriptions or ethograms, even of the most commonly used laboratory monkey, the rhesus, have been lacking until recently (but see now Hinde and Rowell, 1962, and Rowell and Hinde, 1962, for a detailed account of the behavior of *M. mulatta* in captivity). The profuse literature on the chimpanzee in captivity could no doubt yield worthwhile descriptive data for comparison with the behavior of these animals in the wild, but it is only now in the *Macaca* and *Papio* genera that a sufficiency of data from several species is becoming available for cross-checking and validating of field and captivity sources (for baboon data see Zuckerman, 1932; Bolwig, 1959; Washburn and DeVore, 1961, etc.; Kummer, 1956, 1957; Hall, 1962b).

Observation, description and terminology are interdependent, as Schneirla (1950) and Emlen (1958) have emphasized. There is no need to revive any of the classics of the anecdotal literature nowadays to bring this home, but other aspects of the terminological issue are just as important. So, while evading conscientiously the use of such human sociological terms, when making comparisons amongst the primates, as "leadership," "property-owning," and so on, one alternative, that of coining altogether new categorial terms such as allelomimetic, etepimeletic, etc. (Scott, 1956), may merely add to the semantic difficulties, not reduce them. Terminology is a very central problem of method in dealing with primate behavior because investigators converge upon it from disciplines between which there is not at present much overlap of concepts—social anthropology, comparative psychology, and zoology being the obvious examples. To some extent, the discipline background of the investigator may affect his choice of levels in making his observations, because his terminological-conceptual framework is such as to attune him to the ordering and perceiving of a certain class of data.

Scientific economy will necessitate a great deal of cross checking and analyzing of the data from field and laboratory and intermediate studies so as to improve the whole conceptual scheme and to establish criteria for behavioral comparison. It is possible, for example, that concepts derived from sub-primate animal studies such as that of "intention movement" and "displacement activity" should be tested by applying them to the primate observational data before new terms are adduced from the latter with the implication of discontinuity between the "lower" and the "higher" animals. The same sort of problem now arises in selecting and classifying comparable data on the behavior repertoires of the monkeys and apes as has long been known in the discussions of comparative learning processes. The concepts of learning "theory" or "theories" should be measured up to the naturalistic behavior of the animals, for such concepts or their revised versions are certainly necessary in arriving at the characteristic forms of social organization achieved by any wild species or genus.

The task of accurate description of behavior in the wild is one involving considerable practical and observational technique. Until recently, field studies had never been sufficiently prolonged or concentrated to allow the observer to get on really close-range terms with wild groups. In order to work out the behavior repertoire of a species, it seems that it is necessary for the observer to be able to identify all or most of the individuals in the group, or at the very least to know them by sex and age-category, for the significance of a particular behavior pattern or sequence may only be understood through observing and describing occurrences within the characteristic social framework. As Carpenter (1958, p. 75) noted: "The stimulus value of any given gesture or social pattern depends, to a very considerable extent, upon the social status and role of the individual producing it. Furthermore, vocal patterns have various functional or stimulus values depending on the characteristics of the context in which the behavior occurs."

The significance of context will be illustrated later, and the difficulty of distinguishing the "meaning" of certain emotional expressions in different species, and even in the same species, was stressed by Darwin (1872) in discussing the "smile" indicative of pleasure and the "scowl" indicative of fear or anger. Mood can change very rapidly indeed in these animals, so that the mere noting of behavior consequent upon a particular set of intention-movements or a particular expression is not reliable unless it has been followed out on a sufficient number of animals in a sufficient number of instances. The so-called threat-yawn, described in baboons and other monkeys, is an example of this. Likewise, the function and derivation of so-called "primitive instrumental acts" (Carpenter, 1935) such as the throwing down of sticks or stones can only be established satisfactorily for comparative purposes if the occurrences are ordered in the behavioral context and very accurately 'placed' in terms of the factors eliciting the behavior.

3. REPRESENTATIVE SAMPLING

As the inferences for a morphological taxonomy are derived from a sample of specimens considered to represent the species or genus, so must the behavior-samples be representative of the appropriate categories throughout their geographical distribution. "Representativeness" is, in terms of behavior, complicated by the variety of sources from which observations are obtained, such as laboratory, zoo and field studies in the natural habitat.

This can be illustrated by considering the possible effects upon breeding behavior of removing the animals from one climatic region to another, so that one might doubt the relevance of northern hemisphere zoo data on baboon births as providing anything representative of the genus. But, even within the natural geographical range of the animals, there may be considerable variations geared, it seems, to climatic factors and the availability of food. Thus, Washburn and DeVore (1961) report that births in the Kenya baboon groups tend to reach a peak with the onset of the rainy season, while Zuckerman's (1932) sample of foetuses and infants from a group of Chacma baboons in the Eastern Cape indicated to him that, for that particular region, births occurred in any month of the year. Seasonal variation in baboon births is likewise reported by Wingfield (personal communication) in Southern Rhodesia, and Stevenson-Hamilton (1947) stated that, in the Eastern Transvaal, the majority of infants are born from November onwards for two or three months. Data are also available as to birth seasonality in *M. fuscata* (Imanishi, 1957), the rutting season beginning usually at the end of December and lasting until the end of March. Pregnancy being of about 5 months' duration births occur from mid-May to mid-August.

Reliable representative statements about the mating behavior and sexual reproductive physiology of these animals can only be derived from long-term studies which have adequately sampled the different climatic zones of the total distribution area. Further, comparative statements such as Grassé's (1952) to the effect that sexuality is much less pronounced in gibbons and howlers than in baboons are based on non-comparable data, namely the zoo observations of Zuckerman (1932) and the field data of Carpenter (1934, 1940). Further, no account may have been taken of general factors that might grossly affect the conclusions. Thus, not only are long-term seasonal differences likely to operate, but Hall (1962b) has noted that sexual and social behavior of all kinds is much more frequent, in *P. ursinus*, in the first two hours or so after dawn than at any other period of the day. To achieve representativeness, therefore, for behavior data the natural history of the animal must be accurately known so that observations can be ordered to allow for known sources of variation.

A similar problem regarding "typicality" of behavior concerns the nature of aggressiveness found in baboon groups in zoos (Zuckerman, 1932) and in the wild. Contrary to zoo data, Washburn and DeVore (1961) and Hall (1962b) report that, in wild groups, there tends to be a balancing of agressive and cohesive forces, aggression being usually of the threat or brief chastisement kind, without visible injury to the victim. Thus zoo data here is, by definition, unrepresentative of what has so far been observed in the wild groups, as also was the whole captivity environment, but may be of subsidiary interest in showing how stress situations can elicit an abnormal degree of aggressiveness.

The problem of regional variations in the behavior of groups assigned to the same species has already been mentioned, and the question of the representative behavior sample can again be illustrated by this. Chacma baboons are "typically" described as using a special technique in eating scorpions. Thus, Morgan (1890), to cite one source, describes the baboon as pouncing upon the live scorpion, rapidly extracting its sting, and then putting the creature in its mouth. A wild baboon was fed with live scorpions twenty-four hours after its capture in Southwest Africa. It simply grabbed the scorpion by its head, bit it, chewed it up, and swallowed it, sting and all (Hall, 1962a). When live scorpions were presented in paper bags and under stones for the baboons of a Cape group to find, the invariable reaction of all the twenty animals that investigated was a quick startle-response and avoidance (Hall, in press).

Finally, representativeness is very much a function of the observational methods used by the investigator. This is already apparent from some of the examples, but it goes without saying that the observer-interference factor must be eliminated except in so far as one is interested in flight-distances and in agonistic behavior relating to human beings. Fortunately it seems that some monkeys and apes, when for long periods undisturbed, become so tolerant of the observer that he becomes, unless he changes his behavior, a neutral part of their environment. This has been the case with one group of baboons in the Cape (Hall, 1962a), and recent observations by Wingfield (personal communication) of baboons in Southern Rhodesia and of a second large group of baboons in the Cape (Hall, in press) show that progressive familiarization, without necessarily introducing food, can lead to a very great reduction of distance. The same condition was found by Emlen and Schaller (1960) for the Mountain gorilla, and it is doubtful whether the Japanese technique of provisioning is necessary in order to achieve the desirable observation-distance.

4. QUANTIFICATION

It is well established that differences between closely-related species in birds, and also in *Drosophila*, are often largely quantitative (Mayr, 1958), and it is possible that quantitative data may be useful for behavior comparisons between taxa of monkeys and apes. However, meaningful quantification must stem from adequate solution of the previous problems of observation, description, terminology and representative sampling. Implicitly quantitative statements of a general kind are fairly common in the primate literature, as where grooming is described as "frequent" in one species, "rare" in another, and non-occurring in a third, or mating is likewise given a rather dubious frequency formulation regardless of factors which may nullify the comparability of the evidence. Provided that the terms are clearly defined, however, there is every reason to use increasingly frequency-data and time-samples in order to compare the behavior-repertoires or the components of them.

Such timing and frequency-quantifying is arduous in the field, but seems particularly appropriate where the significance of any particular behavior-component is in question. Thus, Hall (1962b) has attempted to work out, in terms of frequencies, the pattern of mating behavior in *P. ursinus*, noting the occurrences of components such as the "copulation-call" of the female, the frequency of apparent initiation of mating by male and female sex partners, and also the correlation between mating-frequency by the dominant male and the number of aggressive episodes amongst the females. Although such statistical treatment of behavior observations is of value only in proportion to the reliability of the observations on which it is working, it is quite clearly an important aid to defining relationships and setting up hypotheses for observational and experimental study which could not have been derived from even the most thorough qualitative descriptions. Similar data on mating behavior is given in Carpenter's (1942c) study of *M. mulatta*.

It would further seem to be essential to use such an approach in objectively defining the dominance characteristics in a group or in a species. Thus, the frequency with which certain individuals groom, are presented to, mate with fully-oestrous females, show aggressive behavior, and so on, provides, as it were, the behavioral profile of the individual, and such data may be used ultimately to define certain social characteristics of the species. Other quantitative information, such as the spacial lay-out of a group of *M. fuscata* or baboons, provides a comparative framework of the sociometric kind.

Lastly, there is little doubt that, as the behavioral data accumulates on these animals in sufficient detail, and as the scope of naturalistic experiments upon their behavior increases, quantification, combined with the technical aids of multi-channel event-recording, will become progressively more valuable. It would be premature to suggest that factorial techniques may be applicable, but this seems likely from the present trend in ethological studies, and analysis of variance techniques may clearly aid in the preliminary search for significant hypotheses for more precise testing.

COMPARATIVE BEHAVIOR EXAMPLES

Having discussed some of the problems in establishing any reliable behavior characters that may aid in the task of primate classification, it remains to point out some of the chief needs for future behavior research by selecting, for a more detailed examination, two broad classes of behavior, the agonistic and the affectional. The problems we have reviewed are common to all comparative animal behavior work, but tend to be magnified because of the apparent social complexity of the monkeys and apes, because they have been essentially regarded as "learning" animals, and because they seem especially to elicit observational and methodological error.

1. Agonistic Behavior

Generally to be included under this heading are a wide variety of behavior patterns supposedly expressing the emotions of "fear" or "anger," these "moods" sometimes leading on to active avoidance or actual attack. The repertoire of stances, movements, and gestures that the human observer may record includes such diverse behavior as defecation and erection of fur, and "primitive" types of instrumental act. Further the behavior is complicated by the occurrence of seemingly "irrelevant" acts, i.e., acts apparently not directed at attaining any goal in the environmental situation, but the performance of which seems to express a conflict of incompatible tendencies which have been very recently, or are currently, aroused.

All such behavior patterns are, in probably varying degress, communicative, in that they may elicit some sort of preparedness, or expectation, in animals of the same group that observe them. For, as Nissen (1951, p. 425) has, I believe, correctly generalized: "When we study social behavior . . . we become intimately concerned with the emotions; when we investigate emotions, we find them primarily in a social setting." This does not, of course, mean that each time a member of a group expresses an emotion, for example by giving the alarm-bark, others in the group will necessarily be alerted or will take any different action from that in which they are presently engaged. For it depends, in baboons at any rate, which animal does the barking. Thus, when a low-status female of a group in the Cape gave an alarm-bark on discovering a snake, none of the rest of the group, so far as the observer could see, paid any attention. That the bark was, however, given with reference to the rest of the group was indicated by the fact that, immediately on giving it, she looked back in the direction of the other animals.

Often in monkeys intention-movements or other expressions of threat may be intermingled with opposite tendencies of moving-away, so that alternating sequences of behavior appropriate to the attack and escape tendencies are evident. Such ambivalent behavior is easy enough to analyze into its respective categories when it involves obvious intention-movements, but it is much harder to deal with when it comprises minor expressions, such as ear-flattening or movements of eyes, lids or brows. The behavior-patterns are not, in the Chacma baboon at any rate, particularly stereotyped, though a certain sequence is usually apparent when, for example, distance-threat develops into actual contact, i.e., attack.

Distance-threat may "typically" begin with staring, flattening of the ears against the head, raising of the brows and a rapid lowering-raising of the light-colored eyelids, raising of the muzzle and barking. The whole head may be jerked forward and down-up in the direction of the opponent, at any rate when a dominant male is preparing to chase and attack a lesser animal. Sometimes the next, sometimes the accompanying, stage is for the animal to rise up on its hindlegs, but this seems chiefly to be a posture enabling the animal to see more clearly, whether the sequel to it is attack or escape. In other words, this is essentially an "observing-response" that may precede either outcome. The mane-fur of adult males appears to stand up somewhat, and, as Darwin (1872) pointed out, this makes the animal appear larger than normal. An animal thus charging after its victim, and uttering repeatedly its loud two-phase bark, will usually, if it catches up with its victim, seize it in his mouth anywhere between the nape of the neck and the base of the tail. A large male attacking an adult female has been observed to lift her briefly in his jaws completely off the ground. However, biting is far from invariable, and seizing with the hands and beating upon the ground is the commoner result of contact (Bolwig, 1959; Hall, 1962b).

Threat-intention is also expressed by a rapid striking-forward action of one or both forelimbs, from sitting or quadrupedal standing or bipedal standing positions. It sometimes seems that the forward-thrusting movement is linked with a rapid scraping-back, along the ground, of the hand, and this is probably a first phase of the action of actually seizing a victim. Slapping, or striking with one hand, is not uncommon in minor attacks, as, for example, when an adult female cuffs a young animal.

A rapid turning of the head from side to side is fairly common amongst the females when a squabble amongst them occurs, but it does not seem clear whether this reaction is a part of threat or of uncertainty or of watching for the intention-movements of either antagonists or allies.

Such is a fairly broad sample of the threat-attack behavior components of this species, corresponding, it would seem, closely with that recorded of the Kenya baboons by Washburn and DeVore (film, "Baboon behavior"), and in many respects with Zuckerman's (1932) account of the Hamadryas. However, both Zuckerman for this species, and Haddow (1952) for *P. anubis*, include in the threat-display grinding of the teeth, and this has not yet been observed in our close-range observations of the wild Chacma.

The need for very precise observation and detailed description not only of the behavior-component but also of the context is illustrated by another component reported as a part of the threat-display of babbons, and of other monkeys, namely yawning. According to Darwin (1872, p. 138): "Baboons often show their passion and threaten their enemies in a very odd manner, namely, by opening their mouths widely as in the act of yawning. Mr. Bartlett (a keeper at the London Zoo) has often seen two baboons, when first placed in the same compartment, sitting opposite to each other and thus alternately opening their mouths, and this action seems frequently to end in a real yawn. Mr. Bartlett believes that both animals wish to show to each other that they are provided with a formidable set of teeth, as is undoubtedly the case."

Darwin quotes examples of threat-yawning in some species of Macaque and *Cercopithecus* monkeys. Zuckerman (1932) likewise includes yawning in the threat-behavior of *P. hamadryas* at the London Zoo, and notes that this is a gesture performed by many monkeys and shows little interspecific variation. However, the "yawning threat" of the Gelada baboon is said by him to be accompanied by a complete eversion of the upper lip, which slaps back as the animal opens its mouth. Haddow (1952) likewise describes yawning as a part of the threat-display of *P. anubis* in the wild.

These statements about this particular behavior-character, yawning, are quoted because they indicate the possibility of confusion of expression and function. Since Darwin's account, it seems to have been assumed that yawning, whenever it occurs in a nervous or disturbed animal, has a particular function, namely of exposing the large canine teeth to the view of the opponent. If this were indeed the case, then we would expect to find that the yawner usually yawns facing its potential adversary, and, further, that yawning has a fairly frequent place amongst the other more straightforward components of threat-attack behavior. DeVore's (1962) detailed description of what he calls the "harassment" sequence of behavior amongst the adult males of one unusual group (in terms of constitution) in Kenya is a clear example where "canine display" is a part of the threat behavior directed at the harassed male by the harasser, but evidence will be cited below which indicates that yawning may also be classed as a displacement activity, occurring in situations where the other threat-components do not occur.

The point that this example seems to make clear is that simple, general descriptions of a particular kind of behavior may sometimes be insufficient for comparative purposes, and, further, that the "place" of such a behavior character in the repertoire can be assigned with greater accuracy by a careful situational analysis.

From the description given so far of threat-attack behavior in *Papio* species, it must still be uncertain whether there are any grounds for indicating distinctive characters, either in the quality of an act or in terms of intensity, or frequency-variations. No doubt some distinctions can soon be made between this class of behavior in baboons and in, say, macaques, and that, further, the differences may be even more evident from comparison with representatives of the genus *Cerco-pithecus*, as indicated by Haddow's (1952) description of the threat-behavior of the Redtail monkey in a tree, when defying a human intruder: the animal crouches, often with tail raised high over the back, and rapidly raises and lowers its head and shoulders with a bobbing movement. Observations of the gaping and head-bowing threat movements of captive Patas monkeys, *Erythrocebus patas* (Hall, to be published), also provide evidence of variations in detail around a general scheme of threat behavior common at least to those Old World monkeys on which systematic data is available. Unfortunately behavior data of the standard required is completely lacking for *Mandrillus* species, the Black ape, *Cynopithecus niger*, and others with which comparisons could be interestingly made with the baboon and macaque data as a reference. J. A. R. van Hooff (1962) has given an interesting prelmiinary account of his of general description and classification of facial expressive movements, including threat, in zoo monkeys and apes, but the comparative picture is complicated by the probability that frequency or qualitative differences in the components of behavior sequences will be found as between individuals of a group according to their age, sex, and group status. Thus, "facial-threat" and "prancing-threat" seemed, in one Cape group of baboons, to be primarily female behavior (Hall, 1962b), and DeVore (1962) has similar observations on the Kenya baboons. Similar expected differences are apparent as between captive male and female rhesus monkeys in experimental situations (Mason, Green and Posepanko, 1960), so that the "profile" of threat-behavior characteristic of a species can only be finally established by the study of its variability amongst individuals and in situations as diverse as the laboratory cage, zoo colony, and natural habitat.

Characterizations of a very general kind, a sample of which has already been considered methodologically, have also been made about aggressiveness in monkeys. Baboons, as is well known, have had a reputation for aggressiveness both amongst themselves, and towards human beings (see Stevenson-Hamilton, 1947, for an evaluation of the reports about *P. ursinus*), The problem is an important one because it typifies the differences between inferences based on impression and those based on systematic observation. Washburn and DeVore (1961) and Hall (1962b and in press) are substantially in agreement from field study that baboon groups are:

(1) under most ecological conditions, tolerant of each other's presence to the extent of going to the same waterhole at the same time, although home-range habit tends to keep them otherwise apart;

(2) characterized internally by a balancing of aggressiveness and protective-

ness the dynamics of which are excellently worked out by DeVore (1962). The dynamics of intra-group and inter-group tensions in baboons show variation around a clear basic pattern which closely resembles that of *Macaca fuscata* (Imanishi, 1957), and it is the pattern-plus-variations which needs to be used for comparison with other species or genera on which the data have been obtained under similar environmental conditions.

ALARM AND AVOIDANCE

We have already considered two components which are common to this and to the threat-attack kind of situation, namely the "observing response" of hindleg-standing, and possibly the side-to-side head movement. Some of the components now to be considered are also to be seen in, for example, the social situation of presenting, because the presenting animal may also be on the alert for being struck at, and so may look round at the dominant animal, and be ready to rush away.

In baboons, as in other monkeys, defecation is a common alarm-reaction readily elicited by threat-attacks, and it is unlikely that this, or any frequency data relating to it, will be of much comparative interest. The rest of the repertoire is fairly complex, and the "meaning" of some of the components is not yet clear. For example, I have observed many instances in *P. ursinus* groups where the tail of a frightened animal (the circumstances of fright being clear in the situation) has gone up from the normal relaxed "riding-whip" position to almost vertically upright, except possibly for the distal section. Yet the exact meaning of these tail-changes, and hence their possible relevance for comparison, is difficult to determine. Even in laboratory observations, we have found great difficulty so far in correlating tail-position variations in Patas monkeys with situations in the environment or with the "mood" of the animal. In baboons, the near-vertical position seems to occur rather frequently in mothers carrying infants, either dorsally or ventrally, but the frequency-data (Table 1) obtained from observations of one group in the Cape in 1961 do not seem to link this with any supporting function that the dorsally-riding young might require.

Tail-angle in baboons, according to a zoo study by Bopp (1954), is said to be one indicator of relative social status within a group. In other words, the degree of fear that one animal might have for another might be partly expressed by the tail's verticality in the subordinate when the dominant one came near and the former presented to him or her. Certainly I have observed in this same large Cape group that adult anoestrous females, in a feeding-space near one of the dominant males, moved away from him with their tails going straight upright as his feeding took him near them, but it is not clear, at present, how such a "signal" fits into the general pattern of submissive behavior such as presenting, and it is surely reasonable to suppose that these expressive movements, when taken in conjunction with others in the repertoire, may have some distinguishing value

TABLE 1

FREQUENCIES OF THREE VARIATIONS IN TAIL-ANGLE IN FEMALE BABOONS
WITH INFANTS ATTACHED TO THEM

| Angle of Tail | Infant Position on Female | | |
	Ventral	Dorsal	Total Observations
Ordinary (riding-whip)	14	30	44
Intermediate	6	10	16
Near-vertical straight	13	10	23
Total observations	33	50	83

These figures suggest that the angle of the females' tail, which is, in fact, proportionally more often near-vertical when the infant is ventral (13 occasions out of 33, i.e., 43 per cent, as compared with 10 out of 50, i.e., 20 per cent, when the infants are riding dorsally), may be one quite important form of expression of "nervousness" or "anxiety." It is also possible that the higher proportional frequency of near-vertical tail in the females with ventral infants may be an indicator of their greater nervousness than those females with the rather older young who usually ride dorsally. That near-vertical is an indicator of fear is supported by the following instances:

SITUATION	BEHAVIOR-PATTERN IN WHICH NEAR-VERTICAL TAIL REACTION IS A COMPONENT
Discovery of live scorpion.	General startle-reaction of adult female.
Discovery of large mole-snake.	Two dark-phase infants riding on their mothers' backs have their tails vertically upright as they gaze at the snake.
Chase by dominant male.	Females rushing away shrieking.
Squabble amongst females.	Threatened, and giving-way, females shrieking and tails waving at vertical.
Presenting to dominant male.	Immature male showing generally nervous behavior as he backed up to the dominant animal.

between monkey species and genera. For example, stump-tailed species of the *Mandrillus* and *Macaca* genera might be expected to show some differences in expressive pattern compared with long-tailed species of Patas and *M. radiata*. Indeed, the Japanese studies report one significant tail-gesture difference between *fuscata* and baboons, namely that elevation of the stumpy tail of a *fuscata* adult male is indicative of dominance.

Facial expressions and head-gestures also seem to have an important communicative function in alarm and avoidance situations, but this is complicated apparently, as are the other parts of the repertoire, by the arousing of alternating or even simultaneously-operating tendencies. Van Hooff (loc. cit.) thus refers to four basic facial expressions, to be observed with many variations in different species, namely the "attack" face, the "aggressive threat" face, the "scared threat" face, and the "crouch" face. The "scared threat" face is the expression usually designated "fear grimace." A subordinate baboon may thus grimace when passing a dominant one, and the reverse has never been observed. In the Chacma baboon, the male sometimes, during the act of copulation, bares his teeth in a similar fashion, but the significance of this gesture, and how it relates to or differs from other facial expressions, is not clear.

Most experimental studies of "fear" in non-human primates have been concerned with working out what sort of stimuli, objects, or situations elicit this emotion, the expression of which is, in a general way, assumed to be recognizable in the different species. Examples are the chimpanzee studies by Haslerud (1938), Yerkes (1943), and Hebb (1946). However, it is now fairly clear that most of the habits of avoidance of certain classes of noxious objects, such as snakes or predators, are the product of early, juvenile learning in the monkey and ape group. A field example where this is indicated, but where the objects given were also used to elicit the startle-avoidance repertoire in distinctive form, was carried out by the writer on the S group of baboons in the Cape on which such experimental situations could be imposed at very close range indeed.

When given a live scorpion in an envelope, or when a live snake was placed under a stone in such a way that the baboon did not see it but was led to expect a food-reward there, the usual startle-reactions comprised the following:

(1) if the animal was sitting when it discovered snake or scorpion, it flung its arms wide apart and jumped back a short distance;

(2) if it was standing, it tended to leap clear of the ground with all four limbs simultaneously, and this reaction is also to be observed when a baboon comes upon a snake in the normal course of its searching for food;

(3) On several occasions, the males of this group, on finding the scorpion, at once swiped quickly at it with one hand, thus knocking it away from them.

Three other behavior-components seen during these tests were a quick shaking of the head from side to side immediately following the startle-response of widely outflung arms, a quick "shrugging" movement of the shoulders as one male walked away from the scorpion, and an action of drawing the hand down and over the muzzle. The latter reaction has been seen in a recently-captured male baboon at Kariba, but neither this nor the other two reactions seem common, and they have only so far been observed in males.

The hitting-away reaction may, it seems, be one which could readily develop into the throwing-action seen sometimes in captive baboons, but it is also evident that certain kinds of alarm or defense behavior are only elicited when the animals are, in nature, intensely aroused, as, perhaps, in the presence of a leopard.

An interesting variant of alarm-behavior sometimes occurs where several animals are involved in a squabble. One may appear partly to mount the other, but supporting itself with one hand over the other's neck or shoulders. However, this does not seem to be a "displaced" mounting, because usually the "mounter's" body is alongside, and not over, that of the other animal. The two animals may thus jerkily point their muzzles in the direction of the other quarrelers.

Alarm at the presence of intruders, or predators, is usually expressed by barking, and similar vocalizations are known for many other monkeys, but the first reactions of a group differ considerably from area to area, day to day, and sometimes, if very closely and suddenly encountered, no vocalization at all will occur, the animals just galloping away. It is also of interest to note, though this may well again be a group reaction common to many species, what Hall (1960) has termed the halt-wait reaction when a group is encountered at its sleeping-place in the morning, and, instead of setting off on its day-range, sits and waits near the cliffs sometimes for two or three hours. It is probable that the security of the sleeping-cliffs is a factor operating here. Defense reactions against other kinds of animal have been notably the function of the dominant male in those groups where one is markedly dominant, but several large males are known simultaneously to interpose themselves between the cause of danger and the rest of the group (Washburn and DeVore, 1961).

There are certain characters here which may usefully lead on to distinguish the baboon genus from other genera, such as the hitting-away of objects, though this is known to occur at least in chimpanzees, but more especially the movements and positions of the tail. There is little to go on at present except where rather gross differences have been featured. Thus, Booth (1957) describes "freezing" behavior in the Olive Colobus monkey when frightened, individuals ". . . adopting on the spot an attitude which has been observed in no other wild species of monkey. The animal sits on its haunches among the dense foliage, with the head bowed down on the abdomen . . . From this position, it is quite unable to observe any further movement on the part of the intruder, and it only darts away if there is further noise from the observer" (p. 427). Carpenter (1934, 1935) describes "bluffing" and concealment as a reaction of howlers and Red spider monkeys to human observers, but there is insufficient close-range observation of wild species to know even a small part of the agonistic behavior-repertoire for comparative purposes.

CONFLICT BEHAVIOR

From such evidence as is available on the reactions of wild monkeys in situations of conflict or drive-frustration, basic species, as opposed to individual, differences may not be very marked. In other words, it seems that very similar kinds of displacement activity, redirection of drive onto substitute objects, and ambivalent behavior (using the categories of Bastock, Morris and Moynihan, 1953) have been recorded amongst widely differing genera, so that the evolutionary interest of these behavior-patterns lies more in the possibility of tracing interactions between learning and emotional responses which are not so readily apparent in lower animals or in man himself.

Yawning we have found to be described as a part of threat-behavior, but reliable evidence indicates that it often occurs in a disturbing situation where it looks like a simple tension-relieving reaction that may be classed as a displace-

ment activity (see Bolwig, 1959; and Hall, 1962b, on *P. ursinus*). I have recorded
it recently on fifteen occasions in adult males of a group whose general behavior
suggested that they were "nervous" at the close presence of the observer, but
not sufficiently nervous actually to move away. It is possible here that two or
three, at least, "incompatible tendencies," such as to escape, to continue feeding,
or to attack, are simultaneously aroused in the dominant animals. That it does not
necessarily involve any threat-intention is indicated by the fact that the yawning animal as often as not has his back to, or is sideways on to, the observer.
Male Chacmas have also been observed to yawn while watching a cow Eland coming threateningly towards the group, and prior to running away from the Eland,
and when approaching close up to view a stuffed Serval cat (Hall, 1960) of
which they were very nervous but which they made no attempt to attack.

Support for the displacement activity explanation comes from Carpenter
(1934, p. 27): "When Howlers are approached by an observer they frequently
yawn with a slow inspiration of breath. This behavior occurs when the animals
are greatly excited." Likewise in the Gibbon, according to Carpenter (1940, p.
168), yawning is associated with frustration, and: "The sequence of behavior
patterns is "understood" by gibbon associates and it exercises a kind of immediate
social control."

Scratching is also a not uncommon "nervous" reaction in both monkeys and
apes. Bolwig (1959) and Hall (1962b) have reported it for the Chacma, and
Carpenter (1935, p. 173) describes Red spider monkeys approaching observers
along the tree-branches and shaking the branches while ". . . concurrently and almost invariably vigorous scratching occurs." In our observations of baboon
groups, cases of genuine "nervous" scratching have not been common, but neither
has the kind of situations likely to evoke such a displacement activity. Köhler
(1925, p. 260) describes this kind of behavior in chimpanzees as taking two forms:
(1) "scratching the head when uncertain and in doubt," (2) "to scratch the whole
surface of the body, especially the arms, the breast, the upper portion of the
thighs, and the lower abdomen, and against the direction in which the hair grows,
is expressive of a wide diversity of emotions." Form (2) may well be one general indicator of a certain level of "anxiety," as we find to be the case in our
captive Patas monkeys in whom scratching-frequency increases very much when
they are in a disturbing situation to which there seems no direct way out (Hall,
to be published), but it is very doubtful if, in the chimpanzee or any other
primate, it signalizes any differentiated emotional state.

Social and sexual behavior-patterns are reported in some species of monkey to
occur in circumstances which indicate that they may be classed as displacement
activity. Carpenter (1942b) noted that, in *M. mulatta*, presenting, at times other
than when it is a greeting response, is a response to generalized, non-sexual excitement, ". . . as when animals are too closely approached by an observer or when
an intra-group fight occurs" (p. 132). Mounting, in the rhesus, and even copulation in the howler (Carpenter, 1934) similarly occur out of their normal context.

Thus, when a group of howlers had been frightened by a pistol shot, a male returned to a female trailing behind the clan, and copulation took place almost as soon as he reached her. Homosexual mountings between subordinate young rhesus males were seen by him to occur during the excitement of aggressive action. He suggests that the tension and emotional excitement persistently characterizing these subordinate animals with insecure social status is the cause of this behavior.

In the Chacma baboon, we have observed only a few instances where mounting or actual copulation has occurred in circumstances of general excitement. The alarming situation created in the group by the presence of the stuffed Serval cat certainly elicited rapid mounting by the dominant male in one group, and by what appeared to be actual copulation between an adult male and an oestrous female in another group.

It is likely that other behavior-patterns that form a part of the maintenance and the social repertoire of a species, for example those concerned with feeding and grooming, may also occur out of normal context in these animals. For example, experimental frustration has, following upon a series of yawns, elicited a rapid and nervous fumbling with and nibbling of food pellets in an adult male Patas (Hall, to be published) in a manner quite unlike his normal feeding. Carpenter's (1934) interesting account of "sham feeding" in Howlers carries the problem even further, for he describes this feeding behavior as occurring in an animal that was disguising its intention to do something, such as to approach close to another member of the group. It is, however, possible that the feeding was primarily, as in the Patas instance, indicative of other conflicting tendencies, such as desire to approach and also to keep away from the other animal, and the "purposiveness" of the act can only be assumed.

The behavioral status of different kinds of conflict-induced response is a matter which may well be of phylogenetic significance, although one may discover, after thorough study, that individual differences in this respect within species are very great. "Redirections" of an agressive tendency onto substitute objects or lower-status animals in a group are quite widely reported. Instances of the latter type of redirection are reported in wild baboon groups. Thus, the dominant male of one Cape group was about ten times more aggressive than normal against members of the group on a day when he had been repeatedly frustrated from attacking a strange baboon which sheltered behind the observer as soon as it was threatened (Hall, 1962b). Similarly, DeVore (1962) describes instances of "transferred threat" where the threatened male chases after a subordinate animal, and so on, and he supposes that this kind of process reinforces the hierarchical system. To what extent this process, supposedly tension-relieving for the individual as well as possibly socially reinforcing, has any distinctiveness amongst the primates is not yet known.

Instances of apparent redirection onto inanimate objects sometimes appear in the threat-pattern of a species. Repeated and vigorous pulling upon branches or rocks or cage-bars, while facing towards the observer, is seen in baboons and other monkeys. Köhler (1925) describes one of his chimpanzees seizing handfuls of grass and herbs, and tearing at them till the bits were strewn around her, these gestures always being partially directed toward the enemy. Yerkes and Yerkes (1929) also describe how frustrated chimpanzees will attack inanimate objects or obstructions as though they were living creatures.

The question whether such redirective gestures and the objects associated with them are "aimed" or not towards the intruder or predator has been much debated, because the aiming or throwing of a missile is assumed to be indicative of a higher intellectual organization imposing itself upon the emotional pattern. Zuckerman (1932) was inclined to the view that the many accounts of monkeys hurling branches and fruit from trees could be simply explained as the more or less accidental outcome of such redirective displays, and thus did not imply any perception of the relationship between such acts and the possible consequence of driving away the intruder. The kind of evidence that favors the "higher" form of interpretation is usually not explicit or detailed enough. For example, a certain Professor Bell is quoted as follows in Kinnaman (1902) with respect to such performances in a wild rhesus monkey group: "I distinctly saw one monkey industriously, with both forepaws, and with obvious malice of pretence, pushing the loose shingle off a shoulder of rock," while he was standing below a landslip on which this group of monkeys was located.

The parallel between such cases and "insight" versus "trial-and-error" kinds of status-interpretation in animal problem-solving will be obvious, and the probability is that the overriding factor not yet assessed is the degree to which such performances vary amongst individuals of a species, and between groups in different parts of the distribution area. In baboons, for instance, local conditions and tradition may determine whether or not individuals roll rocks down at predators just as they determine whether or not they include scorpions in their diet. Hence, even the clear statements of "purposiveness" in such acts, as in Wallace's (1902) report on the orangutan, and Carpenter's (1934, 1935, 1940) records for the howler, red spider monkey, and gibbon, must be used comparatively with the same sort of caution that one should bring to bear upon relative estimates of tool-using skills in laboratory primates.

In this section, we have been dealing rather concretely with the two interrelated problems that are fundamental to the taxonomic and evolutionary significance of the behavior data. On the one hand, we have needed to consider how study of a kind of emotionally expressive repertoire, including a set of facial and posture gestures, and a number of "out-of-context" reactions indicative of conflict, might be used as a means of distinguishing species or genera. The eventual scheme will no doubt comprise a set of basic expressive "plans" with a great range of variations, the place of which in the adaptation pattern of the species will be cogently explained in the same sort of way as are, or will be, the physical adaptations. On the other hand, we have also briefly considered one aspect of the emotional adaptability of the animals which may have some evolutionary significance, for the ability or otherwise of individuals or species to steer their emotional energies into the solving of important problems is surely, even in its "primitive" form, a matter worthy of very careful study.

2. AFFECTIONAL BEHAVIOR

As a class of behavior, this could be used to include all behavior-patterns or their components whereby cohesion between individuals of a group might be facilitated, other than through mating or threat-attack. It would thus include grooming, mounting, presenting, and any other special "greetings" behavior that may be recorded. It would also include the affectional relationship between mother and infant. We shall consider, however, two aspects only, grooming and greetings behavior.

GROOMING

It seems that social grooming may occur in all wild monkeys and apes, and therefore that comparison in its forms and frequencies may be useful. However, while there are quite a lot of statements about grooming in quite a number of species, the evidence is most uneven and, in most cases, falls far short of the standards of reliability required for comparison. Nevertheless, certain trends are interesting, and would be valuable if firmly established. Thus, Carpenter (1934) mentions grooming on only two occasions in his study of the howler, once to say that the tail may be used ". . . as a currying or grooming organ" (p. 29), presumably for self-grooming, and once when describing a mother cleaning up her wet, newborn infant. Haddow (1952) describes grooming as the only mutual behavior that he observed in wild Red tail monkeys, but says it is not common, while Nolte (1955) says that, in groups of wild Bonnet monkeys, *M. radiata*, "Mutual grooming occupied a considerable part of the day" (p. 179).

According to Zuckerman (1932), grooming is common in the zoo Hamadryas baboons, and probably functionally identical with that of the chimpanzee in whom it may be most highly developed. There is thus evidence that ". . . the (grooming) pattern increases both in complexity and in social significance between prosimiae and anthropoid ape" (p. 10). Indeed, Zuckerman was much impressed by the diversity in the grooming characteristics of different primates, and questioned what the evolutionary significance of such differences in frequency and value might be. Yerkes (1933) recorded his view that grooming occurs more commonly and is more definitely socialized in the Old World than in the New World monkeys, but none of these statements has been fully substantiated, nor has that of Maslow's (1935) to the effect that the "mutuality" of grooming is so marked in chimpanzees "that they show, for the first time in the primate order, something that may justifiably be called 'social service' "

(p. 715). Nissen (1931) does not mention the occurrence of grooming at all in his field study of chimpanzees.

From these samples of statements about what is recognized to be a most important social behavior pattern in monkeys and apes, we can see that the task, at present, of making worthwhile comparisons along this axis is extremely difficult. And the reason is mainly that there is little or no experimental or naturalistic evidence about the *general* factors which affect frequency or quality of grooming performances in any species. Thus, we do not really know how validly we can compare statements derived from the study of zoo animals, who may have little else to do all day than groom, with statements derived from field studies where conditions of observation may have been difficult and variable, and where, ,anyway, the animals in most species have to spend most of their day searching for food. Nor do we have any information as to the effects of differing climatic or dietary factors upon grooming frequency.

The problem of determining these general factors before embarking on comparative statements is again illustrated by the fact that in wild baboon groups, grooming is much more common in the early part of the day, and again in late afternoon, than in any other periods (Washburn and DeVore, 1961; Hall, 1962b). Further, not only the relative prominence of this behavior must be known in the natural group, but also the extent of its mutuality. Thus, Bolwig (1959) reports that grooming in the Chacma baboon is essentially a female behavior, i.e., females are the active groomers in many more pairs than are the males. Hall (1962b) analyzed 294 occurrences of grooming in a single group of these baboons. Females were the groomers in 162 cases as against 53 cases by adult males, a ratio of about 3:1 which actually corresponds to the numerical ratio of adult female to adult male in the group. However, the female groomers spent far longer in undivided occasions of grooming than did the males, and, further, anoestrous females were the groomers nearly three times more often than they were groomed. It is also probable that females with young infants are especially favored by grooming attentions. The concept of "mutality" can thus, it seems, be made objectively meaningful for this aspect of the behavior of a species, so that adequate comparisons can then be made with other species in natural conditions.

GREETINGS BEHAVIOR

The form in which gestures of limb, face, voice, and postural changes, are used by a species in casual greetings or social exchanges independently of any primary sexual behavior has received little detailed comparative examination. Carpenter (1942a) describes "greetings ceremonies" between monkeys as including particular facial expressions, a momentary embrace, and a special little cry. Köhler (1925) describes embracing between his chimpanzees, and a "friendly form of welcome" when one animal places the hand on the groin of a sitting animal, or the seated ape grasps his companion's hand, placing it between his thigh and his abdomen, and patting it with his own hand. When the animal is in a standing position, a "friend of the same species" may greet it by placing his hand between its thighs. Köhler describes all such gestures as examples of ". . . the 'borderland' between sociability and secondary sexual manifestations."

Similar behavior-sequences have been observed and analyzed by Hall (1962b) for one group of Chacma baboons. Amongst the components noted were lipsmacking, standing on hindlegs, embracing, genital-stomach nuzzling, and touching with one hand. Sometimes only lipsmacking and touching occur, sometimes a complete performance, but, unlike presenting or mounting, this is essentially a mutual performance engaged in by both animals. An analysis has been made to ascertain if there are any significant differences in the form of the greetings between males and females, adults and juveniles, dominant and subordinate animals. Mutuality is suggested by the fact that the dominant of two females is sometimes observed definitely to initiate the greeting, approaching, while lipsmacking, the subordinate female, holding her briefly under the groin and putting her head down to the subordinate's pelvic region. It is also a performance that the dominant male himself will initiate with other animals of his group.

An interesting variant of greeting the dark-phase infants of this group was seen in males and in females other than the mother. The baboon approached the infant and its mother, lipsmacking vigorously. On reaching the infant, the greeter sometimes stood on its hindlegs and picked up the infant by its backlegs and "kissed" its rump. Sometimes the greeter, standing on hindlegs, just picked up the infant and clasped it briefly against its stomach. Differences in all forms of "affectional" behavior of wild monkeys and apes towards the infants of their group are likely to be taxonomically relevant, but far too little close-range observational data is as yet available, and such intimate social performances as these cannot possibly be reliably observed unless the group and the observer are on very close-range terms. Indeed, one of the most striking results of the accumulation of reliable field data obtained under the ideal conditions of close-range observations of undisturbed groups is the vastly different general picture of what is "normal" in the groups' behavior. The intimacies of daily social behavior would not be seen in groups which were not completely habituated to the observer.

A PROVISIONAL ASSESSMENT

The purpose of reviewing briefly and selectively some of the data, adequate and inadequate, on the agonistic and affectional behavior patterns, using mainly baboon data for reference, has been to arrive at some sort of assessment of the stage now reached in systematic and evolutionary behavior research on the non-human primates. Other facets of the total behavior picture might with equal advantage have been chosen for examination, for example the mating pattern or the whole basic dimension of behavior ontogeny, comparatively treated.

Many of the limitations of method and scope of inquiry have been emphasized, and this may be a not unfair reflection upon the state of a field of inquiry which long suffered from scientific neglect or, in the experimental realm, from the intellectualistic bias mentioned in the historical preamble to this paper. Pioneers such as Zuckerman and Carpenter may well wonder how it is that their leads should have taken so long to follow, and there are no doubt a number of historical and practical factors to account for this. As a mere statement of fact, it was generally the case that the most important developments in comparative animal behavior study came from work on the lower vertebrates using inductive methods of observation sometimes followed by experiment, while the psychologists concentrated chiefly upon the learning processes of monkeys and apes that could be revealed in the laboratory, and the naturalistic and the experimental, the field and the laboratory, exponents hardly began to converge upon the non-human primates, or to consider explicitly such matters as training and method as applying to the study of these animals, until very recently indeed.

The rapprochement we can see starting amongst disciplines and methods in tackling the primate behavior data should also lead, it is hoped, to the fitting of both the ecological and behavioral data into the overall comparative animal framework. In this way, it seems probable that the significance of differences and similarities can be more clearly brought out by such comparison in depth than within the narrower confines of comparison amongst the living primates.

It is finally necessary to ask what it is that the systematic accumulation and functional analysis of the behavior data on these animals can tell us that may be relevant to our scientific understanding of the behavior of living human beings. The scientific exercise of prolonged systematic observation, preferably both in field and laboratory situations, of animals as complex as these could scarcely be bettered as a discipline for the scientific investigation of human behavior in all its diversity. Rather than stopping at the brink of the task—by simply extrapolating from the non-human level of discourse to the human—one may hope that the investigator may derive significant hypotheses for experimental and observational testing on the human material.

BIBLIOGRAPHY

BASTOCK, M., D. MORRIS and M. MOYNIHAN
1953. "Some comments on conflict and frustration in animals." *Behaviour*, 6:66–84.
BOLWIG, N.
"A study of the behavior of the Chacma baboon, *Papio ursinus*." *Behaviour*, 14:136–163.

298   CLASSIFICATION AND HUMAN EVOLUTION

Booth, A. H.

1957. "Observations on the natural history of the Olive colobus monkey, *Procolobus verus* (van Beneden)." *Proc. Zool. Soc., London,* 129.421–430.

Bopp, P.

1954. "Schwanzfunktionen bei Wirbeltieren." *Rev. Suiss. Zool.,* 61:83–151.

Carpenter, C. R.

1934. "A field study of the behavior and social relations of Howling monkeys." *Comp. Psychol. Monogr.,* 10, No. 2, Serial No. 48.

1935. "Behavior of Red spider monkeys in Panama." *J. Mammalol.,* 16:171–180.

1940. "A field study in Siam of the behavior and social relations of the gibbon." *Comp. Psychol. Monogr.,* 16:38–206.

1942a. "Societies of monkeys and apes." *Biol. Sympos.,* 8:177–204.

1942b. "Characteristics of social behavior in non-human primates." *Trans. N. Y. Acad. Sci.,* 4:248–258.

1942c. "Sexual behavior of free ranging Rhesus monkeys (*Macaca mulatta*). I: Specimens, procedures, and behavioral characteristics of estrus." *J. Comp. Psychol.,* 33:113–142. "II: Periodicity of estrus, homosexual, autoerotic and non-conformist behavior." *J. Comp. Psychol.,* 33:143–162.

1958. "Soziologie und Verhalten freilebender nichtmenschlicher Primaten." *Handbuch der Zoologie, Band 8 (Mammalia)* Teil 10. Berlin.

Clark, W. E. Le Gros

1959. *The antecedents of man.* Edinburgh: Edinburgh University Press.

Darwin, C.

1872. *The expression of the emotions in man and animals.* London: Murray.

1874. *The descent of man.* 2nd ed. London, Murray.

DeVore, I.

1962. "The social behavior and organization of baboon troops." Ph.D. thesis, University of Chicago.

Emlen, J. T.

1958. "The art of making field notes." *The Jack-pine warbler,* 36:178–181.

Emlen, J. T. and G. Schaller

1960. "In the home of the mountain gorilla." *Animal Kingdom* (Bull. N. Y. Zool. Soc.), 63:98–108.

Grassé, P.-P.

1952. "Le fait social: sees critères biologiques, ses limites. In *Structure et physiologie des sociétés animales.* Paris; CNRS.

Haddow, A. J.

1952. "Field and laboratory studies on an African monkey, *Cercopithecus ascanius* Schmidti." *Proc. Zool. Soc., London,* 122:297–394.

Hall, K. R. L.

1960. "Social vigilance behaviour in the Chacma baboon, *Papio ursinus.*" *Behaviour,* 16:261–294.

1962a. "Numerical data, maintenance activities, and locomotion in the wild Chacma baboon, *Papio ursinus.*" *Proc. Zool. Soc., London,* 139:181–220.

1962b. "Sexual, derived social, and agonistic behavior patterns in the wild Chacma baboon, *Papio ursinus.*" *Proc. Zool. Soc., London,* 139:284–327.

In press. "Variations in the ecology of the Chacma baboon, *Papio ursinus.*"

To be published. "The behaviour of monkeys to mirror-images."

HARLOW, H. F.
1951. "Primate learning." In C. P. Stone, *Comparative psychology*. London: Staples.

HASLERUD, G. M.
1938. "The effect of movement of stimulus objects upon avoidance reactions in chimpanzees." *J. Comp. Psychol.*, 25:507–528.

HEBB, D. O.
1946. "On the nature of fear." *Psychol. Rev.*, 53:259–276.

HINDE, R. A., and N. TINBERGEN
1958. "The comparative study of species-specific behaviour." In A. Roe and G. G. Simpson, *Behavior and evolution*. New Haven: Yale University Press.

HINDE, R. A. and T. E. ROWELL
1962. "Communication by postures and facial expressions in the rhesus monkey (*Macaca mulatta*)." *Proc. Zool. Soc. London*, 138:1–21.

HOBHOUSE, L. T.
1901. *Mind in evolution*. London: Macmillan.

IMANISHI, K.
1957. "Social behavior in Japanese monkeys, *Macaca fuscata*." *Psychologia*, 1:47–54.

KINNAMAN, A. J.
1902. "Mental life of two Macacus rhesus monkeys in captivity II." *Amer. J. Psychol.*, 13:173–218.

KÖHLER, W.
1925. The mentality of apes. London: Kegan Paul.

KLÜVER, H.
1933. *Behavior mechanisms in monkeys*. Chicago: University of Chicago Press.

KUMMER, H.
1956. "Rang-kriterien bei Mantelpavianen." *Rev. Suiss. Zool.*, 63:288–297.
1957. "Sociales verhalten einer Mantelpavian Gruppe." *Schweiz. Zeitschr. Psychol.*, No. 33, 91 pp.

MASLOW, A. H.
1935. The dominance drive as a determiner of social behavior in infra-human primates. *Psychol. Bull.*, 32:714–715.

MASON, W. A., GREEN, P. C. and C. J. POSEPANKO
1960. "Sex differences in affective social responses of rhesus monkeys." *Behaviour*, 16:74–83.

MAYR, E.
1958. Behavior and systematics. In A. Roe and G. G. Simpson, *Behavior and evolution*. New Haven: Yale University Press.

MORGAN, C. L.
1890. *Animal life and intelligence*. London: Arnold.

NISSEN, H. W.
1931. "A field study of the chimpanzee." *Comp. Psychol. Monogr.*, 8, No. 1, serial No. 36.
1951. "Social behavior in primates." In C. P. Stone, *Comparative psychology*. London: Staples.

NOLTE, A.
1955. "Field observations on the daily routine and social behavior of common Indian monkeys, with special reference to the Bonnet monkey (*Macaca radiata* Geoffrey)." *J. Bombay Nat. Hist. Soc.*, 53:177–184.

ROMANES, G. J.
"On the mental faculties of the bald chimpanzee (*Anthropopithecus calvus*)." *Nature*, 40:160–162.

ROWELL, T. E. and R. A. HINDE
1962. "Vocal communication by the Rhesus monkey (Macaca mulatta)." *Proc. Zool. Soc. London*, 138:279–294.

SCHNEIRLA, T. C.
1950. "The relationship between observation and experimentation in the field study of behavior." *Ann. N. Y. Acad. Sci.*, 51:1022–1044.

SCOTT, J. P.
1956. *The analysis of social organization in animals. Ecology*, 37:213–221.

STEVENSON-HAMILTON, J.
1947. *Wild life in South Africa.* London: Cassell.

THORNDIKE, E. L.
1901. "The mental life of the monkeys." *Psychol. Rev. Mongr. Suppl.*, No. 15.

THORPE, W. H.
1956. *Learning and instinct in animals.* London: Methuen.

VAN HOOFF, J. A. R.
In press. *Facial expressions in primates.*

WALLACE, A. R.
1902. *The Malay Archipelago*, 10th ed. London: Macmillan.

WARDEN, C. J., T. N. JENKINS and L. H. WARNER
1936. *Comparative psychology*, vol. 3 "Vertebrates," Ch. 6, "Primates." New York: Ronald.

WASHBURN, S. L. and I. DEVORE
1961. "The social life of baboons." *Sci. Amer.*, 204:62–71.

YERKES, R. M.
1933. "Genetic aspects of grooming, a socially important primate behavior pattern." *J. Soc. Psychol.*, 4:3–23.
1943. *Chimpanzees: a laboratory colony.* New Haven: Yale University Press.

YERKES, R. M. and A. W. YERKES
1929. *The great apes.* New Haven: Yale University Press.
1935. "Social behavior in infrahuman primates." In C. Murchison, *A handbook of social psychology.* Worcester, Mass.: Clark University Press.

ZUCKERMAN, S.
1932. *The social life of monkeys and apes.* London: Kegan Paul.

A COMPARISON OF THE ECOLOGY AND BEHAVIOR

OF MONKEYS AND APES[1]

IRVEN DeVORE

COMPARISONS OF NONHUMAN PRIMATES have traditionally contrasted the behavior patterns of New World and Old World monkeys. The Platyrrhines of the New World are said to live in loosely organized social groups in which individuals are rarely aggressive and dominance behavior almost absent. The social group of the Old World Catarrhines is described as more rigidly organized by social hierarchies, based on dominance-oriented behavior and frequent fighting among adult males. To the extent that this distinction is valid, the behavioral differences being compared are not those between New and Old World monkeys, but between arboreal and terrestrial species. The only systematic field studies of New World monkeys have been on the howler monkey, *Alouatta palliata* (Carpenter, 1934; Collias and Southwick, 1952; Altmann, 1959), and Carpenter's (1935) brief observations on spider monkeys, *Ateles geoffroyi*. Both of these species are highly specialized, morphologically and behaviorally, for living in the tall trees of the South American jungle. They are seldom seen in the lower branches of trees and almost never come to the ground. On the other hand, behavioral studies of Old World monkeys, principally those of Zuckerman (1932) and Carpenter (1942), had until recently included only species in the baboon-macaque group. In both morphology and behavior the baboon-macaques are more terrestrially adapted than any other monkey or ape.

Comparisons between monkey and ape behavior have also been difficult, since Carpenter's gibbon study (1940) is the only long term study of an ape that has previously been available. This Asiatic brachiator is as highly specialized for arboreal life as the baboon-macaques are for life on the ground. If the monkeys and apes are arranged along a continuum with those that are terrestrially adapted at one pole and those with specialized arboreal adaptations at the other (Fig. 1), it is clear that long-term naturalistic observations have been

1. The preparation of this paper was supported by a National Science Foundation grant for the analysis of primate behavior, while the author was a fellow of the Miller Institute for Basic Research in Science, University of California, Berkeley. In addition to the conference participants, George Schaller and Richard Lee read and commented on this paper. The suggestions of all these persons are gratefully acknowledged, though sole responsibility for its final content rests with the author.

confined almost entirely to species lying at the two extremes and that little has been known of the majority of species falling somewhere in between. The species shown in Figure 1 are those for which some field data are available, and the intention is to suggest only the broad outlines of adaptation to life on the ground or in the trees.

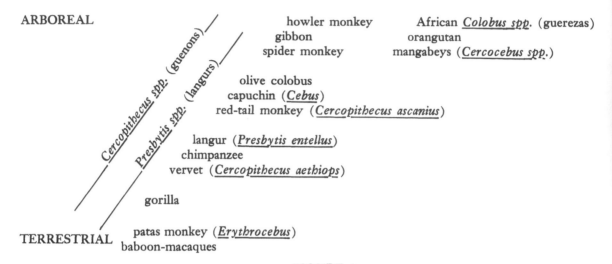

FIGURE 1

Schematic representation of selected species of monkeys and apes along a continuum of relative adaptation to terrestrial or arboreal life.

Napier (1962) has discussed the habitats of monkeys in detail. In brief, spider monkeys, howler monkeys, gibbons, orangutans, and most species of colobus and mangabeys live in the higher levels of mature, tropical forests. The olive colobus lives by preference in the lower forest level, seldom higher than twenty feet from the ground, yet not descending to the ground (Booth, 1957); South American capuchin and African red-tail monkeys, though primarily tree dwellers, may also come to the ground to feed (Carpenter, 1958a, Haddow, 1952). Langurs, vervets, *Cercopithecus l'hoesti*, and chimpanzees seem equally at home on the ground or in trees; and gorillas, patas monkeys, baboons, and macaques are clearly adapted to terrestrial life. Field studies of nonhuman primates are expanding at a rapid rate, with more than fifty workers currently engaged in field research (DeVore and Lee, in press), and in a few years it will be possible to refine greatly the provisional conclusions discussed here. In the following pages the available evidence on home range, intergroup relations, population density, and social behavior is reviewed with respect to adaptation and taxonomy.

HOME RANGE

The size of the area which an organized group of animals customarily occupies, its "home range," varies widely among the primates. Current studies have revealed a high correlation between the size of home range and the degree to which a species is adapted to life on the ground—the more terrestrial the species, the larger its home range (Table 1). Arboreal gibbons occupy a range of only about one tenth of a square mile, and a group of howler monkeys range over about one half of a square mile or less of forest. Since the average gibbon group numbers only four and howler groups average about seventeen, the amount of range needed to support an equivalent amount of body weight in each species is comparable. These home ranges are small, however, by comparison to the home ranges of langur groups, *Presbytis entellus*, studied by Jay (unpub. ms.). Langurs, which frequently feed on the forest floor and raid open fields, range over an average area of three square miles in a year—even though their average group size (twenty-five) is much the same as howler monkeys.

Terrestrial species like the mountain gorilla and the baboon occupy much larger home ranges. Schaller and Emlen (in press) found that gorillas in the Virunga Volcanoes region of Albert National Park, Congo, have an average group size of seventeen, and customarily travel over an area of from ten to fifteen square miles during a year. An average troop of forty baboons, living on the East African savanna in the Royal Nairobi National Park, covers an area of about fifteen square miles during a year. In general, it is clear that a terrestrial adaptation implies a larger annual range; a baboon troop's range is 150 times as large as that of a gibbon group. Even if these figures are corrected to allow for the greater combined body weight of the individuals in a baboon troop, the baboon home range is still at least five times as large as that of gibbons. It is frequently stated that the habitat of arboreal monkeys is actually three-dimensional, and that a vertical dimension must be added to horizontal distances traveled before the true size of a group's range can be accurately determined. Many arboreal species, however, seldom come to the ground, or even to the lower levels of the forest. Howler monkeys are largely confined to the upper levels of primary forest, avoiding secondary growth and scrub forest; gibbons ordinarily stay just beneath the uppermost forest canopy; and most species of African colobus monkeys appear to stay above the shrub layer of primary forest. Limited to arboreal pathways through the forest, such species are far less free to exploit an area completely than are primates that do not hesitate to descend to the ground. In effect these arboreal species occupy a two-dimensional area above the forest floor. Some species which exploit both the arboreal and terrestrial areas of their range with facility, such as South American capuchin monkeys, Indian langurs, some African *Cercopithecus spp.*, and probably chimpanzees, may be accurately described as living in a "three-dimensional range." Even terrestrial species such

TABLE 1
SUMMARY OF AVAILABLE DATA ON GROUP SIZE, POPULATION DENSITY, AND HOME RANGE FOR MONKEYS, APES, WOLVES, AND HUMAN HUNTING GROUPS LIVING IN RELATIVELY ARID, OPEN COUNTRY

Species	Groups		Sample		Population Density (Indiv. per sq. mile)	Group Range (sq. miles)	Source
	Mean	Extreme	Population	No. of Groups			
Orangutan	3	2–5	28	10			Schaller 1961
Gibbon	4	2–6	93	21	11	.1	Carpenter 1940
Black & white colobus	13			1		.06	Ullrich 1961
Olive colobus	12	6–20					Booth 1957
Spider	12	3–17	181	19			Carpenter 1935
Cebidae	15?	5–30					Carpenter 1958
Howler							
1933 census	17	4–35	489	28	82	.5	Carpenter 1934
1951 census	8	2–17	239	30	40		Collias & Southwick 1952
1959 census	18.5	3–45	814	44	136		Carpenter 1962
Langur	25	5–120	665	29	12	3	Jay Unpub. Ms.

Species	Groups		Sample		Population Density	Group Range	Source
	Mean	Extreme	Population	No. of Groups	(Indiv. per sq. mile)	(sq. miles)	
Mountain gorilla (Virunga Volcanoes)	17	5–30	169	10	3	10–15	Schaller 1963
Macaca mulatta							
temples	42	16–78	629	15			Southwick 1961a
forest	50+	32–68	334	7			Southwick 1961b
M. radiata	32		65	2			Nolte 1955
M. assamensis	19		38	2			Carpenter 1958
Baboon							
Nairobi Park	42	12–87	374	9	10	15	DeVore & Washburn (in press)
Amboseli	80	13–185	1203	15			
Human							
bushman	20				.03	440–1250	Shapera 1930; D. Clark pers. comm.
aborigine	35				.08	100–750	Steward 1936
Wolf	12	9–25+				260–1900	Stebler 1944; Murie 1944

as baboons and macaques spend much of their lives in trees, returning to them to sleep each evening and feeding frequently in them at certain seasons.

The ability to exploit both forested and open savanna habitats has had distinct advantages for the baboon-macaque group, enabling it to spread throughout the Old World tropics with very little speciation (DeVore and Washburn, in press). Similarly, the most ground-living Indian langur, *Presbytis entellus,* and the most ground-living *Cercopithecus* monkey, *C. aethiops,* have wider distributions than the other species in their respective genera. In this sense the home range of an organized group of monkeys can be said to be proportional to the geographic distribution of the species. However, this is not always true. The gelada baboon, *Theropithecus gelada,* and the gorilla are geographically restricted, while the gibbon is found throughout Southern Asia. Adaptation of the species to local conditions, the interference of human activities, and competition from other species may explain these contrary examples. Even so, the terrestrial gorilla represents a single, widely separated species while gibbons are divided into a number of well-defined species.

The ability to cross natural barriers and occupy a diversity of environments is reflected in both the size of a group's home range and the amount of speciation that has occurred within its geographic range. This is particularly true when man is considered with the other primates. Speech, tools, and an emphasis on hunting have so altered man's adaptation that direct comparisons are difficult, but if hunter-gatherers living in savanna country with a stone-age technology are selected for comparison, an immense increase in the size of the home range is apparent. Although detailed studies are lacking, the home ranges of two African bushman groups were estimated as 440 and 1250 square miles (Shapera 1930; Clark, personal communication). An average bushman band would number only about 20, or half the size of a baboon troop with a home range of 15 square miles. In Australia, depending on the food resources in a particular area, a band of aborigines has a home range of between 100 and 750 square miles; an average band numbers about 35 (Radcliffe-Brown, 1930; Steward, 1936). Many hunter-gatherer ranges average 100 to 150 square miles, and even the smallest are much larger than the biggest home ranges described for non-human primates. The land area it is necessary for human hunters to control is more comparable to that of group-living predators than it is to that of monkeys and apes. The home range of wolf packs, for example, is between 260 and 1900 square miles (Stebler, 1944; Murie, 1944). Consistent hunting rapidly depletes the available game in a small area, and in the course of human evolution an immense increase in the size of home range would have had to develop concomitantly with the development of hunting.

CORE AREA AND TERRITORIALITY

Within the home range of some animal groups is a locus of intensive occupation, or several such loci separated from each other by areas that are infrequently traversed. These loci, which have been called "core areas" (Kaufmann, 1962), are connected by traditional pathways. A core area for a baboon or langur troop, for example, includes sleeping trees, water, refuge sites, and food sources. Troops of baboons or langurs concentrate their activities in one core area for several days or weeks, then shift to another. Most of a baboon troop's home range is seldom entered except during troop movement from one of these core areas to another; a troop in Nairobi Park with a home range of just over fifteen square miles primarily occupied only three core areas whose combined size was only three square miles. Except that baboon range and core areas are larger, this description would apply equally well to langurs (Jay, unpub. ms.) or, for that matter, to the coati groups, *Nasua narica*, studied by .Kaufmann (1962).

The home ranges of neighboring baboon troops overlap extensively, but the *core areas* of a troop very rarely overlap with those of another troop (DeVore and Washburn, in press). The distinction between a group's home range and the core areas within that range becomes important when the question of territorial defense is considered. Carpenter's review of territoriality in vertebrates concludes that ". . . on the basis of available data territoriality is as characteristic of primates' behavior as it is of other vertebrates" (1958a, p. 242). His description of "territory" in another review (1958b), however, conforms to what has here been described as "home range," following the distinction between "home range" and "territory" (as a defended part of the home range) made by Burt and others (e.g., Burt, 1943, 1949; Bourlière, 1956). Groups of baboons, langurs, and gorillas are frequently in intimate contact with neighboring groups of the same species; yet intergroup aggression was not observed in langurs (Jay, unpub. ms.), tension between baboon troops is rare (DeVore and Washburn, in press), and gorilla males only occasionally show aggression toward a strange group (Schaller and Emlen, in press). Clearly, no definite areal boundaries are defended against groups of conspecifics in these three species. On the other hand, organized groups of these species occupy distinct home ranges, and the fact that a strange baboon troop, for example, is rarely in the core area of another troop indicates that there are spacing mechanisms which ordinarily separate organized groups from each other. Different groups are kept apart, however, not so much by overt aggression and fighting at territorial boundaries, as by the daily routine of a monkey group in its own range, by the rigid social boundaries of organized groups in many monkey species, and, in some species, by loud vocalizations.

Intimate knowledge of the area encompassed in a group's home range is demonstrably advantageous to the group's survival. Knowledge of escape routes from predators, safe refuge sites and sleeping trees, and potential food sources combine to channel a group's activities into daily routines and seasonally patterned movements within a circumscribed area. Beyond the limits of a group's usual range lie unknown dangers and undiscovered food sources, and a baboon troop at the edge of its home range is nervous and ill at ease.

ORGANIZATION OF PRIMATE GROUPS

In addition to the spacing effect of relatively discrete home ranges, most monkeys and some apes live in organized groups which do not easily admit strangers. Gibbons, howler monkeys, langurs, baboons, and macaques all live in such "closed societies." Although Carpenter (1942) described numerous instances of individuals moving from one group to another in his study of the rhesus colony transplanted to Cayo Santiago, Altmann's later study (1962) revealed that almost no individuals changed troops during a two-year period. In retrospect it seems likely that when Carpenter first studied the rhesus colony the six heterosexual groups he described were composed of comparative strangers, and that organized troops had not yet had a chance to stabilize. During Altmann's study approximately the same number of animals were divided into two highly stable troops with almost no individuals' changing troop membership—despite the crowded conditions on the island, where natural spacing mechanisms probably could not operate normally.

Studies by the Japan Monkey Center on Japanese macaques, *Macaca fuscata*, confirm the fact that macaque societies are closed groups (Imanishi, 1960). During more than 1400 hours of observations on more than twenty-five baboon troops, only two individuals changed to new troops (DeVore and Washburn, in press). Jay found that langur troops were similarly conservative, but among langurs individual males or small groups of males may live apart from organized troops. Not all monkey and ape groups have such impermeable social boundaries. Some gorilla groups, for example, can be described as somewhat open or "fluid."

Although different gorilla groups vary considerably in the extent to which they accept individuals into the group, some of them have a high turn-over of adult males (Schaller and Emlen, in press). During a twelve-month study period, some dominant adult males remained with a group of females and young throughout, but other adult males frequently left one group, and either led a solitary existence or joined another group. Males who join a group have access to sexually receptive females, and group-living males usually make little or no attempt to repel them. This "relaxed attitude" toward nongroup members is reflected in intergroup relations. Schaller and Emlen report that "seven different groups were seen in one small section of forest during the period of study." In contrast, it is doubtful if any adult male baboon or macaque can join a new troop without some fighting with the dominant males of the troop.

At present there are not enough detailed studies of the social behavior of monkeys and apes to determine whether organized groups in most species are relatively closed, as in baboons, or relatively open, as in gorillas. A recent study of hamadryas baboons in Ethiopia by Kummer and Kurt (in press), for example, revealed an altogether different kind of group structure. The hamadryas were organized into small groups of from one to four, rarely as many as nine, females and their offspring accompanying only one adult male. These "one-male groups" aggregated into sleeping parties each night which numbered as many as 750 individuals, but during the day the small, one-male units foraged independently, and membership in them remained constant during the observation period, while the number of individuals gathering at sleeping places fluctuated constantly. Kortlandt's (1962) observations of chimpanzees indicate still a different group structure, with anywhere from one to thirty individuals gathered together at one time.

INTERGROUP VOCALIZATIONS

Loud vocalizations seem to aid in spacing groups of some species. Carpenter has suggested that "vocal battles" often substitute for physical aggression in howler monkeys and gibbons. Jay found that langurs also give resounding "whoops" which carry long distances and tend to keep langur troops apart; Ullrich (1961) found the same is true of black and white colobus. Since these loud cries invariably begin at dawn, when no other group is in sight, and often continue as the group leaves its sleeping place and moves to a feeding area, it seems likely that such vocalizations serve more often to identify the *location* of the troop than to issue a challenge to neighboring troops. Altmann's following observations of howling monkeys support this view:

The morning roar, despite its spectacular auditory aspects, was not accompanied by any comparable burst of physical activity. While giving the vocalization, the male stood on four feet or sat. Occasionally, he fed between roars . . . The vocalizations that are given by the males at sunrise are essentially the same as those that are given during territorial disputes, suggesting that these morning howls serve as a "proclamation" of an occupied area (1959, p. 323).

By advertising its position the troop reduces the likelihood that it will meet a neighboring troop. Such location cries can thereby function as spacing mechanisms, but usually without the directly combative connotation which "vocal battle" suggests. When a gibbon or howler group approaches another group, loud vocalizations increase. Two groups of gibbons, however, will mingle without agression after the period of vocalizing has passed (Carpenter, 1940, p. 156), and Jay saw langur troops frequently come together without aggression.

Loud vocalizations of the sort found in gibbons, howlers, colobus and langurs are conspicuously absent in baboons and macaques. The situation on the African savanna, where visibility frequently extends for hundreds or even thousands of yards, is very different from the very limited visibility of dense forests. Baboon ranges are large in Nairobi Park, and troops seldom come near each other. Adjustments in the direction of troop movement can easily be made by sight alone. Even when baboon troops do come together, as they frequently did in that part of the Amboseli Reserve where food and water sources were restricted, no intergroup aggression was observed. In view of the relatively high degree of dominant and aggressive behavior typical of baboons, the fact that intertroop relations are pacific may seem contradictory. If defense of territory is not thought of as fundamental in the behavior pattern of primates, however, the peaceful coexistence of neighboring troops is not inconsistent with a high level of agonistic and dominance behavior between individuals within the group.

Loud vocalizations, as a means of keeping groups apart when they are not in visual contact, are also absent in the gorilla. When two groups come together, or in the presence of man, adult males may give dramatic intimidation gestures (abrupt charges, chest beating, etc.). While some of these gestures may be accompanied by loud vocalizations, their impact is primarily visual. As in the baboons and macaques, visual communication tends to substitute for loud vocalizations, and it is suggested that methods of intergroup communication correlate with the degree to which a species is arboreal or terrestrial. Vocal expression as a spacing mechanism is very important in gibbons and howler monkeys, less important among langurs, and virtually absent in the gorilla, the baboons, and the macaques. It is also reasonable to suppose that vocalizations are more important in intragroup behavior in arboreal species, where foliage interferes with coordination of group activity, than they ordinarily are in terrestrial species. Baboon studies to date have concentrated on troops living in open areas, where troop members are usually in constant visual contact. If baboons become separated from the troop, however, contact is reestablished by vocalizations, and when the troop is feeding in dense vegetation soft vocalizations (e.g., low grunting) are apparently more frequent, and may serve to maintain contact between troop members. Many baboons live in forests and a study of these forest-living baboons could definitely determine whether vocal behavior is more frequent in this habitat than it is on the open plains.

POPULATION DENSITY

Pitelka (1959) has pointed out that the fundamental importance of territoriality lies not in the behavior, such as overt defense, by which an area becomes identified with an individual or group, but in the degree to which the area is used exclusively; that is, functionally, a territory is primarily an ecological, not a behavioral, phenomenon. Recent field studies indicate that if territoriality can be ascribed to most monkeys and apes it is in this more general, functional meaning of the term only. As a means of distributing the population in an available habitat, the core area of a nonhuman primate group's home range may serve the same function as literal, territorial defense does in some vertebrate species. At the human, hunter-gatherer level this is much less true. Although a hunting band may have core areas, around water holes for example, the band tends to consider most or all of its home range an exclusive possession. Human hunting activities require the use of large areas, and the hunting range is usually protected from unwarranted use by strangers. One of the important differences between home ranges which overlap and the exclusive possession of a range is reflected in the population density of the species. Although a howler monkey group ranges over about .5 of a square mile, the 28 troops counted in 1933 could not have exclusively possessed more than an average of .2 of a square mile per group; in 1959 the average per group could have been no higher than .136 of a square mile. In Kenya, about 50 per cent of a baboon troop's home range overlaps with that of adjacent troops, and a troop would seldom possess more than one square mile of its range exclusively. One result of this extensive overlapping is that the population density of the nonhuman primates is far higher than would be expected from a description of home range sizes alone (Table 1).

The population density of howler monkeys on Barro Colorado Island in 1959 was 136 individuals per square mile. Terrestrial species like baboons and gorillas have much lower population densities (10 and 3 per square mile), but they are nevertheless far more densely populated than many hunter-gatherers, who average only .03 to .08 per square mile in savanna country. That hunting man, like other large, group-living carnivores, will inevitably have a lower population density than the other primates can be illustrated by the numbers of lions in Nairobi Park. This area of approximately 40 square miles supports a baboon population of nearly 400, but it supports an average of only 14 lions (extremes: 4–30—i.e., 4 to 30 lions) (Wright, 1960).

In summary, all monkeys studied to date live in organized groups whose membership is conservative and from which strangers are repelled. These groups occupy home ranges which may overlap extensively with those of neighboring groups, but which contain core areas where neighboring groups seldom penetrate. Rather than territorial defense of definite boundaries, monkey groups are spaced by daily routine, tradition, membership in a discrete social group, and the location of adjacent groups. Among the apes these same generalizations would seem to hold for gibbons, but they are much less true of gorillas, and, probably, orang-utans and chimpanzees. The trend toward increasing size of the home range, however, from the very small range of arboreal species to moderately large ranges in terrestrial species, would appear to be true of all monkeys and apes. Means of identifying the position of adjacent groups shifts from loud vocalizations in arboreal species to visual signals in terrestrial ones. Man exemplifies his terrestrial adaptation in his enormously increased home range, his lower population density, and in his reliance on vision in the identification of neighboring groups.

GROUP SIZE

Mention has already been made of the average size of the social group in several species of primates; available data on group size are summarized in Table 1. When the size of the social group is compared to the degree of arboreal or terrestrial adaptation of the species, a trend toward larger groups in the terrestrial species is apparent, in both Old and New World monkeys and among the apes. Although there are no adequate data for the orangutan, the largest group ever reported is five, and the average size of a gibbon troop is four. Schaller found that the average gorilla group was seventeen, temporarily as large as thirty, and Kortlandt (1962) saw as many as thirty chimpanzees together in his study area. The average size of monkey troops varies from 20 or less in the olive colobus, spider, howler, and *Cebidae* group to an average of 25 (but as large as 120) in langurs, and an average of from 40 to 80 in the baboons and macaques (with some troops as large as 200).[2] Present observations suggest three central grouping tendencies in monkeys and apes: five or less for the gibbon and orangutan; from twelve to twenty in arboreal monkeys and the gorilla; and fifty or more in baboons and macaques. This small sample may be misleading, but it is clear at present that, in addition to having a larger home range and lower population density, terrestrial monkeys and apes live in larger organized groups.

SEXUAL DIMORPHISM AND DOMINANCE

Field studies of monkeys indicate that dominance behavior, especially of the adult male, is both more frequent and more intense in ground-living monkeys than in other species. This increase in dominance behavior is accompanied by an increase in sexual dimorphism, particularly in those morphological features which equip the adult male for effective fighting—larger body size, heavier temporal muscles, larger canine teeth, etc. (Washburn and Avis, 1958, p. 431). If the apes are compared with respect to sexual dimorphism, there is a clear trend toward increasing sexual dimorphism from the arboreal gibbon, where the sexes are practically indistinguishable, to the chimpanzee, in which the male is appreciably more robust, to the gorilla, where sexual dimorphism is greatest. Only the orangutan is an exception. The trend toward increased sexual dimorphism in terrestrial species of monkeys is also apparent. Some male characteristics, such as the hyoid bone in howler monkeys and the nose of the proboscis monkey, are more pronounced in arboreal species, but morphological adaptations for *fighting* and *defense* are clearly correlated with adaptation to the ground. The various

---

2. The most thorough census of macaque troops is that of Southwick et al. (1961a, 1961b). Only the figures for forest and temple troops are included in Table 1 because troops in other habitats were subject to frequent trapping. Counts of 399 troops in all habitat categories revealed an over-all average group size of only 17.6.

baboons and macaques all illustrate this tendency. Among *Cercopithecus* species sexual dimorphism is pronounced in *C. aethiops* and the patas monkey (a species closely related to *Cercopithecus*) and decreases in the more arboreal forms.

The trend toward increased fighting ability in the male of terrestrial species is primarily an adaptation for defense of the group. Zuckerman's account of baboon behavior (1932) would indicate that the acquisition and defense of "harems," and concomitant fighting among the males, places a high premium on aggressiveness and fighting ability in intragroup behavior. Behavioral observations on confined animals can be very misleading, however, and no field study of baboons has found that sexual jealousy or fighting is frequent in free-ranging troops (Bolwig, 1959; Hall, 1960, 1961; Washburn and DeVore, 1961a, 1961b). Observations of many baboon troops in close association indicate that intertroop aggression is very rare and that males do not try to defend an area from encroachment by another troop (although baboons were seen trying to keep vervet monkeys away from a fruit tree). The fact that hundreds of hamadryas may gather at one sleeping site would indicate that interindividual tolerance is high in this baboon species as well (Kummer and Kurt, in press). The intergroup fighting of rhesus which Carpenter observed on Cayo Santiago was probably aggravated by unsettled conditions on the island. One index of these conditions is that more infants were killed than were born during his period of study.

Life on the ground exposes a species to far more predators than does life in the trees. Not only are there fewer potential predators in the trees, but also escape is relatively easy. By going beneath the canopy (to escape raptorial birds) or moving across small branches to an adjacent tree (to escape from felines), arboreal species can easily avoid most predators except man. The ultimate safety of all nonhuman primates is in trees, and even the ground-living baboons and macaques will take refuge in trees or on cliffs at the approach of a predator (except man, from whom they escape by running).

Much of the day, however, baboons may be as far as a mile from safe refuge, and on the open plains a troop's only protection is the fighting ability of its adult males. The structure of the baboon troop, particularly when the animals are moving across an open area, surrounds the weaker females and juveniles with adult males. At the approach of a predator, the adult males are quickly interposed between the troop and the source of danger (Washburn and DeVore, 1961b). The structure of a Japanese macaque troop is apparently identical, even though no predators have threatened the Takasakiyama group in many years (Itani, 1954). The ecological basis for sexual dimorphism in baboons has been described elsewhere (DeVore and Washburn, in press). Because only the adult males are morphologically adapted for defense, a baboon troop has twice the reproductive capacity it would have, for the same number of individuals, if males and females were equally large. Adaptation for defense is accompanied by increased agonistic behavior within the troop, but intratroop *fighting* is rare. Stable dominance hierarchies minimize aggression among adults, and male baboons and macaques actively interfere in fights among females and juveniles.

Field studies of other monkeys indicate that intragroup aggressive and agonistic behavior decrease by the degree to which a species is adapted to arboreal life. Among langurs, where sexual dimorphism is less pronounced, females may threaten or attack adult males—behavior that is unparalleled in the baboon-macaques. Langurs are usually in or under trees, where escape is rapid and the need for males to defend the group is much less important. The same argument, with suitable qualifications, holds for the other, more arboreal monkeys. The sexes in the olive colobus, for example, are almost identical in size and form; this species is never seen on the ground (Booth, 1957). This does not imply that the male in other primate species has no protective role. Male colobus, vervets, and howlers have been seen taking direct action against potential predators, and the male in many species is prominent in giving defiant cries and/or alarm calls. With the possible exception of the gibbon, some measure of increased defensive action by adult males is a widespread primate pattern. The evidence does suggest, however, that increased predation pressure on the ground leads to increased morphological specialization in the male with accompanying changes in the behavior of individuals and the social organization of the troop. Although troops of arboreal monkeys may be widely scattered during feeding, a baboon or macaque troop is relatively compact. Some males may live apart from organized groups, either solitarily or in unisexual groups, in less terrestrial species (e.g., langurs), but we discovered no healthy baboons living outside a troop (DeVore and Washburn, in press). Dominant adult males are the focal point for the other troop members in baboons, macaques, and gorillas. When the males eat, the troop eats; when the males move, the others follow. Compared to other monkey species: the baboon-macaques are most dominance-oriented; troop members are more dependent upon adult males and actively seek them out; and the social boundary of the troop is strong.

BEHAVIOR AND TAXONOMY

Species-specific behavior has been valuable in the classification of some vertebrate species, notably the distinct song patterns in birds (e.g., Marler, 1957, 1960; see also Hall, this volume). Spectrographic analyses of primate vocalizations will undoubtedly reveal specific differences in communication patterns, but these are only now being undertaken. Studies of social behavior in monkeys and apes have only begun, and no observations have yet been made in sufficient detail to permit close comparisons. Some general comparisons can be made, however, from studies recently completed.

The African baboons and the Asiatic macaques are very similar in both morphology and general adaptation. Both groups have forty-two chromosomes and their distribution does not overlap, suggesting that they are members of a single radiation of monkeys. Baboons, including drill, mandrill, hamadryas, and savanna forms (but excluding gelada) are probably all species within one genus, *Papio* (DeVore and Washburn, in press). Comparisons between the social behavior of East African baboons and macaques (rhesus and Japanese macaques) have been made elsewhere (DeVore, 1962). At all levels of behavior, from discrete gestures and vocalizations to over-all social structure, baboons and macaques are very much alike. Both groups have an elaborate, and comparable, repertoire of aggressive gestures. In social interactions, the same behavioral sequences occur: an animal who is threatened may redirect the aggression to a third party, or may "enlist the support" of a third party against the aggressor. If support is successfully enlisted, two or more animals then simultaneously threaten the original aggressor. Relations between the adults and the young of both groups are similar.

Play patterns in juvenile groups and the ontogeny of behavior follow the same course. Relationships within the adult dominance hierarchies, and the social structure of the troop are comparable. Some details of gesture and vocalization are certainly distinct, and there is a striking difference, for example, in the form and duration of copulation. Copulation in rhesus monkeys usually involves a series of mounts before ejaculation, as does copulation in the South African baboons ("chacma") studied by Hall (1962), while a single mounting of only a few seconds' duration is typical of East African baboons. Details of gesture and vocalization during copulation also vary between the three groups. On the other hand, most of the behavioral repertoire seems so similar that an infant baboon raised in a macaque troop, or vice versa, would probably have little difficulty in leading the adult life of its adopted group. Behavioral observations clearly confirm the evidence of morphological similarity in this widespread group of monkeys. No other primate, except man, has spread so far with as little morphological change as the baboon-macaque group. With man, these monkeys share the ability to travel long distances, cross water, and live in a wide range of environmental conditions.

On the basis of the present, random studies of monkeys and apes, generalizations regarding trends in behavior must remain speculative, particularly since the majority of the studies are concentrated in the baboon-macaque group. Much more useful statements with regard to the adaptive significance of different behavioral and morphological patterns will be possible when field studies of several species within one genus have been undertaken. The *Cercopithecus* group presents a wide range of ecological adaptations, from swamp-adapted species like *C. talapoin*, through the many forest forms, to the savanna-living *C. aethiops* (see Tappen, 1960). Forms closely related to the *Cercopithecus* group, Allen's swamp monkey, patas, and gelada, would further extend the basis of comparison. A study of patas or gelada monkeys would be particularly useful for cross-generic comparison with baboon and macaque behavior. Although Ullrich has made initial observations, no long term study of an African colobus species is yet available for comparison to Jay's study of Indian langurs. Booth's report (1956, 1957) that olive colobus do not ascend to the upper levels of forests even when these are not occupied by black colobus and red colobus indicate that even brief field observations of the behavior of sympatric primate species would be an immense aid in settling some of the persistent questions in primate taxonomy.

SUMMARY

Field studies of monkeys and apes suggest a close correlation between ecological adaptation and the morphology and behavior of the species. All terrestrial forms occupy a larger home range, and, in monkeys, the geographic distribution of the species increases according to the degree of terrestrial adaptation. Many arboreal species use loud vocalizations in spacing troops; ground-living forms depend more on visual cues. A marked decrease in population density accompanies terrestrial adaptation. Man is part of this continuum, illustrating the extreme of terrestrial adaptation.

Morphological adaptation of the male for defense of the group is more prominent in ground-living species (except man, whose use of tools has removed the selective pressure for this kind of sexual dimorphism), and least prominent in most species that do not come to the ground. The dependence of the other troop members creates a male-focal social organization in terrestrial species. Dominance behavior is much more prominent in terrestrial monkeys, but actual fighting is rare. Terrestrial life and large adult males have not been accompanied by a comparable increase in dominance behavior in the gorilla, however, indicating that defense is more important than intragroup aggression in the development of sexual dimorphism in terrestrial primates. Man's way of life has preserved the division between the male and female roles in adult primate life, but cultural traditions have replaced biological differences in the reinforcement of this distinction.

BIBLIOGRAPHY

ALTMANN, STUART
   1959. "Field observations on a howling monkey society." *J. Mammal.*, 40(3):317–330.
   1962. "A field study of the sociobiology of rhesus monkeys, *Macaca mulatta*." *Ann. New York Acad. Sci.*, 12(2):338–435.
BOLWIG, NIELS
   1959. "A study of the behaviour of the chacma baboon, *Papio ursinus*." *Behaviour*, XIV(1–2):136–163.

BOOTH, A. H.
1956. "The distribution of primates in the Gold Coast." *J. W. Afr. Sci. Ass.*, 2:122.
1957. "Observations on the natural history of the olive colobus monkey, *Procolobus verus* (van Beneden)." *Proc. Zool. Soc. Lond.*, 129:421–431.

BOURLIÈRE, FRANÇOIS
1956. *Natural history of mammals.* (Rev. ed.). New York: A. A. Knopf.

BURT, W. H.
1943. "Territoriality and home range concepts as applied to mammals." *J. Mammal.*, 24:346–352.
1949. "Territoriality." *J. Mammal.*, 30:25–27.

CARPENTER, C. R.
1934. "A field study of the behavior and social relations of howling monkeys, *Alouatta palliata*." *Comp. Psychol. Monogr.*, 10(48).
1935. "Behavior of red spider monkeys in Panama." *J. Mammal.*, 16:171–180.
1940. "A field study in Siam of the behavior and social relations of the gibbon, *(Hylobates lar).*" *Comp. Psychol. Monogr.*, 16:5.
1942. "Sexual behavior of free ranging rhesus monkeys *(Macaca mulatta).*" *J. Comp. Psychol.*, 33:113–162.
1958a. "Territoriality: a review of concepts and problems." See Roe and Simpson, 224–250.
1958b. "Soziologie und Verhalten Frielebendere Nichtmenschlicher Primaten." In *Handbuch der Zoologie*, 8:1–32.

COLLIAS, N. and CHARLES SOUTHWICK
1952. "A field study of population density and social organization in howling monkeys." *Proc. Amer. Phil. Soc.*, 96:143–156.

DEVORE, IRVEN
1962. "The social behavior and organization of baboon troops." Unpublished doctoral dissertation, Univ. of Chicago.

DEVORE, IRVEN and RICHARD B. LEE
In press. "Recent and current field studies of primates." *Folia Primatologia.*

DEVORE, IRVEN and S. L. WASHBURN
In press. "Baboon ecology and human evolution." In *African Ecology and Human Evolution*, F. C. Howell, ed. Chicago: Aldine.

HADDOW, A. J.
1952. "Field and laboratory studies on an African monkey, *Cercopithecus ascanius schmidti* Matchie." *Proc. Zool. Soc. Lond.*, 122, II:297–394.

HALL, K. R. L.
1960. "Social vigilance behaviour of the chacma baboon, *Papio ursinus*." *Behaviour*, 16(3–4):261–294.
1961. "Feeding habits of the chacma baboon." *Advancement Sci.*, 17(70):559–567.
1962. "The sexual, agonistic and derived social behavior patterns of the wild chacma baboon, *Papio ursinus*." *Proc. Zool. Soc. Lond.*, 139:283–327.

IMANISHI, KINJI
1960. "Social organization of subhuman primates in their natural habitat." *Current Anthropology*, 1(5–6):393–407.

ITANI, JUNICHIRO
1954. Japanese monkeys in Takasakiyama. Tokyo, Kobunsha. (in Japanese).

JAY, PHYLLIS
Unpub. Ms. "The social behavior of the langur monkey."

KAUFMAN, JOHN H.
1962. "Ecology and social behavior of the coati, *Nasua narica*, on Barro Colorado Island Panama." *Univ. Calif. Pub. in Zool.*, 60:95–222.

KORTLANDT, ADRIAAN
1962. "Chimpanzees in the wild." *Sci. Amer.*, 206:128–138.

KUMMER, HANS and FRED KURT
In press. "Social units of a free-living population of hamadryas baboons." *Folia Primatologia*.

MARLER, PETER
1957. "Specific distinctiveness in the communication signals of birds." *Behaviour*, XI,1.
1960. "Bird songs and mate selection." *Animal Sounds and Communication, Amer. Institute of Bio. Sci.*, 7.

MURIE, ADOLPH
1944. "The wolves of Mt. McKinley." *Fauna Nat. Parks U.S.*, 5.

NAPIER, JOHN
1962. "Monkeys and their habitats." *New Scientist* 15:88–92.

NOLTE, A.
1955. "Observations of the behavior of free ranging *Macaca radiata* in southern India." *Zeitschrift für Tierpsychologie*, II:77–87.

PITELKA, FRANK A.
1959. "Numbers, breeding schedule, and territoriality in pectoral sandpipers in northern Alaska." *The Condor*, 61(4):233–264.

RADCLIFFE-BROWN, A. R.
1930. "Former numbers and distribution of the Australian aborigines." *Official Year Book of the Commonwealth of Australia*, 23:671–696.

ROE, ANNE and GEORGE GAYLORD SIMPSON (eds.)
1958. *Behavior and evolution.* New Haven: Yale Univ. Press.

SCHALLER, GEORGE
1961. "The orang-utan in Sarawak." *Zoologica, N. Y. Zool. Soc.*, 46:2.

SCHALLER, GEORGE and JOHN T. EMLEN, JR.
In press. "The ecology and social behavior of the mountain gorilla with implications for hominid origins." In *African Ecology and Human Evolution*, F. C. Howell, ed. Chicago: Aldine.

SHAPERA, I.
1930. *The Khoisan peoples of South Africa.* London.

SOUTHWICK, CHARLES H., MIRZA AZHAR BEG and M. RAFIQ SIDDIQI
1961a. "A population survey of rhesus monkeys in villages, towns, and temples of northern India." *Ecology*, 42:538–547.
1961b. "A population survey of rhesus monkeys in northern India: II. transportation routes and forest areas." *Ecology*, 42:698–710.

STARK, D. and H. FRICK
1958. "Beobachtungen an äthiopischen Primaten." *Zoologische Jahrbücher*, 86:41–70.

STEBLER, A. M.
1944. "The status of the wolf in Michigan." *J. Mammal.*, 25:37–43.

STEWARD, JULIAN
1936. "The economic and social basis of primitive bands." *Essays in Honor of A. L. Kroeber*, Berkeley: Univ. of California Press.

TAPPEN, N. C.
1960. "Problems of distribution and adaptation of the African monkeys." *Current Anthropology*, 1:91–120.

ULLRICH, VON WOLFGANG
1961. "Zur Biologie und Soziologie der Colobusaffen." *Der Zoologische Garten*, 25:305–368.

WASHBURN, S. L. and VIRGINIA AVIS
1958. "Evolution of human behavior." See Roe and Simpson, 421–436.

WASHBURN, S. L. and IRVEN DeVORE
1961a. "The social life of baboons." *Sci. Amer.* 204:6.

WASHBURN, S. L. and IRVEN DeVORE
1961b. "Social behavior of baboons and early man." *In Social life of early man*, S. L. Washburn, ed. Viking Fund Pub. in Anthropology, 31:91–104.

WRIGHT, BRUCE S.
1960. "Predation on big game in East Africa." *J. Wild-life Management*, 24:1–15.

ZUCKERMAN, S.
1932. *Social life of monkeys and apes*. London: Kegan Paul.

PSYCHOLOGICAL DEFINITIONS OF MAN

ANNE ROE

A DEFINITION OF MAN, in any terms, may be undertaken in either or both of two ways. Man, as a living species, may be differentiated from all other species presently alive, or one may attempt to view the evolutionary history of this species, and come to a decision as to when a sequence of ancestral (and collateral) populations ceased to be something else, and reached each of the various points of progression signalized respectively by membership in the Hominidae, the Hominae, *Homo*, and *Homo sapiens*. "When" refers to a developmental stage, the appearance of a certain pattern of behavior, rather than to any particular period of time.

Discrimination between man and non-man at the present time presents no difficulties if one looks at it in a superficial fashion. However bitterly anthropologists and others may wrangle over the assignment of names to fossils, I do not think there would be any argument among them or anyone else over the assignment of any living creature to *Homo sapiens*. (Perhaps I should qualify this by saying that there would be full verbal agreement on a conscious level. One of the world's most difficult problems is that this technical agreement does not carry with it emotional agreement. Even among the most sophisticated there are few whose feelings of group association are for the species.)

If there is no argument, why discuss it? Perhaps if we can pinpoint the factors on which the discrimination is based, we will have some useful clues that will help us make the other decision; that is, when did a creature in the past become man? But if we go beyond this quick and easy sorting of living creatures into man and not-man, and ask just how we make the distinction, the matter is not so easy. There is no single attribute which, if carefully examined, applies to all men and to no other creatures, unless of course you define attributes in limited fashion. Certainly, only man has a "language," but if one defines language as a means of communication, or even as symbolic communication, there is no such clear-cut distinction. Tool-using and tool-making; intelligence and reasoning; social forms; culture; all the familiar points of distinction between man and nonman, break down as all-or-none distinctions when they are carefully examined.

What is of particular interest is that practically all of the most crucial criteria that have been called upon at one time or another as distinguishing attributes of man are behavioral criteria. Yet the discontinuities here may be no greater than

320

those for physical criteria. Behavioral analysis is difficult enough when direct observations may be made, but it becomes almost hopelessly complicated when the behavior must be inferred from bones and artifacts whose preservation and discovery are both fortuitous in the extreme. Such inferences, however, can be in part guided by, and in part supplemented by, observational and experimental data on living species.

The diagnosis of *Homo* on morphological grounds developed in this volume includes a combination of a number of morphological differences, several of which are complexes of characters, and at least two of which, locomotion and capacity for manual manipulation, are particularly important for their direct behavioral reference. Two others, prolongation of life periods and the relative increase in size and complexity of the brain also have very important behavioral implications.

Another character, the continuous sexual receptivity of the human female, which is now apparently uniquely human, cannot be inferred from morphological evidence. (It may be somewhat less than continuous. There are certainly rhythmic variations in desire in most women, even though they are masked by behavioral conformity.) Some lengthening of life periods and changes in brain appear to have been necessary conditions for the development of the human way of life, but it is not clear in what ways continuous sexual receptivity may have been involved with these. If the initiation of sexual activity were strictly a male function, this might well have been coordinated with the situation (to be discussed below) of the separation from the rest of a group, of small hunting parties for varying periods of time. But in view of the consistency of the primate pattern of female initiation of sexual activity, it is difficult to see how this might have come about. (There is, of course, some discrepancy of opinion as to the present situation in *Homo sapiens*.) From the standpoint of natural selection, it may, however, be a useful character in a hominid hunting group.

There is fossil evidence that bipedal locomotion, tool-using and tool-making and some degree of prolongation of post-natal life periods preceded the advent of the Hominidae. All of them have, however, been carried farther by the Hominidae. As Washburn and Avis (1958) have pointed out, the human carnivorous habit depends very greatly upon tools and it is probable that this change resulting in an omnivorous diet developed concurrently with tool-use and upright posture. Except in unusual instances, man always walks on two feet. Most modern men include a great deal more animal food in their diet than early man could consistently do. As Schultz (1961) has shown, all of the post-natal life periods have increased in length. Even in recent man, the period of infantile and juvenile dependency has increased culturally if not morphologically. Nevertheless, changes in these characters have been rather minimal and are approaching a limit if they have not yet reached it.

The fantastically great changes in tool-using, tool-making, and communication have come about more in association with development of the central nervous system, particularly the brain, than with changes in peripheral structures such as the hand, although these changes also took place between pre-man and man.

Other elements in the distinctively human adaptation have certainly changed as markedly as tool-using and toolmaking since man appeared. These include symbolic language and thought, self-awareness, and perhaps emotional suscep-tibility. They also include changes in basic motivation, as shown by a marked increase in exploratory, and possibly in sexual behavior, and probably the de-velopment of new basic drives (such as needs for beauty, for understanding and for self-actualization, to use Maslow's classification [1954]) for which we have little, if any, evidence in other forms.

Man developed from such a group as the australopithecines. From the fossil remains, from artifacts, and from what we know of the behavior of living primates, we can make some guesses about the kind of life these creatures were living. They were bipedal and omnivorous, they made and used very crude tools. This much is quite certain. They probably lived in groups of varying size, and it is highly probable that the basis for the form of group living was the protec-tion of the young. In all systematically studied free-living primate groups, the care of the young is the major factor that holds the group together, and primary sexuality is, in this respect, a factor of considerably lesser significance. There is no reason to suppose the australopithecines different in this respect. At what point in evolution the nuclear family became delimited within a larger group there is no way of knowing. But in the earliest men it is extremely likely that sub-groups were composed as they are in the great apes: females with infants, juveniles, and adult males. This is an effective way of life, and the question is why did it change? What situation or situations could have forced a new adaptation?

It seems probable that the first significant step was the development of language, in however crude a form, since, given this, the other characteristics of the human adaptation could follow inevitably. Symbolic language is not necessary for the maintenance of group coherence, or for the kind of life that these creatures were living generally.

Hebb and Thompson propose that "the intellectual development in phylogenesis that eventually makes speech possible is an increasing independence of the con-ceptual activity from the present sensory environment and an increasing capacity for entertaining diverse conceptual processes at the same time. This is the capacity to respond to the present environment in one way and think of responding in another; for perceiving the present situation as it now is, and conceptually adding something else to it; for planning a series of actions of which only the first is immediately feasible. At the lowest level, it is the capacity for delayed response or a simple expectancy; at the highest level, for 'holding' not only a series of words but also of sentences, whose meaning only becomes clear with later words or sentences." But they point out that chimpanzees are entirely capable of a deceitful attack which demonstrates that they can plan a series of actions in ad-vance, and act in one way while thinking about acting in the opposite way.

In the many and elaborate discussions of what language is and how communication differs between animals and man, perhaps the simplest distinction—and the one finally agreeable to most proponents—is that stated by Hebb and Thompson (1954) in these terms: "Man has what neither chimpanzee nor parrot has: the capacity to combine and readily recombine representative or symbolic noises (words), movements (gestures), or modifications of inert things (writing, carving). We propose therefore that the minimal criterion of language as distinct from other purposive communication, is twofold. First, language combines two or more representative gestures or noises purposefully, for a single effect; and secondly, it uses the same gestures in different combinations for different effects, changing readily with circumstances."

We must ask not only how language developed but also why it developed. Apparently the how is more concerned with the development of central processes than with the development of peripheral structures. The problem is not with the muscular and structural properties of the vocal organs of chimpanzees or other apes but with the cortical and subcortical tracts. The speech areas in the brain of man are apparently unique, and recent investigations have mapped them in some detail. The techniques of electrical stimulation of exposed cortical areas in conscious subjects have added greatly to our knowledge of these.

There are several reports of stimulation of some part of the brain of an animal producing some form of vocalization, but these seem to have been unconfirmed. Penfield and Roberts state that ". . . . there are four cortical areas in which a gentle electrical current causes a patient, who is lying fully conscious on the operating table, to utter a long-drawn vowel sound which he is quite helpless to stop until he runs out of breath. Then, after he has taken a breath, he continues helplessly as before. Other animals lack this inborn vocalization transmitting mechanism in the motor cortex . . . . Curious as it may seem, this is the most striking difference between the cortical motor responses of man and other mammals, and it seems likely that it bears some relationship to man's ability to talk. Another striking peculiarity of the human motor cortex, which may also bear some relationship to speaking and writing, is the relatively large area devoted to mouth . . . and the relatively large area of hand as well."

It is clear that thinking is not dependent upon speech, but it is also clear that it is enormously facilitated by speech, and in fact that it is changed by the introduction of language, both phylogenetically and ontogenetically. Vygotsky insists that thought and speech have different genetic roots, and that they develop independently of each other, but that in man these two separate lines of development eventually meet and that thereupon "thought becomes verbal and speech rational." Such a fusion does not develop in other primates.

Let us suppose that this pre-human group lived for a considerable period in a situation which provided them with a degree of continuing security from enemies, that there was adequate vegetable and perhaps marine food in the near vicinity and that this diet was supplemented by hunting. Whether only some of the adult males went on hunting expeditions, or all of them is not of great importance (except that if all went we must assume even greater security since none were then present to protect the females and young). That small groups of young males may split off from the group for short periods is not uncharacteristic of other primates, and an easy extension of such behavior would permit them to go on hunting forays. If the group continued to live in the same spot, such forays would inevitably become increased in length of time, since they would have to go farther and farther afield. They would be limited in the amount that could be transported at any one time. When they returned to the group, it is clear that to be able to communicate their observations to others of the group would be of the greatest advantage. Bees can do as much, and for the same reason. Sharing of the experience of those who went different places and those who stayed at home would have many advantages other than just the acquisition of food— greater group cohesiveness, greater skills, etc.

It should be noted that by this time, some other developments must have taken place. The most necessary of these would be the custom of sharing a kill. This is not characteristic of free-living primates at the present time. Beyond the nursing of infants, and perhaps the mothers' seeing in a vague sort of way that the infants get other food, sharing food seems to occur relatively rarely and then largely between temporary consort pairs, although the observation of sharing as a response to "begging" shown by the chimpanzees observed by Goodall is highly suggestive of the way in which this could gradually have come about.

All of this means that there was already some division of labor, in a sense, beyond that of infant care and group protection which is already seen in other primate groups. Vygotsky (1962) has suggested that speech was "born of the need of intercourse during work," but the reconstruction offered above is better phrased as "born of the need of intercourse as a result of the division of labor."

At the same time that this living situation could have provoked the development of speech, other changes must have been taking place. Change in posture must have introduced a marked perceptual change. The same eye can see considerably farther from five feet above the ground than it can from two feet, which is not only defensively useful in a plains living animal, but introduces a spatial extension of the world which may come to be associated with a temporal one. To see ahead is not only a figure of speech.

One of the striking aspects of man is the temporal extent of the world he lives in —that is, he not only remembers the past but he anticipates the future (sometimes even correctly) to a greater degree than any other animal. All animals have some sort of memory. It is obvious that memory has enormous selective advantage for any organism, and it is specifically developed in many. The return to a burrow after extended foraging is one of the simplest forms, which becomes much less simple when it is seen in migrating herds, who return year after year to the same pastures. To what extent foresight is involved in such behavior is debatable, but it is not conceptually necessary to adduce it by way of explanation.

Change from gathering to hunting involves considerably more. To be efficient, the hunter must not only remember where game has been seen before, he must also remember how the game behaves, and from this be able to predict the behavior of the game. Many carnivores do this to some extent (I am not suggesting it is done consciously, at least not in our terms), but the length of time over which they seem to anticipate the behavior of their prey is extremely limited. Any increase in accuracy of perception, memory or prediction would have enormous selective advantage, and would certainly depend in part upon changes in the central nervous system.

Some discussion of neurophysiological theories of thought may be helpful here. Brain size is a diagnostic character for man, and the marked increase apparently came about rather rapidly, geologically speaking, and was perhaps the consequence of such behavioral changes as tool-use, speech, and increase in other perceptual and cognitive activities. It is entirely probable that tool-using and tool-making stimulated imagery development—a form of thinking not dependent upon speech. At the same time the development of speech would give thinking an entirely new cast, and enormously facilitate it.

The relation of absolute brain size to intelligence has long been a matter of debate. Among living men it is clear that brain size is not a reliable measure of the individual's capacity (Dobzhansky, 1962). Rensch, working with a range of animals from mice to elephants, with birds and with fish, has concluded that memory retention is about proportional to brain size. Hebb has suggested that intelligence may be associated, not with amount of tissue, but with the ratio between the area of the cortex with presumed associative function and the area of the cortex concerned directly with either receptor input or motor output, the A/S ratio. He also assumed the existence of autonomous central processes, underlying perception, attention, perceptual learning, and thought. From review of animal studies he concluded that (Hebb, 1949, p. 116),

"(1) more complex relationships can be learned by higher species with large A/S ratios at maturity; (2) simple relationships are learned about as promptly by lower as by higher species; and (3) the first learning is slower in higher than lower species."

In order to explain the slower early learning of higher species he suggested that autonomous central processes must be established within the associative areas before receptor inputs can gain extensive control over motor action, and therefore he distinguished between what he called primary learning and later learning, considering the former to be primarily in terms of perceptual experience. More recent developments in neurophysiological theory expand but do not refute Hebb's suggestions. They have been reviewed by Pribram (1960).

In latest theory the "association areas" now appear to be intrinsic systems in which are stored the approximate counterparts of the computer's memory with its representations and strategies for processing information. The reflex arc is replaced by a Test-Operate-Test-Exist sequence which operates somewhat after the fashion of an analogue computer. This concept presumes two reciprocally connected neuronal systems. One, analogous to the intrinsic system of the brain, performs the test functions, i.e., when receptor inputs arrive, a search is made among the hierarchically arranged store of representations of action systems or of perceptual data to find one that matches the input. If no match is found, control is shifted to the extrinsic system which then operates on either receptor mechanisms or environment or both until the incongruities of the test are resolved. "The general pattern of reflex action, therefore, is to test the input energies against some criteria established in the organism, to respond if the result of the test is to show an incongruity, and to continue to respond until the incongruity vanishes, at which time the reflex is terminated. Thus there is 'feedback' from the result of the action to the testing phase, and we are confronted by a recursive loop." (Miller, Galanter and Pribram, 1960, p. 26).

The unit of behavior, then, is the TOTE, a feed-back unit. More complicated behavior is explained in terms of series of TOTE hierarchies. Essentially, it is clear, Miller, Galanter, and Pribram view man as a system for processing information. Behavior is organized and they speak of this organization in terms of Plans, which they consider to be essentially the same as a program for a computer, with this definition: "A Plan is any hierarchical process in the organism that can control the order in which a sequence of operations is to be performed."

The import of this for our discussion will be clear from the following quotation: "Does language introduce new psychological processes, or are all our verbally acquired skills foreshadowed by processes observable in lower animals? To the extent that language relies upon TOTE hierarchies and we have seen TOTE hierarchies in animals, there is nothing new here except a greater degree of complexity. But that comment is about as helpful as the remark that both animals and men are constructed of atoms. A more interesting question concerns the possibility that some new configuration of these basic components may have emerged. In particular, we might ask: Is the capacity to use Plans to construct Plans to guide behavior a new psychological process? A motor Plan may be instinctive, or it may be a skill acquired after long hours of practice—here men and animals are on an equal footing. But in the discussion of memorizing and even more clearly in the discussion of speaking we have found it necessary to believe that a motor Plan could be constructed very quickly and efficiently, not by rote, but by the operation of a higher-level Plan that had the motor Plan as its object. Something more is involved here than the usual discussions of insight versus trial-and-error—a motor Plan could be insightfully selected by an organism that could not execute a Plan to construct a motor Plan. Perhaps some of the apes, even some of the higher mammals, might have the rudiments of this higher-level planning ability. If so, then man may indeed have to retreat into his greater complexity to explain his unique accomplishments. If not, we may have here the key to an evolutionary breakthrough as important as the development of lungs and legs."

The implication of a sort of quantum shift in intellectual processes seems to be reinforced by studies of the development of thought in children. The most extensive of these have been carried out by Piaget and his associates, and by Vygotsky and associates. Problem-solving is seen as based on a hierarchical organization of symbolic representations and information-processing strategies deriving to a considerable degree from past experience. Piaget's observations, which have been published in a large series of books and papers, not all of which are readily accessible, have been summarized recently by Hunt (1961), and my discussion is largely taken from him.

The first one and a half or two years are called by Piaget the sensorimotor period, during which he thinks the developing mind goes through six stages, each of which incorporates the reactions of the preceding stage in a hierarchical fashion. These stages are outlined below:

1. Exercising the ready-made schemata (age range roughly 0 to one month).
2. Primary circular reactions: coordination of ready-made schemata to form motor habits and perceptions, no response to vanished object (age range roughly one to 4.5 months).
3. Secondary circular reactions: coordination of motor habits and perceptions to form intentional acts, development of prehension, beginning search for vanished objects, magicophenomenalistic causality, and subjective time series (roughly 4.5 months to 8 or 9 months).
4. Coordination of secondary schemata: application of familiar schemata in new situations as means and as ends, active search for vanished object but no account of the sequence of visible displacements, elementary objectification of causality and time, beginnings of imitation of auditory and visual models (age roughly 8 or 9 months to 11 or 12 months).
5. Tertiary circular reactions: discovery of new means through active experimentation, interest in novelty, object permanence through sequences of visible displacements, systematic imitation of new models, appreciation of objective, spatial, causal, and temporal sequences (age range roughly 11 or 12 months to between 18 months and two years).
6. Internalization of sensorimotor schemata: invention of new means through mental combination, object permanence through invisible sequences of displacements, representative spatial, causal, and temporal series, beginnings of symbolic imitation (age range roughly 18 months to two years as the new preconceptual phase begins).

In the discussion of each of these stages, Hunt brings in extensive data from animal experimentation and states:

"Although genuine intelligence emerges at the sixth sensorimotor stage, the gap between the sensorimotor intelligence of the infant aged between a year and a half and two and the reflective intelligence of adults is wide indeed. In fact, this gap is approximately equivalent to that between the intelligence of the dog, the monkey, or the chimpanzee and the intelligence of adult man."

In the transition from sensorimotor to reflective intelligence, there are several essential factors. One is an increased speed of thought—required to allow knowledge of successive phases of action to be incorporated into a simultaneous whole. An increase in scope is required, which permits concrete actions affecting real entities to be expanded by symbolic representations in order to carry the individual beyond the role of the immediate and local in perception and action into a universe extended indefinitely in both time and space. Finally, there must be developed a concern not only for the results desired of action but also for the actual mechanisms by which results are obtained so that the search for solutions may be combined with the development of knowledge of their nature.

What of the other characteristics which were mentioned? Hebb and Thompson (1954) suggest that emotional susceptibility apparently increases with intellectual capacity, and derive man's social evolution from a combination of his exceptional intellectual and emotional characteristics. Such susceptibility has two aspects—increases in the number and kinds of stimuli which arouse emotions and greater differentiation of the emotional response itself. Both require increasing complexity of the nervous system, and both must be linked not only in form of expression but also in effective stimuli to the development of language. Hebb and Thompson speculate on the role that emotionality has played in man:

"We have already proposed that the mammal seeks excitement when things are dull. This by itself makes an important change in the theory of 'economic man.' That theory can hold only in an impoverished society and *not* in an economically successful one. With a general increase in wealth and security, the risky venture may be preferred to the sure thing, the interesting occupation to one that pays well.

"But let us now go further. Evidence from species comparison suggests that emotional susceptibility increases with intellectual capacity. Man is the most emotional as well as the most rational animal. But this susceptibility is partly self-concealing: its possessor tends to seek the environment in which a too-strong emotion is least likely to occur, the one in which disturbance is nearer the optimal level. In man, this makes for the establishment of 'civilized' societies, the chief characteristic of which (at least until recently) is not that they improve the economic lot of the average member, but that they provide an environment in which the frequency of acute fear, disgust, anger, and jealousy is decreased. The further a society has advanced along this path, the less subject to strong emotion its members must appear . . . ."

Relative levels of self-awareness in species other than our own are, of course, a matter of speculation. There seems no reason to doubt, however, that the level of consciousness is much higher in man than in any other animal. The evolutionary aspect of self-awareness, as Hallowell (1960) has noted, has hardly been considered, and his own discussion of it is as complete as is possible at the present time. Its central importance for human forms cannot be overestimated: "As a result of self-objectification human societies could function through the commonly shared value-orientations of self-conscious individuals, in contrast with the societies of non-hominid and probable early hominid primates, where ego-centered processes remained undeveloped or rudimentary. In fact, when viewed from the standpoint of this peculiarity of man, culture may be said to be an elaborated and socially transmitted system of meanings and values which, in an animal capable of self-awareness, implements a type of adaptation which makes the role of the human being intelligible to himself, both with reference to an articulated universe and to his fellow men."

It can be speculated that development of self-awareness was the important precursor for several other characteristics which are uniquely human. The conception of possession of personal property can be seen as a natural extension of awareness of self. "Acquisivity," once developed, has had many effects. It may well be the major determinant for the fact that man is the most manifestly aggressive primate. Among other primates, aggression is usually carried no further than threatening and when it is, it seems very rarely to have a lethal outcome. In many of its more complex manifestations, it is linked to the primate pattern of dominance and social hierarchies. It is also at the basis for many patterns of sexual interaction—and probably a source for development of sexual jealousy more important than the sexual drive itself.

The ego functions must also have been a major factor in the increase in exploratory drives in man, and I think also in the development in man of those other drives for beauty, for understanding, and particularly for self-actualization, which are characteristic of him now, although in enormously varying degrees.

Whether all of these changes took place gradually, or whether there were in fact jumps, can only be matter for speculation. It is not inconceivable that the changes in man were not just increases in complexity and diversification, but took place in terms of hierarchical reordering of cognitive and emotional and other processes whenever a new function developed. This could readily have been the case with the introduction of speech, for example, which must have completely changed the nature of man's relation to the world, to other men and to himself.

Finally, let us ask Howells' (1959) question, "Supposing, in a moment of idiot progress, we really killed ourselves off. Would *Homo* rise again?" There can be no doubt that the answer is "No."

"Man came from an australopithecine, or some simpler hominid . . . . Those animals are gone; man has competed them into the grave. There are still apes. They might do for a fresh start, but I strongly suspect they are too specialized, and too busy looking for fruit in the forest, to turn to freer use of hands. Monkeys? Just possibly, if something made it worthwhile for a species to stand up. The new men might then have tails. But, in fact, the monkeys have made no move to mimic the hominoids, or human ancestors, during about thirty-five million years.

"No other higher mammals of this earth will serve. Horses, dogs, elephants, all are deeply committed to being what they are. The next try would have to come from a tree shrew, laboriously repeating all of primate history. And before little

Tupaia could put forth progressive descendants now, the world would have to be swept clean of the kind of competition which might overwhelm them on the way up. This means: get rid of most higher mammals, above all, rats, cats, and monkeys.

"If he fails us, we (or rather our carbon copies) are done for. The remaining links of progress are now missing links; the good chances are gone. The mammal-like reptiles gave out long ago; and getting something human from the specialized creatures in the next ranks is hopeless: birds, snakes, frogs. The fishes? Lobefins, with the makings of lungs and limbs, were put out of business eons ago by the ray-fins, who can never leave the sea. The main army of fishes has gone well past the fork that once led to the land. Only the lungfish remain, waiting in mud for the rain to come again, and the coelacanth, so deep in the ocean that he dies in shallow water.

"We might need brand-new 'vertebrates.' Well, then, eradicate the fish, who rule the seas as we rule the land and who are not likely to stand aside while nature experiments with ridiculously crude forerunners of ostracoderms once more. Conceivably life would have to start afresh. In that case, wipe out everything that moves, to keep the necessary simple molecules from being eaten as they form. So all in all our hopes for repetition are not good, and we had better stay the hand that drops the bomb."

But it would seem that the only way to stay the hand that drops the bomb is to exploit more effectively man's capacity for awareness and understanding. It is only through understanding of man's nature in the light of his past history that we can have any hope of controlling his future history.

BIBLIOGRAPHY

DOBZHANSKY, THEODOSIUS
    1962. *Mankind Evolving*. New Haven and London: Yale University Press.
FREEDMAN, L. Z. and A. ROE
    1958. "Evolution and Human Behavior." In Roe, A., and G. G. Simpson (eds.) *Behavior and Evolution*. New Haven: Yale University Press.
HALLOWELL, A. I.
    1960. "Self, society and culture in phylogenetic perspective." In Tax, S. (ed.) *The evolution of Man*, vol. 2. Chicago: University of Chicago Press.
HEBB, D. O.
    1949. *The organization of behavior*. New York: Wiley.

HEBB, D. O. and W. R. THOMPSON
1954. "The social significance of animal studies," in Lindzey, G. (ed.) *Handbook of social psychology*, Vol. 1. Cambridge, Mass.: Addison-Wesley.

HOWELLS, WILLIAM
1959. *Mankind in the Making*. Garden City, New York: Doubleday.

HUNT, J. McV.
1961. *Intelligence and Experience*. New York: The Ronald Press.

MASLOW, A. H.
1954. *Motivation and Personality*. New York: Harper.

MILLER, G. A., E. GALANTER and K. H. PRIBRAM
1960. *Plans and the structure of behavior*. New York: Henry Holt.

PENFIELD, W. and L. ROBERTS
1959. *Speech and Brain-mechanisms*. Princeton, N. J.: Princeton University Press.

PRIBRAM, K. H.
1960. "A review of theory in physiological psychology," *Ann. Rev. Psychol.*, 11, 1–40.

ROE, A.
1959. "Man's forgotten weapon." *The American Psychologist*, 14, 261–266.

SCHULTZ, A. H.
1961. "Some factors influencing the social life of primates in general and of early man in particular." In Washburn, S. L. (ed.) *Social life of early man*. Viking Fund publications in anthropology 31. New York: Wenner-Gren Foundation.

WASHBURN, S. L. and VIRGINIA AVIS
1958. "Evolution of human behavior," In Roe, A., and G. G. Simpson (eds.) *Behavior and Evolution*. New Haven: Yale University Press.

VYGOTSKY, L. S.
1962. *Thought and Language*. Edited and translated by Eugenia Hanfmann and Gertrude Vakar. Published jointly by MIT Press, Massachusetts Institute of Technology, and John Wiley and Sons, Inc., New York and London.

THE TAXONOMIC EVALUATION

OF FOSSIL HOMINIDS

ERNST MAYR

INTRODUCTION

THE CONCEPTS AND METHODS on which the classification of hominid taxa is based do not differ in principle from those used for other zoological taxa. Indeed, the classification of living human populations or of samples of fossil hominids is a branch of animal taxonomy. It can only lead to confusion if different standards and terminologies are adopted in the two fields. The reasons for the adoption of a single, uniform language for both fields, and the nature of this language, have been excellently stated by G. G. Simpson in his contribution to this volume.

There is, perhaps, one practical difference between animal and hominid taxonomy. Hominid remains are of such significance that even rather incomplete specimens may be of vital importance. An attempt must sometimes be made to evaluate fragments that a student of dinosaurs or fossil bovids would simply ignore. But, of course, even a rather complete specimen is only a very inadequate representation of the population to which it belongs, and most specimens are separated by large intervals of space and time. Yet, it is the task of the taxonomist to derive from these specimens an internally consistent classification.

The non-taxonomist must be fully aware of two aspects of such a classification: first, that it is usually by no means the only possible classification to be based on the available evidence, so that a taxonomist with a different viewpoint might arrive at a different classification; and second, that every classification based on inadequate material is provisional. A single new discovery may change the picture rather drastically and lead to a considerable revision.

The material of taxonomy consists of zoological objects. These objects are individuals or parts of individuals who, in nature, were members of populations. Our ultimate objective, then, is the classification of populations as represented by the available samples.

OBJECTS VERSUS POPULATIONS

The statement that we must classify populations rather than objects sounds almost like a platitude in this year 1962. Yet, it is not so many years ago that the study of fossil man was in the hands of strict morphologists who arranged specimens in morphological series and based their classification almost entirely on an interpretation of similarities and differences without regard to any other factor. He who classifies specimens as representatives of populations knows that populations have a concrete distribution in space and time and that this provides a source of information that is not available to the strict morphologist. Any classification that is inconsistent with the known distribution of populations is of lowered validity.

THE APPLICATION OF TAXONOMIC PRINCIPLES IN CONCRETE CASES

There have been several previous attempts (Dobzhansky, 1944; Mayr, 1944, 1950) to apply the principles of systematic zoology to some of the open problems of hominid classification. A great deal of new evidence has since accumulated and there has been some further clarification of our concepts. The time would seem proper for a new look at some of these problems.

SUBSPECIES OR SPECIES

The decision whether to rank a given taxon in the category "subspecies" or in the category "species" is often exceedingly difficult in the absence of conclusive evidence. This is as true in the classification of living populations (geographical isolates) as it is for fossils. A typical example as far as the hominids are concerned is the ranking of Neanderthal Man. I have found three interpretations of Neanderthal in the literature.

(1) "Neanderthal Man is a more primitive ancestral stage through which *sapiens* has passed." This we might call the classical hypothesis, defended particularly during the period when the interpretation of human evolution was based primarily on the evaluation of morphological series. This classical hypothesis had to be abandoned when it was found to be in conflict with the distribution of classical and primitive Neanderthals and of *sapiens* in space and time.

(2) "Neanderthal is an aberrant separate species, a contemporary of early *sapiens* but reproductively isolated from him."

(3) "Neanderthal is a subspecies, a geographic race, of early *sapiens*."

What evidence is there that would permit us to come to a decision as to the relative merits of alternatives (2) and (3)? We must begin by defining rigidly what a species is and what a subspecies. As clearly stated by Simpson (1961) and

Mayr (1957), degree of morphological difference per se is not a decisive primary criterion. A species is reproductively isolated from other species coexisting in time and space, while a subspecies is a geographic subdivision of a species actually or potentially (in the case of geographical isolates) in gene exchange with other similar subdivisions of the species.

The difficulty in applying these concepts to fossil material is obvious. It can be established only by inference whether two fossil taxa formed a single reproductive community or two reproductively isolated ones. In order to draw the correct inference, we must ask certain questions:

Does the distribution of Neanderthal and *sapiens* indicate that they were reproductively isolated? Not so many years ago Neanderthal was considered by many as a Würm "eskimo," but he is now known to have had an enormous distribution, extending south as far as Gibraltar and North Africa and east as far as Iran and Turkestan. There is no evidence (but see below) that Neanderthal coexisted with *sapiens* anywhere in this wide area. Where did *sapiens* live during the Riss-Würm Interglacial and during the first Würm stadial? No one knows. Ethiopia, India and southeast Asia have been suggested, but these will remain wild guesses until some properly dated new finds are made. All we know is that at the time of the first Würm interstadial Cro-Magnon Man suddenly appeared in Europe and overran it in a relatively short time.

THE *SAPIENS* PROBLEM

There has been much talk in the past of the "Neanderthal problem." Now, since the average morphological differences between the classical Neanderthal of the first Würm stadial and the earlier Neanderthals of the Riss-Würm Interglacial have been worked out, and since the distribution of Neanderthal has been mapped, *sapiens sapiens* has become the real problem. Where did he originate and how long did it take for pre-*sapiens* to change into *sapiens?* Where did this change occur?

All we really know is that *s. sapiens*, as Cro-Magnon, suddenly appeared in Europe. Sufficient remains from the preceding period of unmixed Neanderthal in Europe and adjacent parts of Africa and Asia (and a complete absence of any blade culture) prove conclusively that *s. sapiens* did not originate in Europe. The rather wide distribution of types with a strong supraorbital torus (e.g., Rhodesia, Solo-Java) suggests that *s. sapiens* must have originated in a localized area. The sharpness of distinction between Neanderthal and *s. sapiens* (except at Mt. Carmel) further indicates that Neanderthal, as a whole, did not gradually change into *sapiens*, but was replaced by an invader.

There is a suspicion that evolutionary change can occur the faster (up to certain limits), the smaller and more isolated the evolving population is. If *s. sapiens* lost his supraorbital torus very quickly (and the various other characters it had before becoming *sapiens*) then it can be postulated with a good deal of assurance that *sapiens* evolved in a rather small, peripheral, and presumably well isolated population. Even if we assume that the rate of change was slow and the evolving population large, we must still assume that *sapiens* was rather isolated. Otherwise, one would expect to find more evidence for intergradation with late Neanderthal.

It is obvious that the available evidence is meager. Let us assume, however, for the sake of the argument, that Neanderthal and *sapiens* were strictly allopatric, that is that they replaced each other geographically. Zoologists have interpreted allopatry in the past usually as evidence for conspecificity, because subspecies are always allopatric. We are now a little more cautious because we have discovered in recent years a number of cases where closely related species are allopatric because competitive intolerance seems to preclude their geographical coexistence. The rapidity with which Neanderthal disappeared at the time Cro-Magnon Man appeared on the scene would seem to strengthen the claim for competitive intolerance and consequently for species status of these two entities. Yet, here is clearly a case where it is perhaps not legitimate to apply zoological generalizations to man. The Australian Aborigines and most of the North American Indians disappeared equally or perhaps even more rapidly, and yet no one except for a few racists would consider them different species.

We are thus forced to fall back on the two time-honored criteria of species status, degree of morphological difference and presence or absence of interbreeding. Our inference on the taxonomic ranking of Neanderthal will be based largely on these two sets of criteria, supplemented by a third, available only for man.

1. DEGREE OF MORPHOLOGICAL DIFFERENCE

The amount of difference between the skulls of Neanderthal and *sapiens* is most impressive. There are no two races of modern man that are nearly as different as classical Neanderthal and *sapiens*. And yet one has a feeling that the differences are mostly of a rather superficial nature, such as the size of the supra-orbital and occipital torus and the general shape of the skull. The cranial capacity, on the other hand, is remarkably similar in the two forms. The gap between Neanderthal and *sapiens* is to some extent bridged by two populations, Rhodesian Man and Solo Man, which are widely separated geographically from Neanderthal. Although sharing the large supraorbital torus with Neanderthal, these two other populations differ in many details of skull shape and cranial capacity from Neanderthal as well as from *s. sapiens*. Whether or not these peripheral African and Asiatic types acquired their Neanderthaloid features independently, can be established only after a far more thorough study and the investigation of additional material. It seems quite improbable that they are directly related to Neanderthal. In view of their small cranial capacity they may have to be classified with *H. erectus*.

As it now stands, one must admit that the inference to be drawn from the degree of morphological difference between Neanderthal and *sapiens* is inconclusive.

2. INTERBREEDING BETWEEN NEANDERTHAL AND *sapiens*

Cro-Magnon Man, on his arrival in western Europe, seems to have been re-markably free from admixture with the immediately preceding Neanderthal. There is, however, some evidence of mixture in the material from the two caves of Mt. Carmel in Palestine. Both caves were inhabited early in the Würm glacia-tion. The older cave (Tabun) was inhabited by almost typical Neanderthals with a slight admixture of modern characters, the younger cave (Skhul) by an essentially *sapiens* population but with distinct Neanderthaloid characters. The date is too late to consider these populations to have belonged to the ancestral stock that gave rise both to Neanderthal and modern man. It seems to me that the differences between Tabun and Skhul are too great to permit us to consider them as samples from a single population coming from the area of geographical intergradation between Neanderthal and modern man, although this could be true for the Skhul population. Hybridization between invading Cro-Magnon Man and Neanderthal remnants is perhaps a more plausible interpretation for the Skhul population, while there is no good reason not to consider Tabun essentially an eastern Neanderthal, particularly in view of its similarity to the Shanidar speci-mens (Stewart, 1960).

Repeated re-examinations of the Mt. Carmel material have thus substantiated the long-standing claims that this material is evidence for interbreeding between Neanderthal and *sapiens*.

3. THE CULTURAL EVIDENCE

As our knowledge of human and hominid artifacts increases, it becomes neces-sary to include this source of evidence in our considerations. My own personal knowledge of this field is exceedingly slight, but when I look at the implements assigned to Neanderthal and those assigned to Cro-Magnon, I feel the differences are so small that I can not make myself believe they were produced by two different biological species. I realize that the history of human or hominid arti-facts goes back much further than we used to think, yet this is not in conflict with my hunch that there was no opportunity for the simultaneous existence of two separate hominid species of advanced tool makers.

I would like to add some incidental comments on tools and human evolution. The history of peoples and tribes is full of incidences of a secondary cultural deterioration, *vide* the Mayas and their modern descendants! Most of the modern native populations with rudimentary material cultures (e.g., certain New Guinea mountain natives) are almost surely the descendants of culturally more advanced ancestors. This must be kept in mind when paleolithic cultures from Africa and western Eurasia are compared with those of southern and eastern Asia. Stone tools and the hunting of large mammals seem to be closely correlated. Could such peoples have lost their tool cultures after they had emigrated into areas poor in large game? Could this be the reason for the absence of stone tools in Javan *Homo erectus?*

Conclusion.—The facts that are so far available do not permit a clear-cut decision on the question whether Neanderthal was a subspecies or a separate species. It seems to me, however, that on the whole they are in better agreement with the subspecies hypothesis. It would seem best for the time being to postulate that Neanderthal (*sensu stricto*) was a northern and western subspecies of *Homo sapiens* (*sensu lato*), which was an incipient species but probably never reached species level prior to its extinction.

POLYTYPIC SPECIES AND EVOLUTION

All attempts to trace hominid phylogeny still deal with typological models. "*Australopithecus* gave rise to *Homo erectus,*" etc. In reality there were widespread polytypic species with more advanced and more conservative races. One or several of the advanced races gave rise to the next higher grade. It may happen in such a case that the descendant species lives simultaneously (but allopatrically) with the more conservative races of the ancestral species. This is often interpreted to indicate that the ancestral species could not have given rise to the descendant species. *True,* as far as the ancestral species in a typological sense is concerned, but *not* true for the ancestral species as a polytypic whole.

The concept of most polytypic species being descendants of ancestral polytypic species creates at once two formidable difficulties. One of these is caused by unequal rates of evolution of the different races. Let us say that there was an ancestral species 1 with races 1a, 1b, 1c and 1d. Race 1a evolved into species 2, absorbing in the process much of race 1b, and now forms races 2a and 2b. Race 1c became extinct and race 1d persisted in a relic area without changing very drastically. We now have 2 (a and b) and 1 (d) existing at the same time level, even though they represent different evolutionary stages (morphological grades). It is thinkable, for instance, (in part after Coon, 1962) that Heidelberg Man was the first population of *Homo erectus* to reach the *sapiens* level, and that as *Homo sapiens heidelbergensis* it was contemporary with *Homo erectus* of Java and China. (This is purely a thought model, as long as only a single mandible of Heidelberg Man is available.) It is possible that *Homo erectus* persisted in Africa as *rhodesiensis* and in Java as *soloensis* at a time when European populations clearly had reached *Homo sapiens* level. Such a possibility is by no means remote, in view of the many polytypic species of Recent animals in which some races are highly advanced and others very primitive.

I am calling attention to this situation to prevent too far a swing of the pendulum. The late Weidenreich arranged fossil hominids into morphological series strictly on the basis of morphology without regard to distribution in space and time (e.g., Neanderthal—Steinheim—*H. sapiens*). Some modern authors tend to swing to the other extreme by classifying fossil hominids entirely on the basis of geological dating without paying any attention to morphology. The unequal rates of evolutionary change in widely dispersed and partially isolated races of polytypic species make it, however, necessary to take morphology and distribution equally into consideration. Even though *Homo sapiens* unquestionably descended from *Homo erectus*, it is quite possible, indeed probable, that some races of *Homo erectus* still persisted when other parts of the earth were already populated by *Homo sapiens*.

The same argument is even more true for genera. The fact that *Homo* and *Australopithecus* have been found to be contemporaries does not in the least invalidate the generally accepted assumption that *Homo* passed through an *Australopithecus* stage. The Australopithecines consisted of several species (or genera, if we recognize the generic distinction of *Paranthropus*) and each of these species, in turn, was polytypic. Only a segment of this assemblage gave rise to *Homo*. Much of the remainder persisted contemporaneously with *Homo*, for a longer or shorter period, without rising above the Australopithecine grade. The modern concepts of taxonomy and speciation do not require an archetypal transformation (*in toto*) of *Australopithecus* into *Homo*.

The second great difficulty caused by the evolution of polytypic species is a consequence of the first one. It is the difficulty to determine what part (which races) of the ancestral species has contributed to the gene pool of the descendant species. This in turn depends on the amount of gene flow between the races of the ancestral species while it passed from the level of species 1 to the level of species 2. The amount of gene flow is determined by the nature of the interaction between populations in zones of contact. Unfortunately, the situation in the near-human hominids (*Homo erectus* level) was probably different from both the anthropoid condition and the condition in modern man. A number of possibilities are evident in an area of contact between races:

(1) Avoidance
(2) Extermination of one by the other
(3) Killing of the men and absorption of the women
(4) Free interbreeding

There is much evidence that all four processes have occurred and it becomes necessary to determine their relative importance in individual cases. The Congo pygmies, the bushmen, and various negritoid pygmies in the eastern tropics illustrate avoidance by retreating into inferior environments. The Tasmanians and some Indian tribes illustrate extermination. The white invaders in North America and Australia absorbed extremely few genes of the native peoples. The frequently made assertion that invaders kill off the men and take the women is often contradicted by the facts. The sharpness of the difference between classical western European Neanderthal and invading Cro-Magnon indicates to me that Cro-

Magnon did not absorb many Neanderthal genes (some contrary opinions notwithstanding). Language and cultural differences must have militated at the *Homo erectus* level against too active a gene exchange between different races. The distinctness of the negro, mongoloid, and caucasian races supports this assumption. Gene flow obviously occurred, but against considerable obstacles.

SYMPATRIC SPECIES OF HOMINIDS

When one reads the older anthropological literature with its rich proliferation of generic names, one has the impression of large numbers of species of fossil man and other hominids coexisting with each other. When these finds were properly placed into a multi-dimensional framework of space and time, the extreme rarity of the coexistence of two hominids became at once apparent. We have already discussed the case of Neanderthal and *sapiens*, but there are others in the Middle and early Pleistocene.

At the *Homo erectus* level, we have Java Man and Pekin Man, originally described as two different genera, but so strikingly similar that most current authors agree in treating them as subspecies. Ternifine Man in North Africa may be another representative of this same polytypic species. The existing material is, however, rather fragmentary. A further contemporary is Heidelberg Man, whose massive mandible contains teeth that appear smaller and more "modern" than those of a typical *erectus*. Was this a second species or merely a deviant peripheral isolate? This can only be settled by additional discoveries.

In Africa we find incontrovertible evidence of contemporaneity of several species of hominids. *Australopithecus* and *Paranthropus* apparently differed considerably in their adaptations. Perhaps this is the reason why they are not found together in most South African deposits. Yet the degree of difference between them and the time span of their occurrence leaves no doubt that they must have been contemporaries. Here, then, we have a clear case of the contemporaneity of two species of hominids. The fragments of the small hominid (*Telanthropus*) found at Swartkrans with *A. robustus*, which may belong to an Australopithecine or *Homo*, supply additional proof for the coexistence of two hominids in South Africa. In Java, in the Djetis layers, there is also the possibility of the coexistence of two hominids, "*Meganthropus*" and *Homo erectus*.

By far the most exciting instance of the coexistence of two hominids is that established by Leakey in East Africa. In layer 1 of Olduvai "*Zinjanthropus*," an unmistakable Australopithecine of the *Paranthropus* type, is associated with remains of an advanced hominid, "co-Zinjanthropus," that—when better known—may well turn out to be closer to *Homo* than to *Australopithecus*. Whether the tools of this layer were made by both hominids or only the more advanced, can be determined only when the two types are found unassociated at other sites. This will also influence the decision on the identity of the maker of the Sterkfontein tools in South Africa.

The picture that emerges from all these new discoveries is that only one species of hominids seems to have been in existence during the Upper Pleistocene, but that there is much evidence from the Middle and Lower Pleistocene of several independent lines. Some of these gave rise to descendant types, others became extinct. The coexisting types were, so far as known, rather distinct from each other. This is what one would expect on the basis of *a priori* ecological considerations. The principle of "competitive exclusion" would prevent sympatry if there were not considerable ecological divergence. *Australopithecus* and *Paranthropus*, or *Zinjanthropus* and the associated hominid, "co-Zinjanthropus," were able to co-exist only because they utilized the resources of the environment differently. Whether one of them was more of a hunter, the other more of a gatherer (or hunted), whether one was more carnivorous, the other more of a vegetarian, whether one was more of a forest creature, the other a savanna inhabitant, all this still remains to be investigated, when better evidence becomes available.

It is important to emphasize that nothing helped more to make us aware of these problems and to assist in the reconstruction of evolutionary pathways than an improvement of the classification of fossil hominids both on the generic and on the specific levels. It is here that the application of principles of zoological taxonomy has been particularly fruitful. Indeed, the earlier morphologists never appreciated the biological significance of the problem of coexistence or replacement of closely related species.

GENERIC PROBLEMS

The category "genus" presents even greater difficulties than that of the species. There is no non-arbitrary yardstick available for the genus as reproductive isolation is for the species. The genus is normally a collective category, consisting of a group of species believed to be more closely related to each other than they are to other species. Yet, every large genus includes several groups of species that are more closely related to each other than to species of other species groups within the same genus. For instance, in the genus *Drosophila* the species belonging to the *virilis* group are more closely related to each other than to those belonging to the *repleta* group, yet both are included in *Drosophila*. They are not separated in different genera because the species groups have not yet reached the degree of evolutionary divergence usually associated with generic rank. As Simpson (1961) has pointed out, the genus usually has also a definite biological significance, indicating or signifying occupation of a somewhat different adaptive niche. Again, this is not an ironclad criterion because even every species occupies a somewhat different niche, and sometimes different genera may occupy the same adaptive zone.

It is particularly important to emphasize again and again that the function of the generic and the specific names in the scientific binomen are different. The specific name stresses the singularity of the species and its unique distinctness.

The generic name emphasizes not a greater degree of difference but rather the belonging-together of the species included in the genus. To place every species in a separate genus, as was done by so many of the physical anthropologists of former generations, completely stultifies the advantages of binomial nomenclature. As Simpson has stated correctly in this volume, the recognition of a monotypic genus is justified only when a single isolated known species is so distinctive that there is a high probability that it belongs to a generic group with no other known ancestral, collateral or descendant species. The isolated nature of *bamboli*, the type species of *Oreopithecus*, justifies the recognition of this monotypic genus.

Of the literally scores of generic names proposed for fossil hominids, very few deserve recognition. More and more students admit, for instance, that the degree of difference between *Homo erectus* and *H. sapiens* is not sufficient to justify the recognition of *Pithecanthropus*.

There are a number of reasons why it would seem unwise to recognize the genus *Pithecanthropus* in formal taxonomy. First of all, *Homo* would then become a monotypic genus and *Pithecanthropus* contain at most two or three species. This is contrary to the concept of the genus as a collective category. More importantly, the name *Pithecanthropus* was first applied to an actual fossil hominid when only a skull cap was known and the reconstruction envisioned a far more anthropoid creature than *erectus* really is. When the teeth and other body parts were discovered (or accepted, like the femur) it was realized that the total difference between *erectus* and *sapiens* was really rather small and certainly less than is normally required for the recognition of a zoological genus. The recognition of *Pithecanthropus* as a genus would lead to an undesirable heterogeneity of the genus category.

The genus *Australopithecus* has already many of the essential characters of *Homo*, such as a largely upright posture, bicuspid premolars, and reduced canines. For this reason I suggested previously (Mayr, 1950) that "not even *Australopithecus* has unequivocal claims for generic separation." I now agree with those authors who have since pointed out not only that the upright locomotion was still incomplete and inefficient, but also that the tremendous evolution of the brain since *Australopithecus* permitted man to enter a new niche so completely different that generic separation is fully justified. The extraordinary brain evolution between *Australopithecus* and *Homo* justifies the generic separation of these two taxa, no matter how similar they might be in many other morphological characters. Here, as in other cases, it is important not merely to count characters but to weight them.

Whether or not one wants to recognize only a single genus for all the known Australopithecines or admit a second genus, *Paranthropus*, is largely a matter of taste. The species (*robustus*) found at Swartkrans and Kromdraai is larger and seems to have more pronounced sexual dimorphism than *A. africanus*. Incisors and canines are relatively small, while the molars are very large and there are pronounced bony crests on the skull, particularly in adult males. These differences are no greater than among species in other groups of mammals. *Zinjanthropus* in East Africa seems to belong to the more massive *Paranthropus* group. The two Australopithecines (*africanus* and *robustus*) seem to represent the same "grade" as far as brain evolution is concerned, but the differences in their dental equipment and facial muscles indicate that they may have occupied different food niches. It may well depend on future finds whether or not we want to recognize *Paranthropus*. The more genuinely different genera of hominids are discovered, the more important it may become to emphasize the close relationship of *Australopithecus* and *Paranthropus* by combining them in a single genus. It depends in each case to what extent one wants to stress relationships. We have a similar situation among the pongids. I have pointed out earlier (Mayr, 1950) that gorilla and chimpanzee seem to me so much nearer to each other than either is to man or to the orang or to the gibbons, that degree of relationship would seem to be expressed better if the gorilla were included in the genus *Pan* rather than to be recognized as a separate genus. The decision on generic status is as always based on somewhat arbitrary and subjective criteria. One cannot prove that gorilla and chimpanzee belong to the same genus, but neither can one prove that they belong to different genera.

DIAGNOSTIC CHARACTERS

The collective nature of the categories above the species level show clearly why it is often so difficult to provide an unequivocal diagnosis for taxa belonging to the higher categories. Those who think that "the characters make the genus" have little difficulty in characterizing differences between species and calling them generic differences. It is much easier to characterize the species "chimpanzee" and the species "gorilla" than to find diagnostic characters that clearly distinguish the chimpanzee-gorilla group from man, the orang and the gibbons. Higher categories often can be diagnosed only by a combination of characters, not by a single diagnostic character. The definition of the genus *Homo* presented in this volume is an example of such a combinational diagnosis.

The problem of the relation between taxonomic ranking and diagnostic characters will become increasingly acute as new Pliocene and Miocene fossils are found. Nothing would be more short-sighted than to base the classification of such finds on isolated "diagnostic" characters. We must ask ourselves each time whether relationship will be expressed better by including such new taxa in previously established ones, or by separating them as new taxa. If we combine them with previously established taxa, that is, if we include them in previously recognized genera, subfamilies, or families, we may have to modify the diagnosis of such taxa. We must always remember that the categories above the species are collective categories and subjectively delimited. The pronouncement made by Linnaeus, "It is the genus that gives the characters, and not the characters that make the genus," is true not only for the genus but for the categories at the family level.

Diagnostic characters are a convenient tool of the working taxonomist, they should never become a strait jacket.

NOMENCLATURE AND COMMUNICATION

Superimposed on all the taxonomic difficulties are some purely nomenclatural ones. Simpson, in this volume, has already pointed out that it is altogether inadmissible to change a scientific name because it is considered inappropriate. A scientific name is, so to speak, merely a formula and its etymological meaning is quite irrelevant. What is important is to avoid arbitrary changes, because the words of a language lose all usefulness if they are shifted around or replaced by new ones. If an anthropologist wants to play around with names, let him concentrate on the vernacular names. No one will care whether he talks of Heidelberg Man or Man of Mauer, or of Pekin Man rather than Man of Choukoutien.

As soon as an anthropologist employs zoological nomenclature that has very definite rules, he must obey these rules. In particular, I would like to call the attention of anthropologists to Article 35b of the Rules of Zoological Nomenclature, which states that the names of families and subfamilies must be based on the name of an included genus. Since there is no genus Euhomo, there can be no family name Euhominidae. If a subfamily is recognized for the Australopithecines, it can only be Australopithecinae. Not only is this system the only valid one, but it also has the advantage of being simple and unambiguous. I should hope that such confusing terms as Praehominidae would soon disappear from the literature.

Those who give names to fossil Hominidae might also be more careful in the choice of specific names. To have several *africanus* and *robustus* in this family is confusing, particularly during this period of rearranging of genera. It would not seem an impossible demand that only such new specific names be given that had not been given previously to other species in the family Hominidae.

THE CLASSIFICATION OF THE MISSING LINK

Nothing characterized the early study of human evolution as much as the search for the missing link. When one looks at early reconstructions of the missing link, one realizes how strongly the concept was dominated by the ancient idea of the *scala naturae*. If evolution were limited to a single lineage, as thought for instance by Lamarck, the missing link would simply be the halfway stage between the anthropoids and man. Now as we realize that there is no single line of descent but a richly branching phylogenetic tree, the search for the missing link has become somewhat illusory. It is now evident that there is not just one missing link but a whole series of missing links. There is first the species which was at the branching point between the Pongidae and the Hominidae. On the hominid branch there were those species that first acquired such essentially human traits as making tools, making fire, and possessing speech. There is the first species to be referred to the genus *Homo*, and there is the species that acquired a brain capacity about halfway between the anthropoids and modern man. We may already have representatives satisfying most of these qualifications, and rather than searching for *the* missing link we are now beginning to classify kinds of missing links.

It is now clear that we must distinguish between two essential phenomena. There is on one hand the phylogenetic branching of the hominid line from the pongid line. Yet even after this branching had taken place, which presumably was sometime during the Miocene, there was no sign of Man on the new hominid line. The hominids throughout the Miocene and Pliocene were still apes, and even the Australopithecines of the early Middle Pleistocene can hardly be classified as human. It does appear that *Homo erectus* qualifies better as representative of the stage between prehuman hominids and Man than any other form. It is almost certainly the stage at which the hominids became Man.

THE HIGHER CATEGORIES

The Pleistocene hominids present no problem at the family level. They all clearly belong to the Hominidae. Whether or not to separate the Australopithecines in a subfamily Australopithecinae is essentially a matter of taste. Matters are more difficult when it comes to Pliocene and Miocene fossils. Not only are most of them known from insufficient fragments, but the criteria on which to base the decision pongid or hominid become increasingly elusive as we go back in time. Furthermore, there is considerable probability of the existence of additional equivalent branches of anthropoids or near-anthropoids which have since become extinct. *Oreopithecus* seems to represent such a branch.

The evidence concerning the branching-off point of the hominid from the pongid line seems on first sight contradictory. Schultz finds that all the great apes agree with each other in very numerous characters of general morphology, in which they differ from Man. Yet, the African apes (*Pan* sensu lato) are closer to Man than to orang or gibbon in hemoglobin structure (Zuckerkandl, 1963), in serum proteins (Goodman, 1963) and in chromosomal morphology (Klinger, 1963). What can be the explanation of this apparent conflict in the evidence? Perhaps the simplest interpretation would be to assume that Man's shift into the niche of the bipedal, tool-making, and speech-using hominid necessitated a drastic reconstruction of his morphology, but that this reconstruction did not, in turn, require a complete revamping of his biochemical system. Different characters and character complexes thus diverged at very different rates. If one assumes this to be correct, one will conclude that the *Homo*-line branched from the *Pan*-line well after the line of their common ancestor had separated from the orang (*Pongo*)-line.

Full awareness of mosaic evolution is particularly important for the correct placing of early hominid fossils. As Le Gros Clark (1950) and I (Mayr, 1950) have emphasized, the Hominidae are a classical example of mosaic evolution.

Every character or set of characters evolved at a different rate. Even *Australopithecus* is still essentially anthropoid in some characters while having considerably advanced toward the hominid condition in other characters, for instance with respect to upright posture and the general shape of the tooth row. We must furthermore be aware of the fact that evolution is not necessarily irreversible and that temporary specializations may secondarily be lost again. The Pliocene forms *Ramapithecus* from India (Simons, 1961) and *Kenyapithecus* from Africa may well belong to the hominid line. Even more difficult is the allocation of Miocene genera. To place them correctly one may have to use "prophetic characters" (a preevolutionary term of Agassiz), that is, characters which foreshadow future evolutionary trends. It is quite certain that the Miocene hominoids lacked some of the characteristic specializations of both the pongid and the hominid lines. They were not the extreme brachiators that some of the modern pongids are, nor had they reached the completeness of upright posture and the special features of dentition of the later hominids. Such seemingly irrelevant characters as the shape of tooth cusps may be more revealing in such forms than the relative size of the canines or the development of a simian shelf. The recent arguments about *Oreopithecus* show how difficult it is to reach an objective evaluation of the evidence.

CONCLUSIONS

Fossil hominids are samples of formerly existing populations distributed in space and time. Their classification must be consistent with the generalizations derived from the study of polytypic species in animals. Whenever the anthropologist uses the terminology of subspecies, species, and genera, such terminology must be consistent with the meaning of these categories as developed in modern systematics (Simpson, 1961). Application of the principles of systematics has helped to clarify the formerly bewildering diversity of morphological types. Pleistocene hominids display much geographic variation, but the number of full species coexisting at any one time is not known to have exceeded two. The relatively wide distribution of some of the fossil hominids indicates that there has been a considerable amount of gene flow as early as the lowest Middle Pleistocene.

BIBLIOGRAPHY

CLARK, W. E. LE GROS
1950. "New palaeontological evidence bearing on the evolution of the Hominoidea." *Quart. J. Geol. Soc. London,* 105:225–264.

COON, C.
1962. *The Origin of Races.* New York: A. A. Knopf.

DOBZHANSKY, TH.
1944. "On species and races of living and fossil man." *Amer. J. Phys. Anthr.,* 2 (n.s.):251–265.

GOODMAN, M.
1963. "Man's place in the phylogeny of the primates as reflected in serum proteins." In *Classification and Human Evolution.* S. L. Washburn, ed. Chicago: Aldine.

"INTERNATIONAL CODE OF ZOOLOGICAL NOMENCLATURE"
1961. London: *Int. Trust Zool. Nomenclat.*

KLINGER, H. P., J. L. HAMERTON and D. MUTTON
1963. "The chromosomes of the Hominoidea." In *Classification and Human Evolution.* S. L. Washburn, ed. Chicago: Aldine.

MAYR, E.
1944. [Horizontal and vertical subspecies]. N.R.C. Committee on Common Problems of Genetics, Paleontology and Systematics. *Bull.* No. 2 (June):11–16.
1950. "Taxonomic categories in fossil hominids." *Cold Spring Harbor Symp. Quant. Biol.* 25:109–118.
1957. "Species concepts and definitions." In *The Species Problem,* E. Mayr, ed., *Amer. Assoc. Adv. Sci. Publ.* No. 50:1–22.

SIMONS, E. L.
1961. "The phyletic position of *Ramapithecus. Postilla.*" Yale Peabody Mus. No. 57:1–9.

SIMPSON, G. G.
1961. *Principles of animal taxonomy.* New York: Columbia Univ. Press.
1963. "The meaning of taxonomic statements." In *Classification and Human Evolution.* S. L. Washburn, ed. Chicago: Aldine.

STEWART, T. D.
1960. "Form of the pubic bone in Neanderthal man." *Science,* 131:1437–1438.

ZUCKERKANDL, E.
1963. "Perspectives in molecular anthropology." In *Classification and Human Evolution.* S. L. Washburn, ed. Chicago: Aldine.

GENETIC ENTITIES IN HOMINID EVOLUTION

THEODOSIUS DOBZHANSKY

No two humans are ever identical. We take so completely for granted that every person differs from every other that the similarity of identical twins (who are never really "identical") strikes us as startling and peculiar. Individual variation exists, of course, in all biological species, but the realization that two identical flies are no more likely to exist than two identical humans comes as something of a surprise to non-biologists. Humans perceive differences among humans more easily than differences among flies. This legitimate anthropocentrism becomes troublesome in scientific studies of human and organic diversity. We tend to overvalue the magnitude of the differences in human materials relative to the differences found elsewhere.

Need we be much concerned about such overvaluation? Does it, for example, matter whether mankind is taken to be a single species composed of races or an assemblage of several species? Did the mid-Pleistocene hominids belong to one or to several species? The basic issue here is evidently the nature of races, species, and other categories. Are these categories purely arbitrary labels, or do they correspond to real biological phenomena? I wish to present arguments in favor of the latter view.

LEVELS OF INTEGRATION OF ORGANIC SYSTEMS

Biology is a study of life. The lowest, or most elementary, or most fundamental living unit is a gene. A gene is a bit of matter, as far as known always containing, or composed of, deoxyribonucleic or ribonucleic acid. This remarkable substance can engender, in proper environments, synthesis of copies of itself. Except in the simplest viruses, the self-reproducing genes come in integrated systems, chromosomes. Chromosomes are carried in cells. Cells usually form integrated aggregates, individuals.

Individuals are unquestionably real biological units. It can be argued that individuals are even more fundamental units than genes, but such arguments are idle because most scientists regard as fundamental whatever they happen to be interested in. There is no question, however, that individuality has increased as evolution progressed. In asexual form, the entire progeny of an individual has, unless mutation intervenes, the same genotype as the progenitor. In sexual outbreeding species, including man, no two individuals, except identical twins, are likely to carry the same genotypes. Every individual is, then, unique. Evolutionary changes are sequences of unique events, unprecedented and non-recurrent.

Since individuals are innumerable and seemingly endlessly diversified, to study them a scientist must classify them. Groups of more or less similar individuals are treated as taxa, to which formal or informal names are customarily assigned. Groups of taxa constitute taxa of higher and higher orders, making the classification hierarchical. Race, species, genus, etc. are categories of taxa in the hierarchical classification of organisms. Do these categories correspond to some supra-individual levels of biological integration?

Taxonomic categories would have to be invented if they did not exist in nature. The problem of their objectivity vs. arbitrariness has had a long history, which need not be considered here. This problem may be viewed as a reflection in biology of the philosophical dispute between nominalists and realists. To a nominalist only individuals are real, and all taxa are mere group concepts set up for the convenience of the classifier or in deference to his linguistic habits. Classification of organisms is like that of postage stamps, to be arranged as effectively as possible in an album. To a realist, some of the supra-individual categories correspond however to observable levels of biological integration. Thus, the species in sexual organisms is not only a group concept but also a biological phenomenon which exists regardless of whether we classify organisms or not. However, taxonomic species will be most meaningful if they are made to correspond to the biological species.

SPECIES AND TYPES

A historically important variant of the realist position considered species to be the primordial created entities. In its classical, Linnean, version this view negates evolution; but it nevertheless persists in modern biology, more often implicitly than explicitly, in the form of typological modes of thought. Species, or race, or some other category is declared the archetypal entity, and individuals its transitory and imperfect incarnations. Typological approaches of this sort were adopted by such prominent anthropologists as Topinard, Deniker, Hooton, and most recently by the Polish school. According to Topinard, an anthropologist ". . . seeks out the types and multiplies them and supposes that they have been perpetuated without change throughout the upsets and the mix-ups of history and pre-history." Virchow defined races "as acquired deviations from the original type." To Deniker, a race was "a sum total of somatological characteristics once met with in real union of individuals, now scattered in fragments of varying proportions among several 'ethnic groups,' from which it can no longer be differentiated except by the process of delicate analysis." Hooton held that "one must conceive of race not as the combination of features which gives to each person his individual appearance, but rather as a vague physical background, usually more or less obscured or overlaid by individual variations in single subjects, and realized best in a composite picture" (all citations from the anthology of Count, 1950). And Wiercinski, in 1962, paraphrases Czekanowski's definition of race as follows: "A given set of racial features, characterizing a given racial type, is inherited as it would be if determined by one pair of alleles."

Methodological objections against typological approaches to studies of the nature of organic diversity have been stated by Mayr (1959) and Simpson (1961). I can add only that the notion of racial types perpetuated without change or determined by single genes is flatly contradicted by genetic data. Structural race differences are in general complexly polygenic, in fact so much so that the numbers of the genes involved can only rarely be estimated. Genetically simple traits concern usually variables such as blood antigens and malformations which are constituents of the genetic loads which populations carry. It is possible, though not proven, that human populations are polymorphic for some genes with relatively major and pleiotropic effects which determine different constitutional types or somatotypes. The complexes of morphological traits conditioned by these genes may have been misinterpreted as persistent racial types or racial components. Even if confirmed, this would in no way vindicate typology as a method of race study.

TAXA AND THE GENETIC BASIS OF DISCONTINUITY

Excepting the simplest viruses, the genotypes of all organisms are compounded of many genes. Thousands or tens of thousands can be taken as orders of magnitude anywhere above the level of microorganisms. The figure 30,000 sometimes given for man is to my knowledge not substantiated by any data but may be adopted for the purpose of argument. Now, with n variable genes, 2^n kinds of homozygotes, or 3^n homozygous and heterozygous gene combinations, are potentially possible in diploid organisms. Provided that all or a substantial fraction of the gene loci undergo mutation, the number of possible gene combinations is practically infinite.

However, only a very minute fraction of the potentially possible gene combinations can be realized. This is so for two reasons. One reason is trivial—the numbers of individuals alive, or having lived, is very much smaller than that of possible gene combinations. The other reason goes to the heart of the problem we are considering. Only a minority, probably only a tiny minority of the possible genotypes, make their carriers fit to survive and to reproduce in any existing environments. The genotypes actually observable occur in clusters of variants which are adapted to dwell in certain ecological niches. Sewall Wright designates these clusters as occupants of the "adaptive peaks." The gaps between the "peaks" are the "adaptive valleys," corresponding to the unrealized and mostly unfit or inviable gene combinations. Thus, living mankind is a cluster of variant genotypes only slightly smaller in number than the number of the persons alive (taking identical twins to be genetically identical). The species chimpanzee is another cluster of genotypes. Although there is no way to prove this experimentally, there is every reason to suspect that many mixtures of human and chimpanzee genes would be unfit or inviable, so that an adaptive valley separates the two species.

The clusters of genotypes and the gaps between them may be small or large; wholly discontinuous or joined by a few exceptional or hybrid genotypes; increasing or decreasing with time. The clustering of the realized gene combinations is, it must be emphasized, universal in the living world. Clusters, and clusters of clusters, are observed in animals as well as in plants, in higher and in lower, in sexual and in asexual organisms. Indeed, clustering is one of the devices whereby life masters the environments of our planet.

The clusters and the gaps between them are matters of observation. The clustering is, however, exploited by systematists for their own ends. The clusters are treated as taxa and are given names. Consider for example the taxa *Homo sapiens sapiens*, *Homo sapiens neanderthalensis*, *Homo erectus*, *Hominidae*, *Pongidae*, etc. Are these taxa "natural"? For our present purpose we can dodge the controversial issue as to whether a "natural" taxon must necessarily be monophyletic; a taxon is natural when it corresponds to an observable cluster of genotypes, and unnatural if it does not.

Some authors tried to challenge the biological validity of taxa by claiming that discontinuous clusters of organic forms do not in fact exist, and are projections on nature of our linguistic habits. This argument may as well be inverted—our linguistic habits may have been influenced by the observation that the organic variation is so generally and so strikingly discontinuous. Continuity on any one time level would indeed be a biological absurdity—it would demand the production of masses of gene combinations filling the adaptive valleys with corpses. On the other hand, no discontinuity is generally found between generations succeeding each other in time. This is why paleontologists make a virtue of necessity, and utilize gaps in the fossil record to cut what was an essentially continuous series of forms into convenient temporal species, genera, etc. It should then be emphasized that contemporaneous species (whether living or fossil) are quite different biological phenomena from temporal species which succeed each other in time. This matter is discussed in more detail in the papers of E. Mayr and of G. G. Simpson in the present volume.

CATEGORIES AND TAXA

The problem of the biological reality of taxonomic categories is entirely separate from that of the taxa. As pointed out above, a taxon is "natural" if it comprises a cluster of related genotypes separated from other clusters by more or less pronounced genetic discontinuities. Natural taxa occur everywhere, from viruses to primates. But it does not follow that what is traditionally regarded the same category of taxa, say the category of species, is everywhere the same biological phenomenon. Moreover, some categories may, and do, have a firm biological basis while others rest only on convention. Linnaeus used fewer categories than do modern systematists; this only means that with more and more organisms to classify, additional categories became desirable. Whether the gibbons are only generically distinct from other apes, or whether a subfamily, family, or a super-family should be set up for them, can be argued only on grounds of convenience and consistency. There is no known biological phenomenon that would make an array of forms a family rather than a genus.

The category of species has a particular and unique biological meaning because speciation is one of the basic evolutionary processes. In sexually reproducing and contemporaneous organisms species are more than class concepts; they are cor-porate bodies organized for the function of procreation. Species can be defined biologically as inclusive Mendelian populations, breeding communities, set apart from other breeding communities by reproductive isolating barriers. Reproduc-tive isolation acts to circumscribe the array of potential mates. Individuals of a species are interdependent members of an organic system, in basically the same way, though obviously not to the same extent, as cells of a multicellular body.

Species are genetically closed systems. This means that a mutation, no matter how favorable, arising in, for example, the species chimpanzee, cannot benefit the species gorilla or the species man. Genetic closure is, of course, also a property of all higher categories, which are groups of more or less similar species. Races, breeds, or subspecies are, on the contrary, genetically open systems. They do exchange genes along the geographic boundaries where their distributions come in contact. A favorable mutant or a gene combination arising in one race may, if propelled by natural selection, diffuse to other races and thus become a com-mon property of the species as a whole.

Gene diffusion between races often finds visible expression in geographic char-acter gradients, or clines. Such gradients may make it difficult or impossible to draw clear lines of demarcation between the races, especially when the gradients in different characters are uncorrelated. This difficulty has been used by some authors as a basis to argue that man has no races, or even that races in general do not exist. This is about as logical as it would be to argue that youth is not dif-ferent from old age because the gradient between the two is almost completely smooth.

Races are genetically different but not reproductively isolated populations. Race as a biological phenomenon has a genetic basis that is less clear-cut than that of species. The genetic differences between races may be of different orders of magnitude. Populations may differ in the incidence of one or of few genes; such populations are racially distinct, but it is not expedient to treat them as name-able taxa. On the other hand, races may differ in many genes, and some of the differences may be qualitative, certain gene alleles reaching fixation in one and other alleles in other races. The extreme situation, not found in living man but known in many biological materials, are races which are developing reproductive isolation and thus are becoming transformed into species. The difficulty of rigorously defining races is not that races are unreal or arbitrary entities, but rather that there exist so many different kinds of minor and major races and races which are incipient species.

ANAGENESIS AND CLADOGENESIS IN HOMINID EVOLUTION

Although I have made honest efforts to familiarize myself with paleoanthropology, I realize the risks of trespassing outside the field of one's scientific specialization. Despite the obvious incompleteness and inadequacies of the available data concerning human evolution, it may nevertheless be helpful to point out some possible biological implications of these data. Which of these implications and inferences are really warranted will have to be decided by future studies; the present rate of discovery of hominid fossils is so high that this future is hopefully not very distant.

Two types of evolutionary changes, anagenesis and cladogenesis, may usefully be distinguished (see especially Rensch, 1960 and Simpson, 1953). Anagenesis is change in time without diversification, cladogenesis is splitting and branching of an ancestral form into two or several derived ones. A species is anagenetically transformed in time into more and more distinct forms, but remains a single species at any one time level. A cladogenetic change splits the ancestral species, so that on some later time levels two or more distinct species exist contemporaneously. Both anagenetic and cladogenetic changes occur in most evolutionary histories, and that of the hominids is not an exception. The accumulating data suggest, however, more and more clearly that in the hominid evolution anagenesis predominates.

Let us consider the evolutionary situation of the hominid stock from mid-Pleistocene on. The remains of four distinct hominids having lived some 400,000 to 500,000 years ago have been described and named as follows (all the dates taken from Oakley, 1962; I am obligated to Dr. B. G. Campbell for the names of the fossil hominids given below):

Pithecanthropus erectus, Trinil beds, Java.
Sinanthropus pekinensis, North China.
Atlanthropus mauritanicus, Algeria.
Palaeanthropus heidelbergensis, Germany.

Moving from mid-Pleistocene across a time gap, thus far scarcely represented by fossil remains, to a time level 70,000–30,000 years ago, we find what appears to be a completely changed situation. Hominid remains are now available from a fair number of localities in the Old World, and the main ones have been classified as follows:

Protanthropus neanderthalensis, from Western Europe to Western and Central Asia.

Cyphanthropus rhodesiensis, South Africa, Rhodesia.

Javanthropus soloensis, Java.

Homo spelaeus, Europe (Cro-Magnon).

Four genera are thus represented on each of the two time levels, and none of these genera are common to both. Moreover, the genus *Homo,* to which the now living mankind belongs, appears only relatively recently (some 30,000 years ago according to Oakley's reckoning of the age of the Cro-Magnoid skeleton from Combe Capelle), and even then a specific name, *H. spelaeus,* different from the living species, *H. sapiens,* has been suggested. To be sure, so great a proliferation of generic names is endorsed by only few extreme "splitters," and more conservative authors place *neanderthalensis, rhodesiensis* and *soloensis* in the genus *Homo,* and regard *H. spelaeus* as no more than a fossil representative of *H. sapiens.* It is also usual to regard *erectus* and *pekinensis* as species of the same genus, *Pithecanthropus.*

But even taking this relatively "conservative" subdivision, one would have to conclude that a great deal of cladogenetic evolution on the specific and the generic levels has come to pass in the hominid family during the Pleistocene. Moreover, the living *H. sapiens* could have arisen from only one of the four species recorded on the older time levels. At least six of the species listed above must be regarded not as ancestors but as only collateral relatives of *H. sapiens,* all having become extinct without contributing any of their genes to the gene pool of the now living mankind. It is, of course, also possible that the real ancestor of *H. sapiens* is yet to be discovered; the known fossils are then all remains of the extinct collateral relatives.

Assume, for the sake of argument, that *H. sapiens* arose from *Pithecanthropus pekinensis,* or from *Atlanthropus mauritanicus,* or from an as yet undiscovered mid-Pleistocene species. *H. sapiens* has, then, inherited its gene pool from its mid-Pleistocene ancestral species (or genus). It obviously makes no sense to ask whether gene exchange could or could not have taken place between these species, because these species lived at different times. The situation is different with contemporaneous species. If they were reproductively isolated, then the species concerned would represent independent evolutionary lines, most of which must have eventually become extinct. Only one species at any one time level could have been ancestral to *H. sapiens.* (This reasoning would not apply to plants, among which new species may arise by hybridization of preexisting species, followed by a doubling of the chromosome complement in the hybrid).

Another conception of the evolution of the hominid stock assumes a predominance of anagenesis over cladogenesis. Despite the numerous specific and generic names proposed, the facts are not incompatible with the assumption that hominid species were few in number, only a single one having lived at any one time level since the mid-Pleistocene. The now extant species, *Homo sapiens,* is polytypic, i.e., composed of a number of races. *H. sapiens* existed already in late Pleistocene times, when it was composed of races (or sub-species, the terms race and sub-species being near synonyms), among which the principal ones were:

Homo sapiens neanderthalensis
Homo sapiens rhodesiensis
Homo sapiens soloensis
Homo sapiens sapiens (or *spelaeus*)

The late Pleistocene *H. sapiens* descended in turn from a single polytypic mid-Pleistocene *H. erectus*, subdivided into at least the following subspecies:

Homo erectus erectus
Homo erectus pekinensis
Homo erectus heidelbergenis
Homo erectus mauritanicus

MECHANISMS OF ANAGENETIC AND CLADOGENETIC EVOLUTIONARY PATTERNS COMPARED

The two interpretations of hominid evolution outlined above assume, respectively, cladogenesis on both the specific and generic levels, or race differentiation on the intra-specific level only. The point worth stressing is that what is here involved is not a matter of arbitrary labeling or of the priority in naming this or that fossil find, but two markedly different conceptions of hominid evolution. If four or more distinct reproductively isolated species or genera were living in mid-Pleistocene times, then the living mankind has inherited its gene pool from only one of them. If, on the other hand, the mid-Pleistocene forms all belonged to a single polytypic species *Home erectus*, then *H. sapiens* of the upper Pleistocene and of modern times may have genes which were carried in all mid-Pleistocene and upper Pleistocene subspecies. Anagenetic evolution can be compared to a river which flows as a single stream, with possibly some meandering. Cladogenic evolution is then a stream subdividing into branches, most of which are lost in desert sands.

It should be made clear that if *H. erectus* became transformed as a whole into *H. sapiens*, then the contributions of the several subspecies of *H. erectus* to the gene pool of *H. sapiens* may have been far from equal. Consider that the proportions of the genes of, for example, Amerindian, White, Mongoloid and Australian Aboriginal races in the gene pool of the now living mankind are surely different from what they were 1,000 or 10,000 years ago. It would be grotesque to classify these races as species because some of them expanded while others contracted. The race *Homo sapiens neanderthalensis* was "suddenly" replaced in Europe some 30,000 years ago by people which on the basis of their skeletal remains were like the living race *H. sapiens sapiens*. The "sudden" replacement may have taken time of the order of 1,000 years or more, and may well have involved both some massacre and some interbreeding between the invaders and the indigenous populations. According to Coon (1939), the presence of the genes of *H. sapiens neanderthalensis* is discernible in some European populations.

Which of the subspecies of *H. erectus* contributed more and which fewer genes to the gene pool of *H. sapiens* cannot be decided on the present evidence. This evidence is certainly far from sufficient to establish the relative roles of anagenesis and cladogenesis in hominid evolution. New evidence is, however, accumulating rapidly, and it is essential to keep an open mind for its interpretation. The riot of specific and generic names of which paleoanthropology has been the victim can only obfuscate the significance of the valuable findings of the past and of those now being made. A new specific name should not be given to a newly discovered form unless it can be proven beyond reasonable doubt that it cannot belong to a race of a previously known species. Inventing a new generic name for every species stultifies the basic purpose of the binary nomenclature. Latin trinomials, or simple geographic designations of the specimens found, are preferable for the purposes of communication to specific or generic names, since misuses of the latter lead to misunderstanding of the basic character of the hominid evolution.

If mankind has developed anagenetically from a polytypic mid-Pleistocene ancestral species, and that species in turn from a polytypic early Pleistocene species, then the genes of some very ancient local races may now be cosmopolitan in distribution. And conversely, the ancestral genes which proved disadvantageous may have been eliminated by natural selection. An anagenetic descent of a more recent from a more ancient species is best envisaged as a gradual reconstruction of the gene pool of the species as a whole. The question where *H. sapiens* arose from *H. erectus* has different meanings on the assumption of anagenetic or of cladagenetic descent. In the former instance a new species may be said to have arisen almost everywhere where the ancestral species lived, while in the latter the territory of origin was probably much more limited. With anagenetic evolution, every one of us may conceivably carry genes which have first appeared by mutation in ancestors who lived over a territory coextensive with the distribution areas of the ancestral species.

Mutations are, however, only the raw materials of evolution. Neither the speed nor the direction of the evolutionary changes is determined by how frequently mutations arise. Far more important are the processes of natural selection, crossing and recombination, and genetic drift in their manifold interactions. Anagenetic evolution may, then, involve genetic events in different parts of the geographic area occupied by a species. Adaptively favorable gene patterns may arise in several or many places; the populations in which these "evolutionary inventions" have turned up may then spread, come in contact, hybridize, and form new, even superior gene patterns, which then spread from the new centers and continue the process.

Even if the adaptive key events, the paramount genetic improvements, would arise in only one race of the ancestral polytypic species, these improvements may then spread to the species as a whole by migration, hybridization, and natural selection. What is here crucial is whether the improved race before it spreads does or does not become a reproductively isolated new species. If it does, then it can only supplant its former relatives by driving them to extinction. Clado-genetic evolution on the species level has then occurred. But if the improved race remains a race, i.e., continues capable of interbreeding with other races of its species, then it inherits the earth by just as certain but, in a sense, subtler means of intercrossing and genetic replacement.

Weidenreich (1946) believed that the Mongoloid race of man has inherited cer-tain morphological traits from the *pekinensis* race of *H. erectus*. He also thought it possible to trace a direct line of descent from *H. erectus modjokertensis* (Djetis beds, Java, estimated age 600,000 years), through *H. erectus erectus* (Trinil beds, Java, 500,000 years), *H. sapiens soloensis* (upper Pleistocene, Java), to the living Australian Aboriginal race of *H. sapiens*. The apparent survival of some of the features of *H. sapiens neanderthalensis* in Europe has been mentioned above. The situation in Africa is quite obscure, but also most hopeful because the rate of dis-covery of new fossil material on this continent is the highest. The remains of the so-called *Telanthropus* are too fragmentary for classification, but they may represent a form perhaps intermediate between an australopithecine and *H. erec-tus*. The skull from Olduvai Bed II, as yet undescribed in detail, may, according to Oakley (1962), be related to *H. erectus*. *H. sapiens rhodesiensis*, and the Kan-jera and Florisbad remains, may conceivably help to bridge the gap between the *erectus* and the *sapiens* stages. All this is, to be sure, very much disputed ground, and more data must be obtained before any of these conjectures are either con-firmed or invalidated. The point which must be emphasized is that the extreme models of the hominid evolution, which assume respectively a prevalence of cladogenesis on specific and generic levels, or else an anagenetic evolution of a single species, are proposed only as working hypotheses. Neither of these can be established or rejected on the present evidence, and, furthermore, the truth quite possibly lies somewhere between the two. That more than a single hominid species existed contemporaneously and in part also sympatrically in Villafranchian and possible in mid-Pleistocene times seems pretty clearly established.

Australopithecus africanus and *A. robustus* (with its more ancient race *boisei*) are perhaps the best examples (Robinson, 1954) despite much dispute about their ages. The so-called "*Meganthropus*" fragments from Djetis beds in Java which may belong to a race of *A. robustus* (subsp. *paleojavanicus*) seem to have been contemporaneous and sympatric with *Homo erectus modjokertensis*. At present, we can only suspend judgment concerning their taxonomic status and concern-ing the possibility that some of these forms may have been interbreeding.

It has been known for some time (e.g., Movius, 1948), and stressed by Dr. Leakey in the present volume, that two rather distinct cultural traditions, re-flected in different types of stone tools, existed in eastern Asia on the one hand and in Africa, Europe, and Western Asia on the other. These cultural traditions preserved their distinction for a long time, embracing parts of middle and late Pleistocene. It is possible that the carrier of the eastern Asiatic tradition was *H. erectus* (of which the races *erectus* and *pekinensis* are so far known), and that the carrier of the other tradition was another species of *Homo* represented by the races *mauritanicus*, perhaps *heidelbergensis*, and may be by some recently discovered African forms as yet undescribed in detail. This possibility is supported by the fact that *erectus* and *pekinensis* share some morphological traits not found in the western forms. If so, *H. erectus* has not evolved into *H. sapiens*, but became extinct without issue. For example, the Javanese *H. sapiens soloensis* would then be, contrary to Weidenreich, not a descendant of *H. erectus erectus*, but rather a race of the invading *H. sapiens* which supplanted its predecessor.

The problem is really how to discriminate between these two possibilities, a single species of *Homo*, or two species, an eastern and a western one in mid-Pleistocene? One would like to know whether anywhere and at any time races of *H. erectus* and of *H. sapiens*, or of a western species ancestral to the latter, lived sympatrically but without production of intermediate or hybrid types. One may also hope that populations bridging the gap between the supposed species may be discovered. It is evidence of this nature, sympatric occurrence of distinct forms, that leads us to assume the existence of at least two distinct species of *Australopithecus*. No mid-Pleistocene hominid fossils are known from extensive territories in Asia, particularly from India and the surrounding countries, which may well provide the critical evidence. In the absence of such evidence, there is simply no foundation for a decision whether the mid-Pleistocene forms of eastern Asia and the western forms belonged to a single or to two or perhaps more species.

Morphological evidence, the structural differences between the known eastern and western representatives of the mid-Pleistocene *Homo*, is by itself not necessarily decisive for attributing to them either a subspecific or a full specific status. Opinions have fluctuated widely among paleoanthropologists as to what magnitude of morphological difference warrants generic, specific, or racial ranks. We have seen above that a generic status was at one time or another attributed to at least eight forms which the consensus of the participants in the present volume places in a single genus *Homo*. It was Weidenreich (1946 and earlier) who first concluded that the existing morphological evidence is compatible with the assumption that only a single species of *Homo* lived at any one time level from mid-Pleistocene on, and his view was supported by Dobzhansky (1944) and by Mayr (1950) on the basis of genetic and evolutionary considerations.

Geographically remote races of a species may differ morphologically about as much as some admittedly good species, and yet may lack the reproductive isolation that would make them specifically distinct. One should not assume that certain forms are or are not distinct species unless this can be proven, and until this is proven it is preferable, in the opinion of many systematists which I fully share, to classify them tentatively as races. It is misleading to assume that certain morphological characters necessarily indicate specific or generic rather than racial differences. An evolutionist cannot fail to infer that any sort of difference observed between species may also appear, albeit in a rudimentary form, between races of one species. Since this Burg Wartenstein symposium is being held in a medieval castle, this milieu tempts me to put the above principle in Latin: Nihil est in specie quod non prius in varietate erat.

SOME PECULIARITIES OF THE HOMINID EVOLUTIONARY PATTERNS

Biologists classify organisms because organisms are very numerous and so diversified that no two individuals are ever alike. The organic diversity is a product of the evolutionary process. How evolution occurs is not immaterial to a classifier who wishes to understand the nature of the taxa he describes. The basic causes of evolution are the same everywhere, from viruses to hominids, since mutations occur in all organisms and all organisms are subject to natural selection. And yet the evolutionary patterns are varied in different groups of creatures. Comparative study of evolutionary patterns is as legitimate an endeavor as comparative anatomy, comparative physiology, or comparative biochemistry.

Anagenetic as well as cladogenetic changes have been taking place in hominid evolution. But anagenesis has decidedly predominated. Although there were two, and perhaps more than two, contemporaneous species of *Australopithecus*, the genus *Homo* seems to have been represented since its emergence by a single polytypic species at any one time level. A question that inevitably arises is what favored the conservation of the specific unity and impeded the differentiation of man into several species.

Speciation, i.e., subdivision of a single ancestral species into two or more reproductively isolated derived ones, is not merely a function of the number of gene differences accumulated between diverging races, and not even of the time elapsed since their origin. Speciation is a form of genetic specialization of organisms living in different environments. Suppose that two Mendelian populations become races having different gene pools, each internally balanced and adapted to a certain mode of life, but such that many of the recombination genotypes formed, if these populations interbreed, are low in fitness. If so, genetic impediments to interbreeding and gene exchange may be favored by natural selection. When the reproductive isolation becomes strong or complete, we have two or more separate species derived from a single original one.

If, on the other hand, race hybrids show no loss of fitness compared to the parental forms, then the main stimulus to speciation is lacking. This, as has been pointed out by many students of man, including Dr. G. A. Harrison in the present volume, is notably the case of the hybrids between human races. Experiments have been made crossing morphologically indistinguishable populations of Drosophila from different localities; at least in some instances the F_1 hybrids showed an apparent hybrid vigor, but in F_2 there was instead a hybrid breakdown, i.e., a loss of fitness (Vetukhiv, 1957 and other publications). Man behaves in this respect differently from Drosophila. Contrary to what racists so stridently proclaim, there is not a detectable trace of a loss of fitness in either the F_1 or in later generations of hybrids between even the most distinctive human races. Nor is there any evidence of heterosis in the F_1.

Our species is able to tolerate, in at least the environments created by culture, a remarkably great genetic variety without a loss of Darwinian fitness. This does not mean that the genetic differences among humans became adaptively neutral when our remote ancestors originated and developed culture. I wish to suggest that the situation is, in a sense, the converse; far from being detrimental or neutral, the genetic variety may be increasing the fitness of human populations. Culture is an adaptive instrument which permits the human species to evolve by fitting its environments to its genes more often than by changing its genes to fit its environments. The culture does not, however, create a uniformity of environments, except in a narrowly physical sense, such as the control of temperature by wearing clothing or by heating or air conditioning. On the contrary, culture originates an ever increasing variety of environments or adaptive or ecological niches.

Civilization makes this variety virtually inexhaustible. Two factors contribute most notably to the fitness of the human species: the ability of the carriers of most human genotypes to select for themselves a suitable niche, and the ability to adjust themselves to niches which they select or which are selected for them. One may say that man is genetically specialized to be unspecialized as to the precise ecological niche which a human individual may have to occupy in his life. Man has remained a single species because subdivision into several adaptively specialized species would not be advantageous. This is really a paraphrase of Mayr's argument (1950) that man has remained a single species because he fills up all the available ecological niches; subdivision into several adaptively specialized species would have proved disadvantageous and would have been selected against.

Just how ancient is this ecological versatility of the human species is a matter of speculation. However, some reasonable inferences can be made. Washburn and DeVore (1961) and DeVore and Washburn (1962) have pointed out that ground living primates tend to range over more extensive territories than tree-living ones. There are more species of tree-living Cercopithecus in the forests of Africa than of the ground living baboons or macaques from Cape Town to Japan. Greater mobility conduces, other things being equal, to more gene exchange between local populations, less geographic differentiation, and less speciation. The australopithecines were certainly ground living, and their mobility was probably no smaller than that of baboons.

When man became a hunter, by mid-Pleistocene if not earlier, his mobility must have increased greatly. A single species then could easily inhabit a territory extending from Java and Peking to Heidelberg and to Olduvai and southward. Birdsell (1957) has made most interesting calculations of the speed with which human populations at the hunting-gathering level can spread to new territory. His estimate for the occupation of the continent of Australia is about 2,200 years, from the initial entry of the first successful migrants from the north to a population "saturation" stage of the continent. Birdsell extrapolates that a successful species of *Australopithecus* may have required only about 23,000 years to spread from South Africa to southeastern Asia.

The genetic consequences of migration are worth more thought and study than they have thus far received. What happens when a race of a polytypic species improves its adaptedness and spreads to the areas of other races? There is no need to ascribe to our remote ancestors a degree of ferocity attained by only some of their descendants; the invaders need not have slaughtered the entire populations of the territories invaded. Genocide is less probable than genosorption, i.e., incorporation of genes of one population into the gene pool of another. Furthermore, even if the genes of the invaders are superior in some respects, the Darwinian fitness of their descendants in the new territory will probably be increased by incorporation of some of the genes of the indigenous populations. The latter may, for example, have a superior resistance to the endemic diseases. Suppose that the territories of *H. erectus erectus* and *H. erectus pekinensis* were invaded by migrants from the west. These westerners may have been superior genetically or culturally or both. However, provided the easterners and the westerners belonged to the same species, it is most likely that a population resulted which had some genes from both. Even if a stage of cultural development was reached at which the invaders were so "advanced" that they massacred all or most of the native males, they may have spared and interbred with the females.

Changes in sexual biology which have taken place in human evolution are relevant at this point. The almost uninterrupted sexual receptivity of the human female, the complete helplessness of the human infant which must be carried and cared for by the mother, males taking upon themselves the role of economic providers, and finally the development of incest taboos, made man very different from non-human primates. Washburn and DeVore infer that the incest taboos and exogamy rules probably existed from mid-Pleistocene on. The outbreeding and the gene diffusion which these rules induced obviously hindered speciation, but not anagenetic transformations of the human species. I fully agree with the view expressed by these authors, that the way of life of the hominids "based on tools, intelligence, walking and hunting, was sufficiently more adaptable and effective so that a single species could occupy an area which ground monkeys could occupy only by evolving into at least a dozen species. This comparison gives some measure of the effectiveness of the human way of life, even at the level of Peking and Ternifine man."

It is a pleasure to acknowledge my obligation to Drs. Bernard Campbell, G. H. Harrison, Ernst Mayr, William L. Straus, Jr., and Irven DeVore for the suggestions, criticisms, and discussions of the facts and ideas with which the present paper is concerned.

ADDENDUM

This article was concluded and sent to the publisher before the appearance of *The Origin of Races* by C. S. Coon. Dr. Coon and I are in agreement that the now living polytypic species, *Homo sapiens*, is descended from the single polytypic *Homo erectus* of the mid-Pleistocene. Dr. Coon has, however, chosen to believe that *H. erectus* was transformed into *H. sapiens* not once but five times, and that this transformation occurred much earlier in some places than in others. This belief, which has made Dr. Coon's work attractive to racist pamphleteers, is neither supported by conclusive evidence nor plausible on theoretical grounds. The specific unity of mankind was maintained throughout its history by gene flow due to migration and the process for which the word "genosorption" is suggested above. Excepting through such gene flow, repeated origins of the same species are so improbable that this conjecture is not worthy of serious consideration; and given a gene flow, it becomes fallacious to say that a species has originated repeatedly, and even more fallacious to contend that it has originated five times, or any other number above one.

BIBLIOGRAPHY

BIRDSELL, J. B.
   1957. "Some population problems involving Pleistocene man." *Cold Spring Harbor Symp. Quant. Biol.*, 22:47–69.
COON, C. S.
   1939. *The races of Europe*, New York: Macmillan.
COUNT, E. W.
   1950. *This is race*. New York: Schuman.
DEVORE, I. and S. L. WASHBURN
   1962. "Baboon ecology and human evolution." In: *African ecology and human evolution*, Viking Fund Publ. in Anthropology (in press).
DOBZHANSKY, TH.
   1944. "On species and races of living and fossil man." *Amer. Jour. Phys. Anthropology*, 2:251–256.

MAYR, E.

1950. "Taxonomic categories in fossil hominids." *Cold Spring Harbor Symp. Quant. Biol.*, 15:109–118.

1959. "Darwin and the evolutionary theory in biology." In: *Evolution and anthropology: A centennial appraisal*, Washington: Anthropological Society of Washington.

MOVIUS, H. L.

1948. "The lower Palaeolithic cultures of southern and eastern Asia." *Trans. Amer. Philosophical Soc.*, 38:329–420.

OAKLEY, K. P.

1962. "Dating the emergence of man." *Advan. Science*, January:415–426.

RENSCH, B.

1960. *Evolution above the species level.* New York: Columbia Univer. Press.

SIMPSON, G. G.

1953. *The major features of evolution.* New York: Columbia Univ. Press.

1961. *Principles of animal taxonomy.* New York: Columbia Univ. Press.

VETUKHIV, M.

1957. "Longevity of hybrids between geographic populations of *Drosophila pseudo-obscura*." *Evolution*, 9:348–360.

WASHBURN, S. L. and I. DEVORE

1961. "Social behavior of baboons and early man." In: *Social life of early man*, Viking Fund Publ. Anthropology, 31:91–104.

WIERCINSKI, A.

1962. The racial analysis of human populations in relation to ethnogeneses. *Current Anthropology*, 1962:2–20.

INDEX

Hallowell, A. I., 328
Hamerton, J. L., 235
Hamilton, W. J., 211
Hand: characteristics of, 132–43; Oldu-
vai, 183; opposition in human, 182
Harlow, H. F., 273
Harrison, G. A., 55, 56, 76, 78, 79, 181,
254, 261, 264, 266, 267
Harrisson, T., 187
Hasek, M., 210
Haslerud, G. M., 289
Healy, J. R., 59, 79
Hebb, D. O., 289, 322, 323, 325, 328
Heinke, F., 55
Hellman, M., 101, 143
Hemmings, W. A., 208
Hemochorial placenta, 212, 215
Hemoglobin: in gorilla, 247; in higher
vertebrates, 245; in horse and man,
255; in human, 246
Hewes, G. W., 186
Hicks, J. H., 182
Hill, J. P., 212, 216
Hill, R., 247
Hill, W. C. O., 101
Himwich, H. E., 213
Hinde, R. A., 275, 278
Hobhouse, L. T., 273
Hofer, H., 111, 118, 125
Home range, of monkeys and apes, 303–
306
Hominidae, 12; characters of, 9; classifi-
cation of, 69–70, at generic level, 39;
evolution of, 64–69, genetic entities
in, 347–61; as example of mosaic
evolution, 344–45; locomotor functions of, 178–87; manual ability in,
183; Pleistocene fossils of, 203; South
African fossils of, 33; taxonomic
evaluation of, 11–12
Hominoidea: chromosomes of, 235–41;
classification of, 22–30, 33–34, at
family level, 37–39; East African
fossils of, 32–49; rate of evolution of,
64; stages of evolution of, 44–45
Homo, 8, 12, 25; chromosome of, 237;
differences between types of, 14

Homo sapiens, see Man
Hopwood, G., 32
Howell, F. C., 132
Howells, W., 329
Huizinga, J., 56
Hunt, J. McV., 327
Hürzeler, J., 21, 33, 38, 146, 147, 149,
150, 152, 153
Hylobatidae, 38
Hypodigm, 2

I

Imanishi, K., 280, 287, 308
Immunological system, ontogeny of, 210
Immunological theory of primate evolu-
tion, 215, 226
Ingram, V. M., 204, 206, 245, 249, 252,
263
Intelligence, 320
Ischial callosities, in higher Primates, 106
Itani, J., 313

J

Jacob, F., 252
Jay, P., 303, 307–309, 316
Jolicoeur, P., 59
Jones, G. I., 57
Jones, R. T., 251
Joseph, J., 182

K

Kälin, J., 147, 173
Kanam mandible, 33, 43
Kaplan, N. O., 254
Kaufman, J. H., 307

Milton Keynes UK
Ingram Content Group UK Ltd.
UKHW051945071024
449327UK00026B/2178